XLV. 2. 395

# DEFORMATION-MECHANISM MAPS

The Plasticity and Creep of Metals and Ceramics

**Other Pergamon Titles of Interest**

| | |
|---|---|
| ASHBY & JONES | Engineering Materials |
| ASHBY *et al.* | Dislocation Modelling of Physical Systems |
| ASHWORTH | Ion Implantation into Metals |
| BARRETT & MASSALSKI | Structure of Metals, 3rd Edition |
| BELY | Friction & Wear in Polymer-based Materials |
| COUDURIER *et al.* | Fundamentals of Metallurgical Processes |
| HARRIS | Mechanical Working of Metals |
| HOPKINS & SEWELL | Mechanics of Solids |
| JOHNSON | Plane Strain Slip Line Fields for Metal Deformation Processes |
| KRAGELSKY *et al.* | Friction & Wear, Calculation Methods |
| KRAGELSKY & ALISIN | Friction, Wear & Lubrication (3 volumes) |
| MASUBUCHI | Analysis of Welded Structures |
| OSGOOD | Fatigue Design, 2nd Edition |
| UPADHYAYA & DUBE | Problems of Metallurgical Thermodynamics and Kinetics |

**Pergamon Related Journals**

ACTA METALLURGICA

CANADIAN METALLURGICAL QUARTERLY

ENGINEERING FRACTURE MECHANICS

FATIGUE OF ENGINEERING MATERIALS AND STRUCTURES

INTERNATIONAL JOURNAL OF ENGINEERING SCIENCE

MATERIALS RESEARCH BULLETIN

METALS FORUM

PROGRESS IN MATERIALS SCIENCE

THE PHYSICS OF METALS AND METALLOGRAPHY

# DEFORMATION-MECHANISM MAPS

## The Plasticity and Creep of Metals and Ceramics

By

## H. J. FROST

Dartmouth College, New Hampshire, USA

and

## M. F. ASHBY

Cambridge University, UK

## PERGAMON PRESS

OXFORD · NEW YORK · TORONTO · SYDNEY · PARIS · FRANKFURT

| | |
|---|---|
| UK | Pergamon Press Ltd, Headington Hill Hall, Oxford OX3 0BW, England |
| USA | Pergamon Press Inc., Maxwell House, Fairview Park, Elmsford, New York 10523, USA |
| CANADA | Pergamon Press Canada Ltd, Suite 104, 150 Consumers Rd, Willowdale, Ontario M2J 1P9, Canada |
| AUSTRALIA | Pergamon Press (Aust.) Pty Ltd, PO Box 544, Potts Point, NSW 2011, Australia |
| FRANCE | Pergamon Press SARL, 24 rue des Ecoles, 75240 Paris, Cedex 05, France |
| FEDERAL REPUBLIC OF GERMANY | Pergamon Press GmbH, 6242 Kronberg-Taunus, Hammerweg 6, Federal Republic of Germany |

First edition 1982

**British Library Cataloguing in Publication Data**

Frost, H. J.
  Deformation-mechanism maps.
  1. Deformations (Mechanics)  2. Materials.
3. Plasticity.  I. Ashby, M. F.  II. Title.
TA417.6.F76 1982        620.1'123        82–5201
ISBN 0–08–029338–7                        AACR2
ISBN 0–08–029337–9 (pbk.)

Filmset in Great Britain by
Northumberland Press Ltd, Gateshead, Tyne and Wear
Printed and bound by A. Wheaton & Co., Ltd., Exeter

# FOREWORD

DURING the last three decades the body of data for the plastic behaviour of solids has enlarged enormously. The physical understanding of plasticity and creep, too, has been transformed by the identification of the underlying atomistic processes. This has stimulated the application of the older sciences of solid mechanics, thermodynamics and kinetics to the modelling of plasticity and has led to the formulation of a mechanics of materials which, though still incomplete, is now accepted as a foundation on which to build.

At this stage it appears desirable to try to summarize and classify as much of the data as possible, using the theoretical understanding as a framework for doing so. In this book an attempt is made to catalogue mechanisms for plasticity and derive approximate constitutive laws for them. Data are assembled for a large number of metals and non-metals, and fitted to the constitutive laws. The data and the predictions of the laws are then presented together as deformation-mechanism diagrams, which summarize both, in a compact but accessible form. The method exposes the broad similarities and differences between groups of materials, and suggests a classification of materials by their mechanical behaviour.

The work on which the book is based was carried out in part in the Division of Applied Sciences at Harvard University, and in part in the University Engineering Laboratories at Cambridge University. We have benefited from numerous discussions with many colleagues; it is a special pleasure to acknowledge the considerable contributions of Prof. A. S. Argon, Dr L. M. Brown, Prof. B. Chalmers, Dr R. J. Fields, Dr U. F. Kocks and Prof. D. Turnbull.

January 1982

H. J. FROST
M. F. ASHBY

# LIST OF SYMBOLS, DEFINITIONS AND UNITS

| Macroscopic variables | | Equation number |
|---|---|---|
| $\varepsilon$ | Normal (tensile or compressive) strain | 17.7 |
| $\varepsilon^T$ | Transient strain | 17.9 |
| $\dot{\varepsilon}$ | Strain rate $(s^{-1})$ | 1.5 |
| $\dot{\varepsilon}_1, \dot{\varepsilon}_2, \dot{\varepsilon}_3$ | Principal strain rates $(s^{-1})$ | 1.3 |
| $\dot{\varepsilon}_{ij}$ | Strain rate tensor $(s^{-1})$ | 1.4 |
| $\gamma$ | Shear strain | 1.3 |
| $\gamma^T$ | Transient shear strain | 17.10 |
| $\dot{\gamma}$ | Shear strain rate $(s^{-1})$ | 1.3, 1.4, 1.5 |
| $\dot{\gamma}_1, \dot{\gamma}_2 \ldots \dot{\gamma}_8$ | Shear strain rates of individual mechanisms $(s^{-1})$ | 2.1, 2.9, 2.12, 2.21, 2.23, 2.26, 2.29 |
| $\dot{\gamma}_{p\ell}$ | Shear strain rate caused by low-temperature plastic mechanisms $(s^{-1})$ | 3.1 |
| $\dot{\gamma}_{PD}$ | Shear strain rate limited by phonon drag $(s^{-1})$ | 2.14 |
| $\dot{\gamma}_t$ | Shear strain rate limited by twinning $(s^{-1})$ | 2.15 |
| $p$ | Hydrostatic pressure $(MN/m^2)$ | 17.37 |
| $\sigma_n$ | Normal (tensile or compressive) stress $(MN/m^2)$ | 2.17 |
| $\sigma_1, \sigma_2, \sigma_3$ | Principal stresses $(MN/m^2)$ | 1.1 |
| $\sigma_{ij}$ | Stress tensor $(MN/m^2)$ | 1.2 |
| $\sigma_s$ | Shear stress $(MN/m^2)$ | 1.5 |
| $S_{ij}$ | Deviatoric part of the stress tensor $(MN/m^2)$ | 1.2 |
| $t$ | Time (s) | 17.3 |
| $T$ | Temperature (K) | 1.6, 1.12 |

| Other variables and constants | | Equation number |
|---|---|---|
| $A, A_1, A_2$ | Power-law creep constants | 2.18, 2.19 |
| $A_{HD}$ | Constant for Harper–Dorn creep | 2.23 |
| $a_c$ | Diffusive section of a dislocation core $(m^2)$ | 2.20 |
| $\alpha$ | Dimensionless ideal strength | 2.1 |
| $\alpha$ | Dimensionless constant for low-temperature plasticity | 2.7 |
| $\alpha'$ | Constant for power-law breakdown $(MN/m^2)$ | 2.26 |
| $B$ | Drag coefficient $(Ns/m^2)$ | 2.13 |
| $b$ | Magnitude of Burgers' Vector (m) | 2.2 |
| $b_b$ | Burgers' Vector of boundary dislocation (m) | 17.25 |
| $\beta$ | Constant for low-temperature plasticity | 2.7 |
| $\beta'$ | Constant for power-law breakdown | 2.26 |
| $C_p$ | Specific heat at constant pressure $(J/m\ K)$ | 17.16 |
| $C$ | Constant for drag-limited glide | 17.12 |
| $c_{11}, c_{12}, c_{44}$, etc. | Single-crystal moduli $(MN/m^2)$ | — |
| $D_b$ | Boundary diffusion coefficient $(m^2/s)$: $D_b = D_{0b}\exp - (Q_b/RT)$ | 2.30 |
| $D_c$ | Core diffusion coefficient $(m^2/s)$: $D_c = D_{0c}\exp - (Q_c/RT)$ | 2.20 |

**Other variables and constants**                                            **Equation number**

| Symbol | Definition | Equation number |
|---|---|---|
| $D_v$ | Lattice diffusion coefficient (m$^2$/s): $D_v = D_{0v}\exp - (Q_v/RT)$ | 2.17 |
| $D_{\text{eff}}$ | Effective diffusion coefficient for power-law creep or diffusional flow (m$^2$/s) | 2.20, 2.30 |
| $D_A, D_B$ | Tracer diffusion coefficients for components $A$ and $B$ (m$^2$/s) | 2.27, 2.28 |
| $\tilde{D}$ | Chemical interdiffusion coefficient (m$^2$/s) | 2.28 |
| $d$ | Grain size (m) | 2.29, 2.30 |
| $\delta$ | Diffusive thickness of a grain boundary (m) | 2.30 |
| $E^{EL}$ | Elastic energy of a dislocation (J/m) | — |
| $\varepsilon\ (\dot{\varepsilon})$ | Normal strain (rate); see Macroscopic variables | — |
| $F, \Delta F$ | Helmholtz free energy (J/mol) | 2.6, 2.7, 2.10 |
| $\Delta F_n$ | Free energy of twin nucleation (J/mol) | 2.15 |
| $F$ | Force per unit length on a dislocation (N/m) | 2.4 |
| $f_v, f_c$ | Fraction of atom sites associated with lattice and core diffusion respectively | 2.20 |
| $G, \Delta G$ | Gibbs free energy (J/mol) | 2.6, 2.10 |
| $\Gamma$ | Grain boundary energy (J/m$^2$) | — |
| $\gamma\ (\dot{\gamma})$ | Shear strain (rate); see Macroscopic variables | — |
| $\gamma_A$ | Activity coefficient of component $A$, etc. | 2.28 |
| $\dot{\gamma}_{\text{LIM}}$ | Upper limiting strain rate (taken as $10^6$/s) | 17.13 |
| $\dot{\gamma}_0, \dot{\gamma}_p, \dot{\gamma}_t$ | Pre-exponential constants (s$^{-1}$) | 2.9, 2.12. 2.15 |
| $K\ (K_0)$ | Bulk modulus (at 300 K and atmospheric pressure) (MN/m$^2$) | 17.37 |
| $K$ | Work-hardening constant in tension (MN/m$^2$) | 17.6 |
| $K_s$ | Work-hardening constant in shear (MN/m$^2$) | 17.6 |
| $k$ | Boltzmann's constant ($1 \cdot 381 \times 10^{-23}$ J/K) | — |
| $k$ | Thermal conductivity (J/s m K) | 17.20 |
| $\ell$ | Obstacle spacing (m) | Table 2.1 |
| $M$ | Dislocation mobility (m$^2$/Ns) | 2.4 |
| $m$ | Strain hardening exponent | 17.6 |
| $\mu\ (\mu_0)$ | Shear modulus (at 300 K and atmospheric pressure) (MN/m$^2$) | 2.1, 2.3, Table 2.1 |
| $n, n'$ | Creep exponent | 2.18, 2.19, 2.26 |
| $v$ | Debye frequency (taken as $10^{12}$/s) | 2.5 |
| $P_j$ | Material property | 1.6 |
| $p$ | Hydrostatic pressure; see Macroscopic variables (MN/m$^2$) | 17.37 |
| $p$ | Dimensionless exponent | 2.7 |
| $p_0$ | Atmospheric pressure ($0 \cdot 1$ MN/m$^2$) | 17.37 |
| $Q_c, Q_b, Q_v$ | Activation energies for lattice, boundary and core diffusion respectively (kJ/mole) | 2.17, 2.20, 2.30 |
| $Q_{cr}$ | Activation energy for creep when different from $Q_v$, $Q_b$ or $Q_c$ (kJ/mole) | 2.24 |
| $q$ | Dimensionless exponent | 2.17 |
| $\dot{q}$ | Heat generation rate (J/m$^3$ s) | 17.14 |
| $R$ | Gas constant ($8 \cdot 314$ J/mol K) | — |
| $\rho$ | Dislocation density (m$^{-2}$) | — |
| $\rho_b$ | Density of boundary dislocations (m$^{-1}$) | 17.29 |
| $\rho_m$ | Mobile dislocation density (m$^{-2}$) | 2.3 |
| $S_i (S_i^0)$ | State variable (constant initial value) | 1.6, 1.10 |
| $\sigma$ | Stress; see Macroscopic variables (MN/m$^2$) | — |
| $\sigma_y$ | Tensile yield strength (MN/m$^2$) | — |
| $T$ | Absolute temperature; see Macroscopic variables (K) | 1.6, 1.12 |
| $T_M$ | Melting temperature (K) | — |

| | | |
|---|---|---|
| $T_s$ | Temperature of surroundings (K) | 17.20 |
| $t$ | Time; see Macroscopic variables (s) | — |
| $\hat{\tau}$ | Flow strength (in shear) for obstacle cutting at 0 K (MN/m$^2$) | 2.6, 2.7 |
| $\hat{\tau}_p$ | Lattice resistance of Peierls stress (in shear) at 0 K (MN/m$^2$) | 2.10, 2.12 |
| $\hat{\tau}$ | Twinning stress at 0 K (MN/m$^2$) | 2.15 |
| $\tau_{tr}$ | Threshold stress for creep (MN/m$^2$) | 17.35 |
| $\phi, \Delta\phi$ | Chemical potential (J/atom) | 17.27 |
| $\varphi_i, \Delta\varphi_i$ | Interface potential barrier (J/atom) | 17.24 |
| $V^*$ | Activation volume (m$^3$) | 17.49 |
| $\bar{v}$ | Average dislocation velocity (m/s) | 2.3, 2.5 |
| $v_c$ | Average climb velocity (m/s) | 2.17 |
| $x_A, x_B$ | Atom fractions of $A$ and $B$ | 2.27, 2.28 |
| $\Omega$ | Atomic or molecular volume (m$^3$) | 2.17 |
| $\Omega_i$ | Ionic volume (m$^3$) | 2.29 |

# CONTENTS

1   Deformation Mechanisms and Deformation-mechanism Maps            1
2   Rate-equations                                                    6
3   Construction of the Maps                                         17
4   The f.c.c. Metals: Ni, Cu, Ag, Al, Pb and $\gamma$-Fe           20
5   The b.c.c. Transition Metals: W, V, Cr, Nb, Mo, Ta, and $\alpha$-Fe   30
6   The Hexagonal Metals: Zn, Cd, Mg and Ti                         43
7   Non-ferrous Alloys: Nichromes, T-D Nickels and Nimonics         53
8   Pure Iron and Ferrous Alloys                                    60
9   The Covalent Elements, Si and Ge                                71
10  The Alkali Halides: NaCl and LiF                                75
11  The Transition-metal Carbides: ZrC and TiC                      80
12  Oxides with the Rock-salt Structure: MgO, CoO and FeO           84
13  Oxides with the Fluorite Structure: $UO_2$ and $ThO_2$          93
14  Oxides with the $\alpha$-Alumina Structure: $Al_2O_3$, $Cr_2O_3$ and $Fe_2O_3$   98
15  Olivines and Spinels: $Mg_2SiO_4$ and $MgAl_2O_4$              105
16  Ice, $H_2O$                                                    111
17  Further Refinements: Transient Behaviour; Very High and Very Low Strain Rates; High Pressure   117
18  Scaling Laws and Isomechanical Groups                          133
19  Applications of Deformation-mechanism Maps                     141

Materials Index                                                    163
Subject Index                                                      165

# DEFORMATION MECHANISMS AND DEFORMATION-MECHANISM MAPS

CRYSTALLINE solids deform plastically by a number of alternative, often competing, mechanisms. This book describes the mechanisms, and the construction of maps which show the field of stress, temperature and strain-rate over which each is dominant. It contains maps for more than 40 pure metals, alloys and ceramics. They are constructed from experimental data, fitted to model-based rate-equations which describe the mechanisms. Throughout, we have assumed that fracture is suppressed, if necessary, by applying a sufficiently large hydrostatic confining pressure.

The first part of the book (Chapters 1–3) describes deformation mechanisms and the construction of deformation-mechanism maps. The second part (Chapters 4–16) presents, with extensive documentation, maps for pure metals, ferrous and non-ferrous alloys, covalent elements, alkali halides, carbides, and a large number of oxides. The final section (Chapters 17–19) describes further developments (including transient behaviour, the influence of pressure, behaviour at very low and very high strain rates) and the problem of scaling laws; and it illustrates the use of the maps by a number of simple case studies.

The catalogue of maps given here is, inevitably, incomplete. But the division of materials into iso-mechanical groups (Chapter 18) helps to give information about materials not analysed here. And the method of constructing maps (Chapter 3) is now a well-established one which the reader may wish to apply to new materials for himself.

## 1.1 ATOMIC PROCESSES AND DEFORMATION MECHANISMS

Plastic flow is a *kinetic process*. Although it is often convenient to think of a polycrystalline solid as having a well defined *yield strength*, below which it does not flow and above which flow is rapid, this is true only at absolute zero. In general, the strength of the solid depends on both strain and strain-rate, and on temperature. It is determined by the kinetics of the processes occurring on the atomic scale:

the glide-motion of dislocation lines; their coupled glide and climb; the diffusive flow of individual atoms; the relative displacement of grains by grain boundary sliding (involving diffusion and defect-motion in the boundaries); mechanical twinning (by the motion of twinning dislocations) and so forth. These are the underlying atomistic processes which cause flow. But it is more convenient to describe polycrystal plasticity in terms of the mechanisms to which the atomistic processes contribute. We therefore consider the following *deformation mechanisms*, divided into five groups.

(1) *Collapse at the ideal strength*—(flow when the ideal shear strength is exceeded).
(2) *Low-temperature plasticity by dislocation glide* —(a) limited by a lattice resistance (or Peierls' stress); (b) limited by discrete obstacles; (c) limited by phonon or other drags; and (d) influenced by adiabatic heating.
(3) *Low-temperature plasticity by twinning.*
(4) *Power-law creep by dislocation glide, or glide-plus-climb*—(a) limited by glide processes; (b) limited by lattice-diffusion controlled climb ("high-temperature creep"); (c) limited by core-diffusion controlled climb ("low-temperature creep"); (d) power-law breakdown, (the transition from climb-plus-glide to glide alone); (e) Harper–Dorn creep; (f) creep accompanied by dynamic recrystallization.
(5) *Diffusional Flow*—(a) limited by lattice diffusion ("Nabarro–Herring creep"); (b) limited by grain boundary diffusion ("Coble creep"); and (c) interface-reaction controlled diffusional flow.

The mechanisms may superimpose in complicated ways. Certain *other mechanisms* (such as super-plastic flow) appear to be examples of such combinations.

## 1.2 RATE-EQUATIONS

Plastic flow of fully-dense solids is caused by the shearing, or deviatoric part of the stress field, $\sigma_s$.

In terms of the principal stresses $\sigma_1$, $\sigma_2$ and $\sigma_3$:

$$\sigma_s = (\tfrac{1}{6}((\sigma_1 - \sigma_2)^2 + (\sigma_2 - \sigma_3)^2 + (\sigma_3 - \sigma_1)^2))^{\frac{1}{2}}$$
(1.1)

or in terms of the stress tensor $\sigma_{ij}$:

$$\sigma_s = (\tfrac{1}{2} s_{ij} s_{ij})^{\frac{1}{2}}$$
(1.2)

where
$$s_{ij} = \sigma_{ij} - \tfrac{1}{3}\delta_{ij}\sigma_{kk}$$

(Very large hydrostatic pressures influence plastic flow by changing the material properties in the way described in Chapter 17, Section 17.4, but the flow is still driven by the shear stress $\sigma_s$.)

This shear stress exerts forces on the defects—the dislocations, vacancies, etc.—in the solid, causing them to move. The defects are the *carriers* of deformation, much as an electron or an ion is a carrier of charge. Just as the electric current depends on the density and velocity of the charge carriers, the shear strain-rate, $\dot{\gamma}$, reflects the density and velocity of deformation carriers. In terms of the principal strain-rates $\dot{\varepsilon}_1$, $\dot{\varepsilon}_2$ and $\dot{\varepsilon}_3$, this shear strain-rate is:

$$\dot{\gamma} = (\tfrac{2}{3}((\dot{\varepsilon}_1 - \dot{\varepsilon}_2)^2 + (\dot{\varepsilon}_2 - \dot{\varepsilon}_3)^2 + (\dot{\varepsilon}_3 - \dot{\varepsilon}_1)^2))^{\frac{1}{2}}$$
(1.3)

or, in terms of the strain-rate tensor $\dot{\varepsilon}_{ij}$:

$$\dot{\gamma} = (2\,\dot{\varepsilon}_{ij}\dot{\varepsilon}_{ij})^{\frac{1}{2}}$$
(1.4)

For simple tension, $\sigma_s$ and $\dot{\gamma}$ are related to the tensile stress $\sigma_1$ and strain-rate $\dot{\varepsilon}_1$ by:

$$\left. \begin{array}{l} \sigma_s = \sigma_1/\sqrt{3} \\ \dot{\gamma} = \sqrt{3}\,\dot{\varepsilon}_1 \end{array} \right\}$$
(1.5)

The *macroscopic variables* of plastic deformation are the stress $\sigma_s$, temperature $T$, strain-rate $\dot{\gamma}$ and the strain $\gamma$ or time $t$. If stress and temperature are prescribed (the independent variables), then the consequent strain-rate and strain, typically, have the forms shown in Fig. 1.1a. At low temperatures ($\sim 0.1\ T_M$, where $T_M$ is the melting point) the material work-hardens until the flow strength just equals the applied stress. In doing so, its structure changes: the dislocation density (a *microscopic*, or *state variable*) increases, obstructing further dislocation motion and the strain-rate falls to zero, and the strain tends asymptotically to a fixed value. If, instead, $T$ and $\dot{\gamma}$ are prescribed (Fig. 1.1b), the stress rises as the dislocation density rises. But for a given set of values of this and the other state variables $S_i$ (dislocation density and arrangement, cell size, grain size, precipitate size and spacing, and so forth) the strength is determined by $T$ and $\dot{\gamma}$, or (alternatively), the strain-rate is determined by $\sigma_s$ and $T$.

At higher temperatures ($\sim 0.5\ T_M$), polycrystalline solids creep (Fig. 1.1, centre). After a transient during which the state variables change, a steady

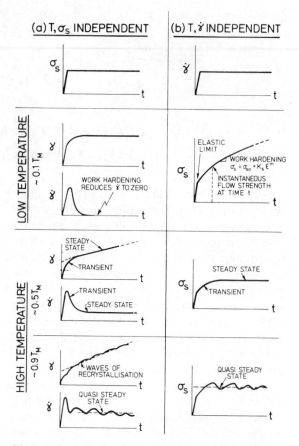

Fig. 1.1. The way in which $\sigma_s$, $T$, $\dot{\gamma}$ and $\gamma$ are related for materials (a) when $\sigma_s$ and $T$ are prescribed and (b) when $\dot{\gamma}$ and $T$ are prescribed, for low temperatures (top), high temperatures (middle) and very high temperatures (bottom).

state may be reached in which the solid continues to deform with no further significant change in $S_i$. Their values depend on the stress, temperature and strain-rate, and a relationship then exists between these three macroscopic variables.

At very high temperatures ($\sim 0.9\ T_M$) the state variables, instead of tending to steady values, may oscillate (because of dynamic recrystallization, for instance: Fig. 1.1, bottom). Often, they oscillate about more or less steady values; then it is possible to define a quasi-steady state, and once more, stress, temperature and strain-rate are (approximately) related.

Obviously, either stress or strain-rate can be treated as the independent variable. In many engineering applications—pressure vessels, for instance—loads (and thus stresses) are prescribed; in others—metal-working operations, for example—it is the strain-rate which is given. To simplify the following discussion, we shall choose the strain-rate $\dot{\gamma}$ as the independent variable. Then each mechanism of deformation can be described by a *rate equation* which relates $\dot{\gamma}$ to the stress $\sigma_s$, the temperature $T$, and to the structure of the material at that instant:

$$\dot{\gamma} = f(\sigma_s, T, S_i, P_j) \qquad (1.6)$$

As already stated, the set of $i$ quantities $S_i$ are the *state variables* which describe the current microstructural state of the materials. The set of $j$ quantities $P_j$ are the *material properties*: lattice parameter, atomic volume, bond energies, moduli, diffusion constants, etc.; these can be regarded as constant except when the plastic properties of different materials are to be compared (Chapter 18).

The state variables $S_i$ generally change as deformation progresses. A second set of equations describes their rate of change, one for each state variable:

$$\frac{dS_i}{dt} = g_i(\sigma_s, T, S_i, P_j) \qquad (1.7)$$

where $t$ is time.

The individual components of strain-rate are recovered from eqn. (1.6) by using the associated flow rule:

$$\frac{\dot{\varepsilon}_1}{\sigma_1 - \frac{1}{2}(\sigma_2 + \sigma_3)} = \frac{\dot{\varepsilon}_2}{\sigma_2 - \frac{1}{2}(\sigma_1 - \sigma_3)} =$$
$$\frac{\dot{\varepsilon}_3}{\sigma_3 - \frac{1}{2}(\sigma_1 - \sigma_2)} = C \qquad (1.8)$$

or, in terms of the stress and strain-rate tensors:

$$\frac{\dot{\varepsilon}_{ij}}{3s_{ij}} = C \qquad (1.9)$$

where $C$ is a constant.

The coupled set of equations (1.6) and (1.7) are the *constitutive law* for a mechanism. They can be integrated over time to give the strain after any loading history. But although we have satisfactory models for the rate-equation (eqn. (1.6)) we do not, at present, understand the evolution of structure with strain or time sufficiently well to formulate expressions for the others (those for $dS_i/dt$). To proceed further, we must make simplifying assumptions about the structure.

Two alternative assumptions are used here. The first, and simplest, is the assumption of *constant structure*:

$$S_i = S_i^0 \qquad (1.10)$$

Then the rate-equation for $\dot{\gamma}$ completely describes plasticity. The alternative assumption is that of *steady state*:

$$\frac{dS_i}{dt} = 0 \qquad (1.11)$$

Then the internal variables (dislocation density and

arrangement, grain size, etc.) no longer appear explicitly in the rate-equations because they are determined by the external variables of stress and temperature. Using eqn. (1.7) we can solve for $S_1$, $S_2$, etc., in terms of $\sigma_s$ and $T$, again obtaining an explicit rate-equation for $\dot{\gamma}$.

Either simplification reduces the constitutive law to a single equation:

$$\dot{\gamma} = f(\sigma_s, T) \qquad (1.12)$$

since, for a given material, the properties $P_j$ are constant and the state variables are either constant or determined by $\sigma_s$ and $T$. In Chapter 2 we assemble constitutive laws, in the form of eqn. (1.12), for each of the mechanisms of deformation. At low temperatures a steady state is rarely achieved, so for the dislocation-glide mechanisms we have used a constant structure formulation: the equations describe flow at a given structure and state of work-hardening. But at high temperatures, deforming materials quickly approach a steady state, and the equations we have used are appropriate for this steady behaviour. Non-steady or transient behaviour is discussed in Chapter 17, Section 17.1; and ways of normalizing the constitutive laws to include change in the material properties $P_j$ are discussed in Chapter 18.

## 1.3 DEFORMATION-MECHANISM MAPS

It is useful to have a way of summarizing, for a given polycrystalline solid, information about the range of dominance of each of the mechanisms of plasticity, and the rates of flow they produce. One way of doing this (Ashby, 1972; Frost and Ashby, 1973; Frost, 1974) is shown in Fig. 1.2. It is a diagram with axes of normalized stress $\sigma_s/\mu$ and temperature, $T/T_M$ (where $\mu$ is the shear modulus and $T_M$ the melting temperature). It is divided into fields which show the regions of stress and temperature over which each of the deformation mechanisms is dominant. Superimposed on the fields are contours of constant strain-rate: these show the net strain-rate (due to an appropriate superposition of all the mechanisms) that a given combination of stress and temperature will produce. The map displays the relationship between the three macroscopic variables: stress $\sigma_s$, temperature $T$ and strain-rate $\dot{\gamma}$. If any pair of these variables are specified, the map can be used to determine the third.

There are, of course, other ways of presenting the same information. One is shown in Fig. 1.3: the axes are shear strain-rate and (normalized) shear stress; the contours are those of temperature. Maps

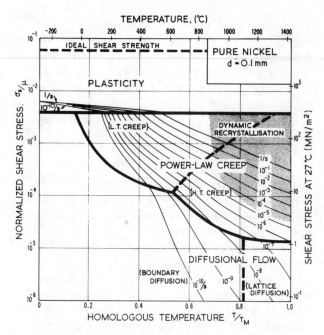

Fig. 1.2. A stress/temperature map for nominally pure nickel with a grain size of 0·1 mm. The equations and data used to construct it are described in Chapters 2 and 4.

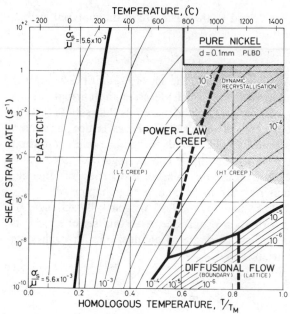

Fig. 1.4. A strain-rate/temperature map for nominally pure nickel, using the same data as Fig. 1.2.

like these are particularly useful in fitting isothermal data to the rate-equations, but because they do not extend to 0 K they contain less information than the first kind of map.

A third type of map is obviously possible: one with axes of strain rate and temperature (or reciprocal temperature) with contours of constant stress (Figs. 1.4 and 1.5). We have used such plots as a way of fitting constant-stress data to the rate-

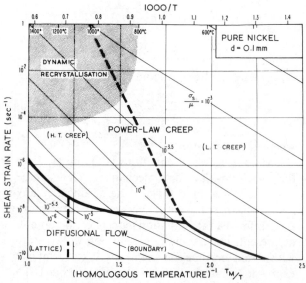

Fig. 1.5. A strain-rate/reciprocal temperature map for nominally pure nickel, using the same data as Fig. 1.2.

equations of Chapter 2, and for examining behaviour at very high strain-rates (Chapter 17, Section 17.2).

Finally, it is possible to present maps with a structure parameter ($S_1$, $S_2$, etc.) such as dislocation density or grain size as one of the axes (see, for example, Mohamed and Langdon, 1974). Occasionally this is useful, but in general it is best to avoid the use of such microscopic structure variables as axes of maps because they cannot be externally

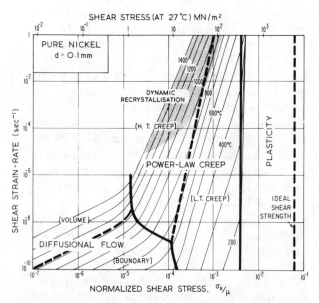

Fig. 1.3. A strain-rate/stress map for nominally pure nickel, using the same data as Fig. 1.2.

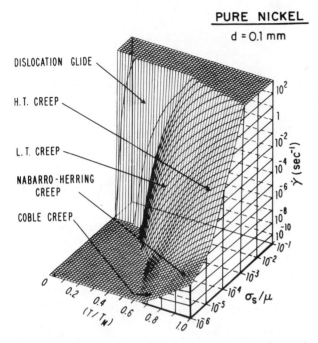

**PURE NICKEL**

d = 0.1 mm

DISLOCATION GLIDE

H.T. CREEP

L.T. CREEP

NABARRO-HERRING CREEP

COBLE CREEP

Fig. 1.6. A three-dimensional map for nominally pure nickel, using the same data as Figs. 1.2, 1.3 and 1.4.

controlled or easily or accurately measured. It is usually better to construct maps either for given, fixed values of these parameters, or for values determined by the assumption of a steady state.

Three of the maps shown above are orthogonal sections through the same three-dimensional space, shown in Fig. 1.6. In general, we have not found such figures useful, and throughout the rest of this book we restrict ourselves to two-dimensional maps of the kind shown in Figs. 1.2, 1.3 and, occasionally, 1.4.

## 1.4   A WARNING

One must be careful not to attribute too much precision to the diagrams. Although they are the best we can do at present, they are far from perfect or complete. *Both the equations in the following sections, and the maps constructed from them, must be regarded as a first approximation only. The maps are no better (and no worse) than the equations and data used to construct them.*

### References for Chapter 1

Ashby, M. F. (1972) *Acta Met.* **20**, 887.
Frost, H. J. and Ashby, M. F. (1973) Division of Applied Physics, Harvard University Report: "A Second Report on Deformation-Mechanism Maps".
Frost, H. J. (1974) Ph.D. thesis, Harvard University.
Mohamed, F. A. and Langdon, T. G. (1974) *Met. Trans.* **5**, 2339.

# CHAPTER 2

# RATE-EQUATIONS

IN THIS chapter we develop, with a brief explanation, the rate-equations used later to construct the maps. We have tried to select, for each mechanism, the simplest equation which is based on a physically sound microscopic model, or family of models. Frequently this equation contains coefficients or exponents for which only *bounds* are known; the model is too imprecise, or the family of models too broad, to predict exact values. Theory gives the form of the equation; but experimental data are necessary to set the constants which enter it. This approach of "model-based phenomenology" is a fruitful one when dealing with phenomena like plasticity, which are too complicated to model exactly. One particular advantage is that an equation obtained in this way can with justification (since it is based on a physical model) be extrapolated beyond the range of the data, whereas a purely empirical equation cannot.

In accordance with this approach we have aimed at a precision which corresponds with the general accuracy of experiments, which is about $\pm 10\%$ for the yield strength (at given $T$ and $\dot{\gamma}$, and state of work-hardening), or by a factor of two for strain-rate at given $\sigma_s$ and $T$). For this reason we have included the temperature-dependence of the elastic moduli but have ignored that of the atomic volume and the Burgers' vector.

Pressure is not treated as a variable in this chapter. The influence of pressure on each mechanism is discussed, with data, in Chapter 17, Section 17.4.

The equations used to construct the maps of later chapters are indicated in a box. Symbols are defined where they first appear in the text, and in the table on pages ix to xi.

## 2.1 ELASTIC COLLAPSE

The *ideal shear strength* defines a stress level above which deformation of a perfect crystal (or of one in which all defects are pinned) ceases to be elastic and becomes catastrophic: the crystal structure becomes mechanically unstable. The instability condition, and hence the ideal strength at 0 K, can be calculated from the crystal structure and an inter-atomic force law by simple statics (Tyson, 1966; Kelly, 1966). Above 0 K the problem becomes a kinetic one: that of calculating the frequency at which dislocation loops nucleate and expand in an initially defect-free crystal. We have ignored the kinetic problem and assumed the temperature-dependence of the ideal strength to be the same as that of the shear modulus, $\mu$, of the polycrystal. Plastic flow by collapse of the crystal structure can then be described by:

$$
\begin{aligned}
\dot{\gamma}_1 &= \infty \quad \text{when} \quad \sigma_s \geqslant \alpha\mu \\
\dot{\gamma}_1 &= 0 \quad \text{when} \quad \sigma_s < \alpha\mu
\end{aligned}
\tag{2.1}
$$

Computations of $\alpha$ lead to values between 0·05 and 0·1, depending on the crystal structure and the force law, and on the instability criterion. For the f.c.c. metals we have used $\alpha = 0·06$, from the computer calculations of Tyson (1966) based on a Lennard–Jones potential. For b.c.c. metals we have used $\alpha = 0·1$, from the analytical calculation of MacKenzie (1959). For all other materials we have used $\alpha = 0·1$.

## 2.2 LOW-TEMPERATURE PLASTICITY: DISLOCATION GLIDE

Below the ideal shear strength, flow by the conservative, or glide, motion of dislocations is possible —provided (since we are here concerned with polycrystals) an adequate number of independent slip systems is available (Figs. 2.1 and 2.2). This motion is almost always *obstacle-limited*: it is the interaction of potentially mobile dislocations with other dislocations, with solute or precipitates, with grain boundaries, or with the periodic friction of the lattice itself which determines the rate of flow, and (at a given rate) the yield strength. The yield strength of many polycrystalline materials does not depend strongly on the rate of straining—a fact which has led to models for yielding which ignore the effect of strain-rate (and of temperature) entirely. But

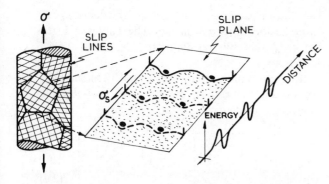

Fig. 2.1. Low-temperature plasticity limited by discrete obstacles. The strain-rate is determined by the kinetics of obstacle cutting.

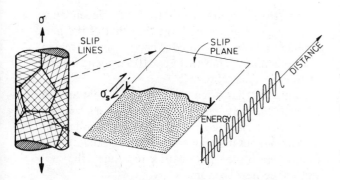

Fig. 2.2. Low-temperature plasticity limited by a lattice resistance. The strain-rate is determined by the kinetics of kink nucleation and propagation.

this is misleading: dislocation glide is a kinetic process. A density $\rho_m$ of mobile dislocations, moving through a field of obstacles with an average velocity $\bar{v}$ determined almost entirely by their waiting time at obstacles, produces a strain-rate (Orowan, 1940) of:

$$\dot{\gamma} = \rho_m b \bar{v} \qquad (2.2)$$

where $b$ is the magnitude of the Burgers' vector of the dislocation. At steady state, $\rho_m$ is a function of stress and temperature only. The simplest function, and one consistent with both theory (Argon, 1970) and experiment is:

$$\rho_m = \alpha \left( \frac{\sigma_s}{\mu b} \right)^2 \qquad (2.3)$$

where $\alpha$ is a constant of order unity. The velocity $\bar{v}$ depends on the force $F = \sigma_s b$ acting, per unit length, on the dislocation, and on its *mobility*, $M$:

$$\bar{v} = MF \qquad (2.4)$$

The kinetic problem is to calculate $M$, and thus $\bar{v}$. In the most interesting range of stress, $M$ is determined by the rate at which dislocation segments are thermally activated through, or round, obstacles. In developing rate-equations for low-temperature plasticity (for reviews, see Evans and Rawlings,

1969; Kocks *et al.*, 1975; de Meester *et al.*, 1973) one immediately encounters a difficulty: the velocity is always an exponential function of stress, but the details of the exponent depend on the shape and nature of the obstacles. At first sight there are as many rate-equations as there are types of obstacle. But on closer examination, obstacles fall into two broad classes: discrete obstacles which are bypassed individually by a moving dislocation (strong dispersoids or precipitates, for example) or cut by it (such as forest dislocations or weak precipitates); and extended, diffuse barriers to dislocation motion which are overcome collectively (a lattice-friction, or a concentrated solid solution). The approach we have used is to select the rate-equation which most nearly describes a given class of obstacles, and to treat certain of the parameters which appear in it as adjustable, to be matched with experiment. This utilizes the most that model-based theory has to offer, while still ensuring an accurate description of experimental data.

## Plasticity limited by discrete obstacles

The velocity of dislocations in a polycrystal is frequently determined by the strength and density of the *discrete obstacles* it contains (Fig. 2.1). If the Gibbs free-energy of activation for the cutting or by-passing of an obstacle is $\Delta G(\sigma_s)$, the mean velocity of a dislocation segment, $\bar{v}$, is given by the kinetic equation (see reviews listed above):

$$\bar{v} = \beta b v \, \exp \, -\frac{\Delta G(\sigma_s)}{kT} \qquad (2.5)$$

where $\beta$ is a dimensionless constant, $b$ is the magnitude of the Burgers' vector, and $v$ is a frequency.

The quantity $\Delta G(\sigma_s)$ depends on the distribution of obstacles and on the pattern of internal stress, or "shape", which characterizes one of them. A regular array of box-shaped obstacles (each one viewed as a circular patch of constant, adverse, internal stress) leads to the simple result:

$$\Delta G(\sigma_s) = \Delta F \left( 1 - \frac{\sigma_s}{\hat{\tau}} \right) \qquad (2.6)$$

where $\Delta F$ is the total free energy (the *activation energy*) required to overcome the obstacle without aid from external stress. The material property $\hat{\tau}$ is the stress which reduced $\Delta G$ to zero, forcing the dislocation through the obstacle with no help from thermal energy. It can be thought of as the flow strength of the solid at 0 K.

But obstacles are seldom box-shaped and regularly spaced. If other obstacle shapes are considered, and allowance is made for a random, rather than a regular, distribution, all the results can be described by the general equation (Kocks *et al.*, 1975):

$$\Delta G(\sigma_s) = \Delta F \left[ 1 - \left( \frac{\sigma_s}{\hat{\tau}} \right)^p \right]^q \qquad (2.7)$$

The quantities $p$, $q$ and $\Delta F$ are bounded: all models lead to values of

$$0 \leqslant p \leqslant 1$$
$$1 \leqslant q \leqslant 2 \qquad (2.8)$$

The importance of $p$ and $q$ depends on the magnitude of $\Delta F$. When $\Delta F$ is large, their influence is small, and the choice is unimportant; for discrete obstacles, we use $p = q = 1$. But when $\Delta F$ is small, the choice of $p$ and $q$ becomes more critical; for diffuse obstacles we use values derived by fitting data to eqn. (2.7) in the way described in the next sub-section.

The strain rate sensitivity of the strength is determined by the activation energy $\Delta F$—it characterizes the strength of a single obstacle. It is helpful to class obstacles by their strength, as shown in Table 2.1; for strong obstacles $\Delta F$ is about $2 \, \mu b^3$; for weak, as low as $0.05 \, \mu b^3$. A value of $\Delta F$ is listed in the Tables of Data for each of the materials analysed in this book. When dealing with pure metals and ceramics in the work-hardened state, we have used $\Delta F = 0.5 \, \mu b^3$.

**TABLE 2.1    Characteristics of obstacles**

| Obstacle strength | $\Delta F$ | $\hat{\tau}$ | Example |
|---|---|---|---|
| Strong | $2 \, \mu b^3$ | $> \dfrac{\mu b}{\ell}$ | Dispersions; large or strong precipitates (spacing $\ell$) |
| Medium | $0.2$–$1.0 \, \mu b^3$ | $\approx \dfrac{\mu b}{\ell}$ | Forest dislocations, radiation damage; small or weak precipitates (spacing $\ell$) |
| Weak | $< 0.2 \, \mu b^3$ | $\ll \dfrac{\mu b}{\ell}$ | Lattice resistance; solution hardening (solute spacing $\ell$) |

The quantity $\hat{\tau}$ is the "athermal flow strength"—the shear strength in the absence of thermal energy. It reflects not only the strength but also the density and arrangement of the obstacles. For widely-spaced, discrete obstacles, $\hat{\tau}$ is proportional to $\mu b / \ell$, where $\ell$ is the obstacle spacing; the constant of proportionality depends on their strength and distribution (Table 2.1). For pure metals strengthened by work-hardening we have simply used $\hat{\tau} = \mu b / \ell$

(which can be expressed in terms of the density $\rho$, of forest dislocations: $\hat{\tau} = \mu b \sqrt{\rho}$). The Tables of Data list $\hat{\tau}/\mu_0$, thereby specifying the degree of work-hardening.

If we now combine eqns (2.3), (2.4), (2.5) and (2.7) we obtain the *rate-equation for discrete-obstacle controlled plasticity*:

where

$$\begin{aligned} \dot{\gamma}_2 &= \dot{\gamma}_0 \exp\left[ -\frac{\Delta F}{kT} \left( 1 - \frac{\sigma_s}{\hat{\tau}} \right) \right] \\ \dot{\gamma}_0 &= \frac{\alpha}{b} \left( \frac{\sigma_s}{\mu} \right)^2 \beta b v \end{aligned} \qquad (2.9)$$

When $\Delta F$ is large (as here), the stress dependence of the exponential is so large that that of the pre-exponential can be ignored. Then $\dot{\gamma}_0$ can be treated as a constant. We have set $\dot{\gamma}_0 = 10^6/\text{s}$, giving a good fit to experimental data.

Eqn. (2.9) has been used in later chapters to describe plasticity when the strength is determined by work-hardening or by a strong precipitate or dispersion; but its influence is often masked by that of a diffuse obstacle: the lattice resistance, the subject of the next section.

## Plasticity limited by a lattice resistance

The velocity of a dislocation in most polycrystalline solids is limited by an additional sort of barrier: that due to its interaction with the atomic structure itself (Fig. 2.2). This *Peierls force* or *lattice resistance* reflects the fact that the energy of the dislocation fluctuates with position; the amplitude and wavelength of the fluctuations are determined by the strength and separation of the interatomic or intermolecular bonds. The crystal lattice presents an array of long, straight barriers to the motion of the dislocation; it advances by throwing forward kink pairs (with help from the applied stress and thermal energy) which subsequently spread apart (for reviews see Guyot and Dorn, 1967; Kocks *et al.*, 1975).

It is usually the nucleation-rate of kink-pairs which limits dislocation velocity. The Gibbs free energy of activation for this event depends on the detailed way in which the dislocation energy fluctuates with distance, and on the applied stress and temperature. Like those for discrete obstacles, the activation energies for all reasonable shapes of lattice resistance form a family described (as before) by:

$$\Delta G(\sigma_s) = \Delta F_p \left[ 1 - \left( \frac{\sigma_s}{\hat{\tau}_p} \right)^p \right]^q \qquad (2.10)$$

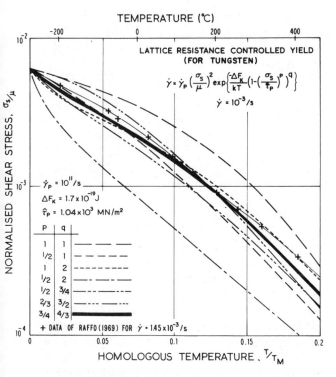

Fig. 2.3. Predicted contours for $\dot{\gamma} = 10^{-3}$/s for various formulations of the lattice-resistance controlled glide equation (eqn. (2.12)), compared with data for tungsten.

where $\Delta F_p$ is the Helmholtz free energy of an isolated pair of kinks and $\hat{\tau}_p$ is, to a sufficient approximation, the flow stress at 0 K*. An analysis of data (of which Fig. 2.3 is an example) allows $p$ and $q$ to be determined. We find the best choice to be:

$$\Delta G(\sigma_s) = \Delta F_p \left[ 1 - \left( \frac{\sigma_s}{\hat{\tau}_p} \right)^{3/4} \right]^{4/3} \quad (2.11)$$

Combining this with eqns. (2.2), (2.3) and (2.5) leads to a model-based *rate-equation for plasticity limited by a lattice resistance*:

$$\dot{\gamma}_3 = \dot{\gamma}_p \left( \frac{\sigma_s}{\mu} \right)^2 \cdot$$

$$\exp - \left\{ \frac{\Delta F_p}{kT} [1 - (\sigma_s/\hat{\tau}_p)^{3/4}]^{4/3} \right\} \quad (2.12)$$

The influence of the choice of $p$ and $q$ is illustrated in Fig. 2.3. It shows how the measured strength of tungsten (Raffo, 1969) varies with temperature, compared with the predictions of eqn. (2.12), with

* The equation of Guyot and Dorn (1967) used in the earlier report (Ashby, 1972a) will be recognized as the special case of $p = 1$, $q = 2$. This equation was misprinted as $(1 - \sigma_p/\sigma_s)^2$ instead of $(1 - \sigma_s/\sigma_p)^2$ in the paper by Ashby (1972a).

various combinations of $p$ and $q$. It justifies the choice of 3/4 and 4/3, although it can be seen that certain other combinations are only slightly less good.

The pre-exponential of eqn. (2.12) contains a term in $\sigma_s^2$ arising from the variation of mobile dislocation density with stress (eqn. (2.3)), which must, when $\Delta F_p$ is small (as here), be retained. In using eqn. (2.12) to describe the low-temperature strength of the b.c.c. metals and of ceramics, we have used $\dot{\gamma}_p = 10^{11}$/s—a mean value obtained by fitting data to eqn. (2.12). If data allow it, $\dot{\gamma}_p$ should be determined from experiment; but its value is not nearly as critical as those of $\Delta F_p$ and $\hat{\tau}_p$, and data of sufficient precision to justify changing it are rarely available. With this value of $\dot{\gamma}_p$, the quantities $\Delta F_p$ (typically 0.1 $\mu b^3$) and $\hat{\tau}_p$ (typically $10^{-2}$ $\mu$) are obtained by fitting eqn. (2.12) to experimental data for the material in question. The results of doing this are listed in the Tables of Data in following chapters.

It should be noted that the values of $\hat{\tau}_p$ (and of $\hat{\tau}$) for single crystals and polycrystals differ. The difference is a Taylor factor: it depends on the crystal structure and on the slip systems which are activated when the polycrystal deforms. For f.c.c. metals the appropriate Taylor factor $M_s$ is 1·77; for the b.c.c. metals it is 1·67 (Kocks, 1970). (They may be more familiar to the reader as the Taylor factors $M = 3·06$ and 2·9, respectively, relating the critical resolved shear strength to tensile strength for f.c.c. and b.c.c. metals. Since $\hat{\tau}_p$ and $\hat{\tau}$ are polycrystal shear strengths, the factors we use are less by the factor $\sqrt{3}$.) For less symmetrical crystals the polycrystal shear strength is again a proper average of the strengths of the active slip systems. These often differ markedly and it is reasonable to identify $\hat{\tau}_p$ for the polycrystal with that of the hardest of the slip systems (that is, we take $M_s = 1$, but use $\hat{\tau}_p$ for the hard system). In calculating $\hat{\tau}_p$ or $\hat{\tau}$ from single crystal data, we have applied the appropriate Taylor factor.

## Plasticity limited by phonon or electron drags

Under conditions of explosive or shock loading, and in certain metal-forming and machining operations, the strain-rate can be large ($> 10^2$/s). Then the interaction of a moving dislocation with phonons or electrons can limit its velocity. The strength of the interaction is measured by the *drag coefficient*, $B$ (the reciprocal of the mobility $M$ of eqn. (2.4)):

$$v = \frac{\sigma_s b}{B}$$

Values of $B$ lie, typically, between $10^{-5}$ and $10^{-4}$ Ns/m$^2$ (Klahn et al., 1970). Combining this with eqn. (2.4) leads to the *rate-equation for drag limited glide*:

$$\dot{\gamma} = \frac{c\rho_m \mu}{B}\left(\frac{\sigma_s}{\mu}\right) \qquad (2.13)$$

where $c$ is a constant which includes the appropriate Taylor factor.

To use this result it is necessary to know how $\rho_m/B$ varies with stress and temperature. For solute-drag, $\rho_m$ is well described by eqn. (2.3); but at the high strain-rates at which phonon drag dominates, $\rho_m$ tends to a constant limiting value (Kumar et al., 1968; Kumar and Kumble, 1969), so that $\rho_m/B$ is almost independent of stress and temperature. Then the strain-rate depends linearly on stress, and the deformation becomes (roughly) Newtonian-viscous:

$$\boxed{\dot{\gamma}_{PD} = C\left(\frac{\sigma_s}{\mu}\right)} \qquad (2.14)$$

where $C(\text{s}^{-1})$ is a constant. Using the data of Kumar et al. (1968) and Kumar and Kumble (1969), we estimate $C \approx 5 \times 10^6$/s.

Drag-controlled plasticity does not appear on most of the maps in this book, which are truncated at strain-rate (1/s) below that at which it becomes important. But when high strain-rates are considered (Chapter 17, Section 17.2), it appears as a dominant mechanism.

**The influence of alloying on dislocation glide**

A *solute* introduces a friction-like resistance to slip. It is caused by the interaction of the moving dislocations with stationary weak obstacles: single solute atoms in very dilute solutions, local concentration fluctuations in solutions which are more concentrated. Their effect can be described by eqn. (2.9) with a larger value of $\hat{\tau}$ and a smaller value of $\Delta F$ (Table 2.1). This effect is superimposed on that of work-hardening. In presenting maps for alloys (Chapter 7) we have often used data for heavily deformed solid solutions; then solution strengthening is masked by forest hardening. This allows us to use eqn. (2.9) unchanged.

A *dispersion of strong particles of a second phase* blocks the glide motion of the dislocations. The particles in materials like SAP (aluminium containing $Al_2O_3$), T-D Nickel (nickel containing $ThO_2$—

Chapter 7), or low-alloy steels (steels containing dispersions of carbides—Chapter 8) are strong and stable. A gliding dislocation can move only by bowing between and by-passing them, giving a contribution to the flow strength which scales as the reciprocal of the particle spacing, and which has a very large activation energy (Table 2.1). Detailed calculations of this *Orowan strength* (e.g. Kocks et al., 1975) lead to eqn. (2.9) with $\hat{\tau} \approx 2\mu b/\ell$ and $\Delta F \gqq 2\mu b^3$. This activation energy is so large that it leads to a flow strength which is almost athermal, although if the alloy is worked sufficiently the yield strength will regain the temperature-dependence which characterizes forest hardening (Table 2.1). We have neglected here the possibility of thermally activated cross-slip at particles (Brown and Stobbs, 1971; Hirsch and Humphreys, 1970), which can relax work-hardening and thus lower the flow strength. To a first approximation, a strong dispersion and a solid solution give additive contributions to the yield stress.

A *precipitate*, when finely dispersed, can be cut by moving dislocations. If the density of particles is high, then the flow strength is high (large $\hat{\tau}$) but strongly temperature-dependent (low $\Delta F$: Table 2.1). If the precipitate is allowed to coarsen it behaves increasingly like a dispersion of strong particles.

## 2.3  MECHANICAL TWINNING

Twinning is an important deformation mechanism at low temperatures in h.c.p. and b.c.c. metals and some ceramics. It is less important in f.c.c. metals, only occurring at very low temperature. Twinning is a variety of dislocation glide (Section 2.2) involving the motion of partial, instead of complete, dislocations. The kinetics of the process, however, often indicate that nucleation, not propagation, determines the rate of flow. When this is so, it may still be possible to describe the strain-rate by a *rate-equation for twinning*, taking the form:

$$\dot{\gamma} \approx \dot{\gamma}_t \exp\left[-\frac{\Delta F_N}{kT}\left(1 - \frac{\sigma_s}{\hat{\tau}_t(T)}\right)\right] \qquad (2.15)$$

Here $\Delta F_N$ is the activation free energy to nucleate a twin without the aid of external stress; $\dot{\gamma}_t$ is a constant with dimensions of strain-rate which includes the density of available nucleation sites and the strain produced per successful nucleation; $\hat{\tau}_t$ is the stress required to nucleate twinning in the absence of thermal activation. The temperature-dependence of $\Delta F_N$ must be included to explain the observation that the twinning stress may decrease with decreasing temperature (Bolling and Richman, 1965).

The rate-equation for twinning is so uncertain that we have not included it in computing the maps shown here. Instead, we have indicated on the data-plots where twinning is observed. The tendency of f.c.c. metals to twin increases with decreasing stacking fault energy, being greatest for silver and completely absent in aluminium. All the b.c.c. and h.c.p. metals discussed below twin at sufficiently low temperature.

## 2.4  HIGH-TEMPERATURE PLASTICITY: POWER-LAW CREEP

At high temperatures, materials show rate-dependent plasticity, or *creep*. Of course, the mechanisms described in the previous sections lead, even at low temperatures, to a flow strength which depends to some extend on strain-rate. But above $0.3\ T_M$ for pure metals, and about $0.4\ T_m$ for alloys and most ceramics, this dependence on strain-rate becomes much stronger. If it is expressed by an equation of the form

$$\dot{\gamma} \propto \left(\frac{\sigma_s}{\mu}\right)^n \tag{2.16}$$

then, in this high-temperature regime, $n$ has a value between 3 and 10, and (because of this) the regime is called *power-law creep*. In this section we consider steady-state creep only; primary creep is discussed in Chapter 17, Section 17.1.

### Power-law creep by glide alone

If the activation energy $\Delta F$ in eqn. (2.9) or $\Delta F_p$ in eqn. (2.12) is small, then thermally-activated glide can lead to creep-like behaviour above $0.3\ T_M$. The activation of dislocation segments over obstacles leads to a drift velocity which, in the limit of very small $\Delta F$, approaches a linear dependence on stress. This, coupled with the $\sigma_s^2$ stress-dependence of the mobile dislocation density (eqn. (2.3)) leads to a behaviour which resembles power-law creep with $n \simeq 3$. This *glide-controlled creep* may be important in ice (Chapter 16) and in certain ceramics (Chapters 9 to 15), and perhaps, too, in metals below $0.5\ T_M$.

The maps of later chapters include the predictions of the glide mechanisms discussed in Section 2.2; to that extent, glide-controlled creep is included in our treatment. But our approach is unsatisfactory in that creep normally occurs at or near steady state, not at constant structure—and eqns. (2.9) and (2.12) describe only this second condition. A proper treatment of glide-controlled creep must include a description of the balanced hardening-and-recovery processes which permit a steady state. Such a treatment is not yet available.

### Power-law creep by climb-plus-glide

At high temperatures, dislocations acquire a new degree of freedom: they can climb as well as glide (Fig. 2.4). If a gliding dislocation is held up by discrete obstacles, a little climb may release it, allowing it to glide to the next set of obstacles where the process is repeated. The glide step is responsible for almost all of the strain, although its average velocity is determined by the climb step. Mechanisms which are based on this climb-plus-glide sequence we refer to as *climb-controlled creep* (Weertman, 1956, 1960, 1963). The important feature which distinguishes these mechanisms from those of earlier sections is that the rate-controlling process, at an atomic level, is the diffusive motion of single ions or vacancies to or from the climbing dislocation, rather than the activated glide of the dislocation itself.

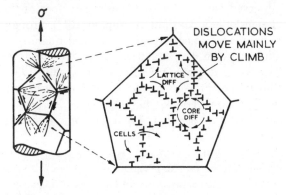

Fig. 2.4. Power-law creep involving cell-formation by climb. Power-law creep limited by glide processes alone is also possible.

Above $0.6\ T_M$ climb is generally *lattice-diffusion controlled*. The velocity $v_c$ at which an edge dislocation climbs under a local normal stress $\sigma_n$ acting parallel to its Burgers' vector is (Hirth and Lothe, 1968):

$$v_c \approx \frac{D_v \sigma_n \Omega}{bkT} \tag{2.17}$$

where $D_v$ is the lattice diffusion coefficient and $\Omega$ the atomic or ionic volume. We obtain the basic climb-controlled creep equation by supposing that $\sigma_n$ is proportional to the applied stress $\sigma_s$, and that the average velocity of the dislocation, $\bar{v}$, is proportional to the rate at which it climbs, $v_c$. Then, combining eqns. (2.2), (2.3) and (2.17) we obtain:

$$\dot{\gamma} = A_1 \frac{D_v \mu b}{kT}\left(\frac{\sigma_s}{\mu}\right)^3 \qquad (2.18)$$

where we have approximated $\Omega$ by $b^3$, and incorporated all the constants of proportionality into the dimensionless constant, $A_1$, of order unity.

Some materials obey this equation: they exhibit proper-law creep with a power of 3 and a constant $A_1$ of about 1 (see Brown and Ashby, 1980a). But they are the exceptions rather than the rule. It appears that the local normal stress, $\sigma_n$, is not necessarily proportional to $\sigma_s$ implying that dislocations may be moving in a cooperative manner which concentrates stress or that the average dislocation velocity or mobile density varies in a more complicated way than that assumed here. Over a limited range of stress, up to roughly $10^{-3}\,\mu$, experiments are well described by a modification of eqn. (2.18) (Mukherjee et al., 1969) with an exponent, $n$, which varies from 3 to about 10:

$$\dot{\gamma} = A_2 \frac{D_v \mu b}{kT}\left(\frac{\sigma_s}{\mu}\right)^n \qquad (2.19)$$

Present theoretical models for this behaviour are unsatisfactory. None can convincingly explain the observed values of $n$; and the large values of the dimensionless constant $A_2$ (up to $10^{15}$) strongly suggest that some important physical quantity is missing from the equation in its present form (Stocker and Ashby, 1973; Brown and Ashby, 1980). But it *does* provide a good description of experimental observations, and in so far as it is a generalization of eqn. (2.18), it has some basis in a physical model.

In this simple form, eqn. (2.19) is incapable of explaining certain experimental facts, notably an increase in the exponent $n$ and a drop in the activation energy for creep at lower temperatures. To do so it is necessary to assume that the transport of matter via *dislocation core diffusion* contributes significantly to the overall diffusive transport of matter, and—under certain circumstances—becomes the dominant transport mechanism (Robinson and Sherby, 1969). The contribution of core diffusion is included by defining an effective diffusion coefficient (following Hart, 1957 and Robinson and Sherby, 1969):

$$D_{\text{eff}} = D_v f_v + D_c f_c$$

where $D_c$ is the core diffusion coefficient, and $f_v$ and $f_c$ are the fractions of atom sites associated with each type of diffusion. The value of $f_v$ is essentially unity. The value of $f_c$ is determined by the dislocation density, $\rho$:

$$f_c = a_c\,\rho,$$

where $a_c$ is the cross-sectional area of the dislocation core in which fast diffusion is taking place. Measurements of the quantity $a_c D_c$ have been reviewed by Balluffi (1970): the diffusion enhancement varies with dislocation orientation (being perhaps 10 times larger for edges than for screws), and with the degree of dissociation and therefore the arrangement of the dislocations. Even the activation energy is not constant. But in general, $D_c$ is about equal to $D_b$ (the grain boundary diffusion coefficient), if $a_c$ is taken to be $2\delta^2$ (where $\delta$ is the effective boundary thickness). By using the common experimental observation* that $\rho \approx 10/b^2\ (\sigma_s/\mu)^2$ (eqn. 2.3), the effective diffusion coefficient becomes:

$$D_{\text{eff}} = D_v\left[1 + \frac{10 a_c}{b^2}(\sigma_s/\mu)^2\frac{D_c}{D_v}\right] \qquad (2.20)$$

When inserted into eqn. (2.19), this gives the *rate-equation for power-law creep*†:

$$\boxed{\dot{\gamma}_4 = \frac{A_2 D_{\text{eff}}\,\mu b}{kT}(\sigma_s/\mu)^n} \qquad (2.21)$$

Eqn. (2.21) is really two rate-equations. At high temperatures and low stresses, lattice diffusion is dominant; we have called the resulting field *high-temperature creep* ("H.T. creep"). At lower temperatures, or higher stresses, core diffusion becomes dominant, and the strain rate varies as $\sigma_s^{n+2}$ instead of $\sigma_s^n$; this field appears on the maps as *low-temperature creep* ("L.T. creep").

### Harper–Dorn creep

There is experimental evidence, at sufficiently low stresses, for a dislocation creep mechanism for which $\dot{\gamma}$ is proportional to $\sigma_s$. The effect was first noted in aluminium by Harper and Dorn (1957) and Harper et al. (1958): they observed linear-viscous creep, at stresses below $5 \times 10^{-6}\,\mu$, but at rates much higher than those possible by diffusional flow (Section 2.5). Similar behaviour has been observed in lead and tin by Mohamed et al. (1973).

---

\* The observations of Vandervoort (1970), for example, show that $\rho \approx \beta/b^2\ (\sigma_s/\mu)^2$ for tungsten in the creep regime.

† Eqn. (2.21) is equally written in terms of the tensile stress and strain-rate. Our constant $A_2$ (which relates shear stress to shear strain-rate) is related to the equivalent constant $A$ which appears in tensile forms of this equation by $A_2 = (\sqrt{3})^{n+1}\,A$. For further discussion see Chapter 4. The later tabulations of data list the *tensile* constant, $A$.

The most plausible explanation is that of climb-controlled creep under conditions such that the dislocation density does not change with stress. Mohamed *et al.* (1973) summarize data showing a constant, low dislocation density of about $10^8/m^2$ in the Harper–Dorn creep range. Given this constant density, we obtain a rate-equation by combining eqns. (2.2) and (2.17), with $\sigma_s \propto \sigma_n$, to give:

$$\dot{\gamma} = \rho_m \frac{D_v \mu \Omega}{kT}(\sigma_s/\mu) \qquad (2.22)$$

This is conveniently rewritten as:

$$\dot{\gamma}_5 = A_{HD}\frac{D_v \mu b}{kT}(\sigma_s/\mu) \qquad (2.23)$$

where $A_{HD} = \rho_m \Omega/b$ is a dimensionless constant.

Harper–Dorn creep is included in the maps for aluminium and lead of Chapter 4, using the experimental value for $A_{HD}$ of $5 \times 10^{-11}$ for aluminium (Harper *et al.*, 1958) and of $1.2 \times 10^{-9}$ for lead (Mohammed *et al.*, 1973). They are consistent with the simple theory given above if $\rho = 10^8–10^9/m^2$. The field only appears when the diffusional creep fields are suppressed by a large grain size. Harper–Dorn creep is not shown for other materials because of lack of data.

**Power-law breakdown**

At high stresses (above about $10^{-3}\mu$), the simple power-law breaks down: the measured strain-rates are greater than eqn. (2.21) predicts. The process is evidently a transition from climb-controlled to glide-controlled flow (Fig. 2.5). There have been a number of attempts to describe it in an empirical way (see, for example, Jonas *et al.*, 1969). Most lead to a rate-equation of the form:

$$\dot{\gamma} \propto \exp{(\beta' \sigma_s)} \exp{-\left(\frac{Q_{cr}}{RT}\right)} \qquad (2.24)$$

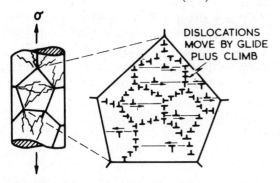

Fig. 2.5. Power-law breakdown: glide contributes increasingly to the overall strain-rate.

or the generalization of it (Sellars and Tegart, 1966; Wong and Jonas, 1968):

$$\dot{\gamma} \propto (\sinh{\beta' \sigma_s})^{n'} \exp{-\left(\frac{Q_{cr}}{RT}\right)} \qquad (2.25)$$

which at low stresses ($\beta' \sigma_s < 0.8$) reduces to a simple power-law, while at high ($\beta' \sigma_s > 1.2$) it becomes an exponential (eqn. (2.4)).

Measurements of the activation energy $Q_{cr}$ in the power-law breakdown regime often give values which exceed that of self-diffusion. This is sometimes taken to indicate that the recovery process differs from that of climb-controlled creep. Some of the difference, however, may simply reflect the temperature-dependence of the shear modulus, which has a greater effect when the stress-dependence is greater (in the exponential region). A better fit to experiment is then found with:

$$\dot{\gamma} = A\left[\sinh{\left(\frac{\alpha' \sigma_s}{\mu}\right)}\right]^{n'} \exp{-\left(\frac{Q_{cr}}{kT}\right)}$$

with $\alpha' = \beta' \mu_0$. In order to have an exact correspondence of this equation with the power-law eqn. (2.21) we propose the following *rate-equation for power-law creep and power-law breakdown*:

$$\dot{\gamma}_6 = A_2'\frac{D_{\text{eff}} \mu b}{kT}\left[\sinh{\left(\alpha'\frac{\sigma_s}{\mu}\right)}\right]^{n'} \qquad (2.26)$$

where $\qquad A_2'\alpha'^n = A_2$

and $\qquad n' = n$

Eqn. (2.26) reduces identically to the power-law creep equation (2.21) at stresses below $\sigma_s \approx \mu/\alpha'$. There are, however, certain difficulties with this formulation. The problem stems from the use of only two parameters, $n'$ and $\alpha'$, to describe three quantities: $n'$ describes the power-law; $\alpha'$ prescribes the stress level at which the power-law breaks down; and $n'\alpha'$ describes the strength of the exponential stress-dependence. Lacking any physical model, it must be considered fortuitous that any set of $n'$ and $\alpha'$ can correctly describe the behaviour over a wide range of stresses.

In spite of these reservations, we have found that eqn. (2.26) gives a good description of hot working (power-law breakdown) data for copper and aluminium. Because we retain our fit to power-law creep, the value of $n'$ is prescribed, and the only new adjustable parameter is $\alpha'$. This will be discussed further in Chapter 4, and Chapter 17, Section 17.2, where values for $\alpha'$ are given.

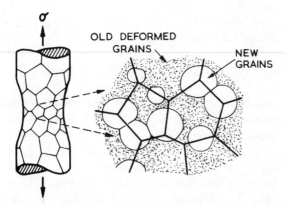

Fig. 2.6. Dynamic recrystallization replaces deformed by undeformed material, permitting a new wave of primary creep, thus accelerating the creep rate.

## Dynamic recrystallization

At high temperatures ($\geqslant 0.6\, T_M$), power-law creep may be accompanied by repeated waves recrystallization as shown in Fig. 2.6 (Hardwick *et al.*, 1961; Nicholls and McCormick, 1970; Hardwick and Tegart, 1961; Stüwe, 1965; Jonas *et al.*, 1969; Luton and Sellars, 1969). Each wave removes or drastically changes the dislocation substructure, allowing a period of primary creep, so that the strain-rate (at constant load) oscillates by up to a factor of 10. The phenomenon has been extensively studied in Ni, Cu, Pb, Al and their alloys (see the references cited above), usually in torsion but it occurs in any mode of loading. It is known to occur in ceramics such as ice and NaCl, and in both metals and ceramics is most pronounced in very pure samples and least pronounced in heavily alloyed samples containing a dispersion of stable particles.

Dynamic recrystallization confuses the high-temperature, high-stress region of the maps. When it occurs the strain-rate is higher than that predicted by the steady-state eqn. (2.21), and the apparent activation energy and creep exponent may change also (Jonas *et al.*, 1969). The simplest physical picture is that of repeated waves of primary creep, occurring with a frequency which depends on temperature and strain-rate, each wave following a primary creep law (Chapter 17, Section 17.1) and having the same activation energy and stress dependence as steady-state creep. But a satisfactory model, even at this level, is not yet available.

Accordingly, we adopt an empirical approach. The maps shown in Chapter 1 and in Chapters 4 to 16, show a shaded region at high temperatures, labelled "dynamic recrystallization". It is not based on a rate-equation (the contours in this region are derived from eqn. (2.21)), but merely shows the field in which dynamic recrystallization has been observed, or in which (by analogy with similar materials) it would be expected.

## The influence of alloying on power-law creep

A *solid solution* influences creep in many ways. The lattice parameter, stacking fault energy, moduli and melting point all change. Diffusive transport now involves two or more atomic species which may move at different rates. Solute atoms interact with stationary and moving dislocations introducing a friction stress for glide-controlled creep and a solute-drag which retards climb-controlled creep (see Hirth and Lothe, 1968 for details of these interactions). If the alloy is ordered, a single moving dislocation generally disrupts the order; if they move in paired groups, order may be preserved, but this introduces new constraints to deformation (see Stoloff and Davies, 1967).

The stress dependence of creep in solid solutions falls into two classes (Sherby and Burke, 1967; Bird *et al.*, 1969): those with a stress dependence of $n = 4$ to 7; and those with $n \approx 3$. The first class resembles the pure metals, and is referred to as "climb-controlled creep". The second is referred to as "viscous-drag-controlled creep" and is believed to result from the limitation on dislocation velocity imposed by the dragging of a solute atmosphere, which moves diffusively to keep up with the dislocation. To a first approximation, both classes of creep behaviour may be described by the power-law creep equation (eqn. (2.21)), with appropriate values of $n$, $A$, and diffusion coefficients. That is what is done here.

The diffusion coefficient for vacancy diffusion, appropriate for climb-controlled creep in a two-component system, is:

$$\bar{D} = \frac{D_A D_B}{D_A x_B + D_B x_A} \qquad (2.27)$$

where $D_A$ and $D_B$ are the tracer diffusion coefficients of components $A$ and $B$, respectively, and $x_A$ and $x_B$ are the respective atomic fractions (Herring, 1950; Burton and Bastow, 1973). The appropriate diffusion coefficient for viscous-drag-controlled creep is the chemical interdiffusivity of the alloy:

$$\tilde{D} = (x_A D_B + x_B D_A)\left(1 + \frac{\partial \ln \gamma_A}{\partial \ln x_A}\right) \qquad (2.28)$$

where $\gamma_A$ is the activity coefficient of the $A$ species. (This average applies because the solute atmosphere' diffuses with the dislocation by exchanging position with solvent atoms in the dislocation path.)

Low-temperature, core-diffusion, limited creep should occur in solid solutions by the same mechanism as in pure metals. The diffusion coefficient must be changed, however, to take into account the solute presence, in the way described by eqn. (2.27). In general, the solute concentration at the dislocation core will differ from that in the matrix, and core diffusion will be accordingly affected; but the lack of data means that diffusion rates have to be estimated (Brown and Ashby, 1980b; see also Chapters 7 and 8). Solid solutions exhibit power-law breakdown behaviour at high stresses, and Harper–Dorn creep at low stresses.

A *dispersion of strong particles of a second phase* blocks dislocation glide and climb, helps to stabilize a dislocation substructure (cells), and may suppress dynamic recrystallization. The stress exponent $n$ is found to be high for dispersion-hardened alloys: typically 7 or more; and the activation energy, too, is often larger than that for self-diffusion. Creep of dispersion-hardened alloys is greatly influenced by thermomechanical history. Cold working introduces dislocation networks which are stabilized by the particles, and which may not recover, or be removed by recrystallization, below 0·8 or 0·9 $T_M$. The creep behaviour of the cold-worked material then differs greatly from that of the recrystallized material, and no steady state may be possible. On recrystallization, too, the particles can stabilize an elongated grain structure which is very resistant to creep in the long direction of the grains.

Early theories of creep in these alloys (Ansell, 1968) were unsatisfactory in not offering a physical explanation for the high values of $n$ and $Q$. The work of Shewfelt and Brown (1974, 1977) has now established that creep in dispersion-hardened single crystals is controlled by climb over the particles. But a satisfactory model for polycrystals (in which grain boundary sliding concentrates stress in a way which helps dislocations overcome the dispersion) is still lacking.

Most *precipitates*, when fine, are not stable at creep temperatures, but may contribute to short-term creep strength: alloys which precipitate continuously during the creep life often have good creep strength. When coarse, a precipitate behaves like a dispersion.

## 2.5 DIFFUSIONAL FLOW

A stress changes the chemical potential, $\phi$, of atoms at the surfaces of grains in a polycrystal. A hydrostatic pressure changes $\phi$ everywhere by the same amount so that no potential gradients appear;

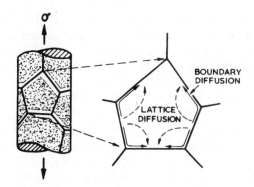

Fig. 2.7. Diffusional flow by diffusional transport through and round the grains. The strain-rate may be limited by the rate of diffusion or by that of an interface reaction.

but a stress field with a deviatoric component changes $\phi$ on some grain surfaces more than on others, introducing a potential gradient, $\Delta\phi$. At high temperatures this gradient induces a diffusive flux of matter through and around the surfaces of the grains (Fig. 2.7), and this flux leads to strain, provided it is coupled with sliding displacements in the plane of the boundaries themselves. Most models of the process (Nabarro, 1948; Herring, 1950; Coble, 1963; Lifshitz, 1963; Gibbs, 1965; Raj and Ashby, 1971) assume that it is *diffusion-controlled*. They are in substantial agreement in predicting a rate-equation: if both lattice and grain boundary diffusion are permitted, the *rate-equation for diffusional flow* is:

$$\dot{\gamma}_7 = \frac{42\sigma_s\Omega}{kTd^2}D_{\text{eff}}$$ (2.29)

where

$$D_{\text{eff}} = D_v\left[1 + \frac{\pi\delta}{d}\frac{D_b}{D_v}\right]$$ (2.30)

Here $d$ is the grain size, $D_b$ the boundary diffusion coefficient and $\delta$ the effective thickness of the boundary.

Like the equation for climb-controlled creep, it is really two equations. At high temperatures, *lattice diffusion* controls the rate; the resulting flow is known as Nabarro–Herring creep and its rate scales as $D_v/d^2$. At lower temperatures, *grain-boundary diffusion* takes over; the flow is then called Coble creep, and scales as $D_b/d^3$.

This equation is an oversimplification; it neglects the kinetics involved in detaching vacancies from grain-boundary sites and reattaching them again, which may be important under certain conditions. Such behaviour can become important in alloys, particularly those containing a finely dispersed second phase. Pure metals are well described by eqn. (2.29), and it is used, in this form, to construct most of the maps of subsequent sections. Interface reaction control, and its influence on the maps, is dealt with further in Chapter 17, Section 17.3.

## INFLUENCE OF ALLOYING ON DIFFUSIONAL FLOW

A *solid solution* may influence diffusional flow by changing the diffusion coefficient (Herring, 1950). When lattice diffusion is dominant, the coefficient $D_v$ should be replaced by $\bar{D}$ (eqn. (2.27)). When boundary diffusion is dominant a similar combined-coefficient should be used, but lack of data makes this refinement impossible at present. More important, the solid solution can impose a drag on boundary dislocations slowing the rate of creep; and solute redistribution during diffusional flow can lead to long transients. These effects are discussed further in Chapter 17, Section 17.3.

There is evidence that a *dispersion of a second phase* influences the way in which a grain boundary acts as a sink and source of vacancies, introducing a large interface-reaction barrier to diffusion and a threshold stress below which creep stops. Its influence is illustrated in Chapter 7, and discussed in Chapter 17, Section 17.3.

The influence of a *precipitate* on diffusional flow is not documented. This sort of creep is normally observed at high temperatures ($> 0.5\ T_M$) when most precipitates will dissolve or coarsen rapidly.

### References for Chapter 2

Ansell, G. S. (1968) In: *Oxide Dispersion Strengthening*. AIME, Gordon & Breach, p. 61.

Argon, A. S. (1970) *Scripta Met.* **4**, 1001.

Ashby, M. F. (1969) *Scripta Met.* **3**, 837.

Ashby, M. F. (1972a) *Acta Met.* **20**, 887.

Ashby, M. F. (1972b) *Surface Sci.* **31**, 498.

Balluffi, R. W. (1970) *Phys. Stat. Sol.* **42**, 11.

Bird, J. E., Mukherjee, A. K. and Dorn, J. E. (1969) In: *Quantitative Relation between Properties and Microstructure*. Israel Univ. Press, Jerusalem, p. 255.

Bolling, G. F. and Richman, R. H. (1965) *Acta Met.* **13**, 709, 723.

Brown, A. M. and Ashby, M. F. (1980a) *Scripta Met.* **14**, 1297.

Brown, A. M. and Ashby, M. F. (1980b) *Acta Met.* **28**, 1085.

Brown, L. M. and Stobbs, W. M. (1971) *Phil. Mag.* **23**, 1185.

Burton, B. (1972) *Mat. Sci. Eng.* **10**, 9.

Burton, B. (1973) *Mat. Sci. Eng.* **11**, 337.

Burton, B. and Bastow, B. D. (1973) *Acta Met.* **21**, 13.

Coble, R. L. (1963) *J. Appl. Phys.* **34**, 1679.

de Meester, B., Yin, C., Doner, M. and Conrad, H. (1973) In: *Rate Processes in Plastic Deformation* (eds J. C. M. Li and A. K. Mukherjee). A.S.M.

Evans, A. G. and Rawlings, R. D. (1969) *Phys. Stat. Sol.* **34**, 9.

Frost, H. J. (1974) Ph.D. Thesis, Division of Applied Sciences, Harvard University.

Gibbs, G. B. (1965) *Mem. Sci. Rev. Met.* **62**, 781.

Guyot, P. and Dorn, J. E. (1967) *Can. J. Phys.* **45**, 983.

Hardwick, D. and Tegart, W. J. McG. (1961) *J. Inst. Met.* **90**, 17.

Hardwick, D., Sellars, C. M. and Tegart, W. J. McG. (1961) *J. Inst. Met.* **90**, 21.

Harper, J. G. and Dorn, J. E. (1957) *Acta Met.* **5**, 654.

Harper, J. G., Shepard, L. A. and Dorn, J. E. (1958) *Acta Met.* **6**, 509.

Hart, E. W. (1957) *Acta Met.* **5**, 597.

Herring, C. (1950) *J. Appl. Phys.* **21**, 437.

Hirsch, P. B. and Humphries, F. (1970) *Proc. R. Soc.* **A318**, 45.

Hirth, J. P. and Lothe, J. (1968) *Theory of Dislocations*. McGraw Hill.

Jonas, J. J., Sellars, C. M. and McTegart, W. J. McG. (1969) *Met. Rev.* **14**, 1.

Kelly, A. (1966) *Strong Solids*. Oxford University Press.

Klahn, D., Mukherjee, A. K. and Dorn, J. E. (1970) *Second International Conference on the Strength of Metals and Alloys*, ASM, p. 951. Asilomar, California.

Kocks, U. F. (1970) *Met. Trans.* **1**, 1121.

Kocks, U. F., Argon, A. S. and Ashby, M. F. (1975) *Prog. Mat. Sci.* **19**.

Kumar, A., Hauser, F. E. and Dorn, J. E. (1968) *Acta Met.* **16**, 1189.

Kumar, A. and Kumble, R. G. (1969) *J. Appl. Phys.* **40**, 3475.

Lifshitz, L. M. (1963) *Soviet Phys. JETP* **17**, 909.

Luton, M. J. and Sellars, C. M. (1969) *Acta Met.* **17**, 1033.

MacKenzie, J. K. (1959) Ph.D. thesis, Bristol University.

Mohammed, F. A., Murty, K. L. and Morris, J. W., Jr. (1973) In: *The John E. Dorn Memorial Symposium*. Cleveland, Ohio, ASM.

Mukherjee, A. K., Bird, J. E. and Dorn, J. E. (1969) *Trans. ASM* **62**, 155.

Nabarro, F. R. N. (1948) Report on a Conference on the Strength of Metals (Phys. Soc. London).

Nicholls, J. H. and McCormick, P. G. (1970) *Met. Trans.* **1**, 3469.

Orowan, E. (1940) *Proc. Phys. Soc.* **52**, 8.

Raffo, P. L. (1969) *J. Less-Common Metals* **17**, 133.

Raj, R. and Ashby, M. F. (1971) *Met. Trans.* **2**, 1113.

Robinson, S. L. and Sherby, O. D. (1969) *Acta Met.* **17**, 109.

Sellars, C. M. and Tegart, W. J. McG. (1966) *Mem. Sci. Rev. Met.* **63**, 731.

Sherby, O. D. and Burke, P. M. (1967) *Prog. Mat. Sci.* **13**, 325.

Shewfelt, R. S. W. and Brown, L. M. (1974) *Phil. Mag.* **30**, 1135.

Shewfelt, R. S. W. and Brown, L. M. (1977) *Phil. Mag.* **35**, 945.

Stocker, R. L. and Ashby, M. F. (1973) *Scripta Met.* **7**, 115.

Stoloff, N. S. and Davies, R. G. (1967) *Prog. Mat. Sci.* **13**, 1.

Stüwe, H. P. (1965) *Acta Met.* **13**, 1337.

Tyson, W. R. (1966) *Phil. Mag.* **14**, 925.

Vandervoort, R. R. (1970) *Met. Trans.* **1**, 857.

Weertman, J. (1956) *J. Mech. Phys. Solids* **4**, 230.

Weertman, J. (1960) *Trans. AIME* **218**, 207.

Weertman, J. (1963) *Trans. AIME* **227**, 1475.

Wong, W. A. and Jonas, J. J. (1968) *Trans. AIME* **242**, 2271.

# CHAPTER 3

# CONSTRUCTION OF THE MAPS

SEVEN or more mechanisms, described in Chapter 2, contribute to the deformation of crystalline solids. The one which is *dominant* (meaning that it contributes most to the total rate of deformation) depends on the stress and temperature to which the solid is exposed, and on its properties. It is helpful, for a given material, to have a way of plotting the field of dominance of each mechanism and of displaying both the experimental data and the predictions that the model-based equations of Section 2 make for it. Such a diagram (or "map") summarizes in a compact form both the experimental and model-based understanding of the materials (Ashby, 1972; Frost and Ashby, 1973).

Ways of doing this were introduced briefly in Chapter 1, and illustrated, for nickel, by Figs. 1.2 to 1.6. For the reasons given there, it is preferable to use as axes, the macroscopic variables $\sigma_s/\mu$, $T/T_M$, $\dot\gamma$, and, when discussing non-steady-state behaviour (Chapter 17, Section 17.1), the strain $\gamma$, or the time $t$. Of the possible combinations, we have found that with axes of $\sigma_s/\mu$ and $T/T_M$ (Fig. 1.2) and that with axes of $\dot\gamma$ and $\sigma_s/\mu$ (Fig. 1.3) are the most useful: the first covers the full range of all the macroscopic variables, and best allows comparison of theory and experiment at low temperatures, while the second displays the creep regime, and permits accurate comparison of theory and experiment at high temperatures. They appear throughout Chapters 4 to 16. The map with axes of $\dot\gamma$ and $T/T_M$ has merit for displaying behaviour at very high strain rates (Chapter 17, Section 17.2).

## 3.1 METHOD OF CONSTRUCTION

A deformation map for a material is constructed by the following procedure. Examples of each step will be found throughout Chapters 4 to 16.

First, data for the material properties are gathered: lattice parameter, molecular volume, and Burger's vector; moduli and their temperature dependencies; and lattice, boundary and core diffusion coefficients (if they exist). It is often necessary to replot data for moduli and diffusion coefficients in order to make a sensible choice of constants, $\mu_0$, $d\mu/dT$, $D_{0v}$, $Q_v$, $\delta D_{0b}$, $Q_b$, etc.

Second, data for the hardness, low-temperature yield, and creep are gathered: flow strength as a function of temperature and strain rate, and creep rate as a function of temperature and stress. These data are plotted on transparent paper with the axes used for the maps themselves: $\log_{10}(\sigma_s/\mu)$ and $T/T_M$, or $\log_{10}(\dot\gamma)$ and $\log_{10}(\sigma_s/\mu)$. Each datum is plotted as a symbol which identifies its source, and is labelled with the value of the third macroscopic variable: $\log_{10}(\dot\gamma)$ or $T$.

Third, an initial estimate is made of the material properties describing glide ($\Delta F$, $\hat\tau$, $\Delta F_p$, $\hat\tau_p$, etc.) and creep ($n$, $A$, etc.) by fitting eqns. (2.9), (2.12) and (2.21) to these data plots. From the plots it is also possible to make an initial estimate of the stress at which the simple power-law for creep breaks down, giving $\alpha'$ of eqn. (2.26).

Fourth, the initial values for the material properties are used to construct a trial map. This is best done by a simple computer program which steps incrementally through the range of $\sigma_s/\mu$ and $T/T_M$ (or $\dot\gamma$ and $\sigma_s/\mu$), evaluating and summing the rate-equations at each step, and which plots the result in the forms shown in Figures which appear in Chapter 1.

All the maps (regardless of the choice of axes) are divided into *fields*, within each of which a given mechanism is dominant. The *field boundaries* are the loci of points at which two mechanisms contribute equally to the overall strain-rate, and are computed by equating pairs (or groups) of rate-equations, and solving for stress as a function of temperature as shown in Fig. 3.1. Superimposed on this are the *contours of constant strain-rate*, obtained by summing the rate-equations in an appropriate way (discussed below) to give a total strain-rate, $\dot\gamma_{net}$ and plotting the loci of points for which $\dot\gamma_{net}$ has given constant values, as shown in Fig. 1.2.

Fifth, the data plots are laid over the trial maps, allowing the data to be divided into blocks according to the dominant flow mechanism. It is then possible to make a detailed comparison between each block of data and the appropriate rate

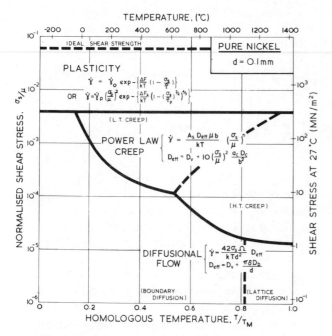

Fig. 3.1. The construction of a deformation-mechanism map. The field boundaries are the loci of points at which two mechanisms (or combinations of mechanisms—see text) have equal rates.

equation. The material properties are now adjusted to give the best fit between theory and experimental data. New maps are now computed and the comparison repeated. Final adjustments are made by constructing maps of the types described in Chapter 1, plotting the data onto them, and examining both goodness-of-fit in individual fields and the precision with which computed and experimental field boundaries coincided. It cannot be emphasized too strongly that for the final map to have any real value, this detailed comparison with data over the entire range $T$, $\sigma_s$ and $\dot\gamma$ is essential.

Finally, the adjusted data are tabulated and the maps redrawn, either with data plotted on them, or on separate data-plots. Such tables and plots for each material discussed in this book will be found in Chapters 4 to 16.

## 3.2  SUPERPOSITION OF RATE-EQUATIONS

The method of combining the rate-equations requires some discussion. Glide plasticity is described by two rate-equations (Section 2.2), one for obstacle-controlled glide ($\dot\gamma_2$, eqn. (2.9)) and one for lattice-resistance-controlled glide ($\dot\gamma_3$, eqn. (2.12)). At the lowest level of approximation, they can be treated as alternatives:

$$\dot\gamma_{\text{plas}} = \text{Least of } \{\dot\gamma_2, \dot\gamma_3\} \qquad (3.1)$$

This is the level adopted here. It is equivalent to assuming that the strongest obstacles control the flow stress, and is entirely adequate for our purposes.

A better approximation is to recognize that, when several strengthening mechanisms (drag, discrete obstacles, lattice resistance) operate at once, their contributions to the flow stress superimpose in a roughly linear way. Even this is an approximation; the superposition is seldom truly linear. The highest precision is possible only by modelling the detailed way in which a given pair of mechanisms interact (see Evans and Rawlings, 1969; Kocks et al., 1975; Frost and Ashby, 1971).

Power-law creep ($\dot\gamma_4$, eqn. (2.21) or $\dot\gamma_6$, eqn. (2.26)) and diffusional flow ($\dot\gamma_7$, eqn. (2.29)) are independent flow mechanisms involving different defects. To a first approximation, their strain-rates add. Power-law creep ($\dot\gamma_4$) and glide ($\dot\gamma_{\text{plas}}$, eqn. (3.1)) do not. Both processes involve the same defect; they describe the same dislocations moving under different conditions. As the stress is raised, the gliding part of the motion of a dislocation becomes more important, and the climbing part less so until, when the boundary between the two fields is reached, power-law climb is not necessary at all. We have solved the problem by treating power-law creep and glide plasticity as alternative mechanisms, choosing always the faster one. This divides the map into two parts, one ("power-law creep") depicting steady-rate flow and one ("plasticity") depicting flow at constant structure—a consequence of the fact that the glide equations do not include recovery, and therefore cannot properly describe the transition from constant-structure to glide-controlled plasticity at steady state. Harper–Dorn Creep ($\dot\gamma_5$, eqn. (2.23)) is treated as an alternative to diffusional flow, again selecting the faster mechanism. Finally, if the ideal strength is exceeded, flow ($\dot\gamma_1$, eqn. (2.1)) becomes catastrophic. In summary, the net strain rate of a polycrystal subject to a stress $\sigma_s$ at a temperature $T$ is:

$$\dot\gamma_{\text{net}} = \dot\gamma_1 + \text{greatest of } (\dot\gamma_{\text{plas}}, \dot\gamma_4 \text{ or } \dot\gamma_6)$$
$$+ \text{greatest of } (\dot\gamma_5, \dot\gamma_7) \qquad (3.2)$$

Within a field, the contribution of one mechanism to $\dot\gamma_{\text{net}}$ is larger than any other. It is separated by field boundaries (heavy, full lines) from fields of dominance of other mechanisms. The power-law creep equation (2.21) and the diffusional flow equation (2.29) each describe the sum of two additive contributions; heavy broken lines separate the regions of dominance of each contribution. The contours of constant strain rate are obtained by solving eqn. (3.2) for $\sigma_s$ as a function of $T$ at constant $\dot\gamma_{\text{net}}$.

## 3.3 TREATMENT OF DATA

The origins and detailed treatment of data are described under the headings of the individual materials. As emphasized earlier, the accuracy of the maps reflects that of the experiments. Different experimenters often report strain-rates that differ by a factor as large as 100 at a given $\sigma_s$ and $T$; it is necessary to judge which experiments more accurately reflect the true material behaviour. These judgements are to some extent subjective because of the large number of variables involved: purity, testing atmosphere, grain size, thermomechanical history, recrystallization effects, type of test, and so forth. The effect of *impurities* is most pronounced in nominally "pure" metals: as little as 0·1% impurity lowers the creep-rate of pure nickel by more than an order of magnitude (Dennison *et al.*, 1966). The low-temperature yield stress of b.c.c. metals is raised substantially by even smaller amounts of interstitial impurities. Grain size (if not extreme) atmosphere, and test type are less important, but, when relevant, we have tried to record them (grain size appears explicitly on all maps). Dynamic recrystallization can be a problem; it can cause the strain-rate to oscillate (at constant stress) or the flow stress to oscillate (at constant strain-rate) and makes it difficult to define steady-state behaviour. If recrystallization (which depends dramatically on purity) occurs only once during a test, it may be neglected in evaluating steady-rate behaviour. But if the test produces repeated recrystallization, as is common in hot torsion tests taken to large strains, the successive waves of recrystallization may overlap to produce another type of steady-state behaviour. This regime is shown as a shaded region on the maps.

As mentioned above, the parameters appearing in the rate-equations are adjusted to give the best description of the experimental data. Most of the adjustment involves the dislocation creep parameters, $n$ and $A$ and the dislocation glide parameters $\hat{\tau}_0$, $\hat{\tau}_p$ and $\Delta F_p$. In some cases the core diffusion coefficients have been adjusted to fit low-temperature creep data.

The data on which each map is based are plotted on the map or as a separate data plot. Points attached by solid lines have the same strain-rate. A dashed line between points indicates a series of intermediate experimental points. Included on the plots are creep, tension, compression, torsion and (occasionally) extrusion tests, all converted to shear stress. Torsion and extrusion data present difficulties because the stress and strain-rate are not constant throughout the specimen; they must be inferred from a flow-field. The data for single crystals are different: they are plotted as critical resolved shear stress whenever possible, and with the standard conversion from tensile to shear stress ($\sigma_s = \sigma/\sqrt{3}$) otherwise. To compare them with the polycrystal data shown on the same map, the reader must multiply the single-crystal stresses by the appropriate factor: 1·77 for f.c.c. metals, for instance, and 1·67 for b.c.c. (see Chapter 2, Section 2.2 for further information on Taylor factors). In arriving at the optimized data of the tables shown in later chapters, single-crystal data were treated in this way.

Maps with normalized stress $\sigma_s/\mu$ and temperature $T/T_M$ have, plotted on them, near the bottom, a line of constant shear stress (usually $\sigma_s = 0·1$ MN/m2). This allows the scale on the right-hand edge of the maps (of shear stress at 300 K) to be translated to give the shear stress at any temperature. All maps with $T/T_M$ as an axis carry, additionally, a scale of temperature in °C.

### References for Chapter 3

Ashby, M. F. (1972) *Acta Met.* **20**, 887.

Dennison, J. P., Llewellyn, R. J. and Wilshire, B. (1966) *J. Inst. Met.* **94**, 130.

Evans, A. G. and Rawlings, R. D. (1969) *Phys. Stat. Sol.* **34**, 9.

Frost, H. J. and Ashby, M. F. (1973) Division of Applied Physics, Harvard University Report.

Frost, H. J. and Ashby, M. F. (1971) *J. Appl. Phys.* **42**, 5273.

Kocks, U. F., Argon, A. and Ashby, M. F. (1975) *Prog. Mat. Sci.* **19**.

# CHAPTER 4

# THE F.C.C. METALS: Ni, Cu, Ag, Al, Pb AND $\gamma$-Fe

THE f.c.c. metals, above all others, are tough and ductile, even at temperatures close to 0 K. When annealed, they are soft and easily worked; but their capacity for work-hardening is sufficiently large that, in the cold-worked state, they have useful strength. Their capacity for alloying, too, is great, leading to ranges of materials, such as the aluminium alloys, the beryllium coppers, the stainless steels, and the nickel-based superalloys, which have remarkable yield and creep strengths. Maps for five pure f.c.c. metals (Ni, Cu, Ag, Al and Pb) are shown in Figs. 4.1 to 4.20. Those for non-ferrous alloys are given in Chapter 7; those for iron and stainless steels in Chapter 8. The maps are based on data plotted on the figures, and the parameters listed in Table 4.1.

## 4.1 GENERAL FEATURES OF MECHANICAL BEHAVIOUR OF F.C.C. METALS

The maps for nickel (Figs. 4.1 to 4.4) typify those for the f.c.c. metals. All show three principal fields: low-temperature plasticity, power-law creep and diffusional flow. Occasionally (aluminium and lead are examples) a field of Harper–Dorn creep replaces that of diffusional flow, and in two instances (copper and aluminium) power-law breakdown is shown.

The f.c.c. metals are remarkable in having an extremely low lattice resistance (certainly less than $10^{-5}\,\mu$); as a result their yield strength is determined by the density of discrete obstacles or defects they contain.* When pure, it is the density and arrangement of dislocations that determines the flow stress, which therefore depends on the state of work-hardening of the metal. Most of the maps describe work-hardened material, and were computed by using a dislocation density of $6.25 \times 10^{14}/\text{m}^2$ (or an obstacle of $\ell = 4 \times 10^{-8}$ m, see Chapter 2, Table 2.1). This choice describes a heavily deformed state,

* The refractory f.c.c. metals Ir and Rh are exceptional in exhibiting a large lattice resistance: roughly $10^{-2}\,\mu$ at 0 K.

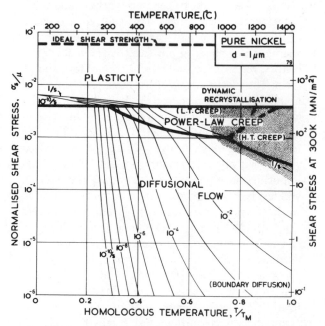

Fig. 4.1. Pure nickel of grain size 1 $\mu$m, work-hardened; the obstacle spacing $l$ is taken as $4 \times 10^{-8}$ m.

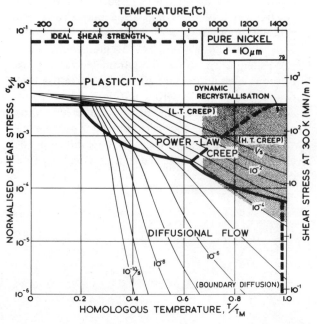

Fig. 4.2. Pure nickel of grain size 1 $\mu$m, work-hardened, as Fig. 4.1.

| Material | Nickel | Copper | Silver | Aluminium | Lead | γ-Iron |
|---|---|---|---|---|---|---|
| *Crystallographic and thermal data* | | | | | | |
| Atomic volume, $\Omega$ (m³) | $1.09 \times 10^{-29}$ | $1.18 \times 10^{-29}$ | $1.71 \times 10^{-29}$ | $1.66 \times 10^{-29}$ | $3.03 \times 10^{-29}$ | $1.21 \times 10^{-29}$ (u) |
| Burgers vector, $b$ (m) | $2.49 \times 10^{-10}$ | $2.56 \times 10^{-10}$ | $2.86 \times 10^{-10}$ | $2.86 \times 10^{-10}$ | $3.49 \times 10^{-10}$ | $2.58 \times 10^{-10}$ |
| Melting temperature, $T_M$ (K) | 1726 | 1356 | 1234 | 933 | 601 | 1810 |
| *Modulus** | | | | | | |
| Shear modulus at 300 K $\mu_0$ (MN/m²) | $7.89 \times 10^4$ (a) | $4.21 \times 10^4$ (f) | $2.64 \times 10^4$ (j) | $2.54 \times 10^4$ (o) | $0.73 \times 10^4$ (r) | $8.1 \times 10^4$ (u) |
| Temperature dependence of modulus, $\dfrac{T_M}{\mu_0}\dfrac{d\mu}{dT}$ | $-0.64$ (a) | $-0.54$ | $-0.54$ (k) | $-0.50$ (o) | $-0.76$ (r) | $-0.91$ |
| *Lattice diffusion†* | | | | | | |
| Pre-exponential, $D_{0v}$ (m²/s) | $1.9 \times 10^{-4}$ (b) | $2.0 \times 10^{-5}$ (g) | $4.4 \times 10^{-5}$ (l) | $1.7 \times 10^{-4}$ (p) | $1.4 \times 10^{-4}$ (s) | $1.8 \times 10^{-5}$ (u) |
| Activation energy, $Q_v$ (kJ/mole) | 284 (b) | 197 (g) | 185 (l) | 142 (p) | 109 (s) | 270 |
| *Boundary diffusion†* | | | | | | |
| Pre-exponential, $\delta D_{0b}$ (m³/s) | $3.5 \times 10^{-15}$ (c) | $5.0 \times 10^{-15}$ (h) | $4.5 \times 10^{-15}$ (m) | $5.0 \times 10^{-14}$ (h) | $8.0 \times 10^{-14}$ (t) | $7.5 \times 10^{-14}$ (u) |
| Activation energy, $Q_b$ (kJ/mole) | 115 (c) | 104 (h) | 90 (m) | 84 (h) | 66 (t) | 159 |
| *Core diffusion†* | | | | | | |
| Pre-exponential, $a_c D_{0c}$ (m⁴/s) | $3.1 \times 10^{-23}$ (d) | $1.0 \times 10^{-24}$ (i) | $2.8 \times 10^{-25}$ (n) | $7.0 \times 10^{-25}$ (q) | $1.0 \times 10^{-22}$ (h) | $1.0 \times 10^{-23}$ (u) |
| Activation energy, $Q_c$ (kJ/mole) | 170. (d) | 117 (i) | 82 (n) | 82 (q) | 66 (h) | 159 |
| *Power-law creep* | | | | | | |
| Exponent, $n$ | 4.6 (e) | 4.8 (e) | 4.3 (e) | 4.4 (e) | 5.0 (e) | 4.5 (u) |
| Dorn constant,‡ $A$ | $3.0 \times 10^6$ (e) | $7.4 \times 10^5$ (e) | $3.2 \times 10^2$ (e) | $3.4 \times 10^6$ (e) | $2.5 \times 10^8$ (e) | $4.3 \times 10^5$ |
| P-L breakdown, $\alpha'$ | — | 794 | — | 1000 | — | — |
| *Obstacle-controlled glide* | | | | | | |
| 0 K flow stress, $\hat{\tau}/\mu_0$ | $6.3 \times 10^{-3}$ (e) | $6.3 \times 10^{-3}$ (e) | $7.2 \times 10^{-3}$ (e) | $7.2 \times 10^{-3}$ (e) | $8.7 \times 10^{-3}$ (e) | $1.7 \times 10^{-3}$ (u) |
| Pre-exponential, $\dot{\gamma}_0$ (s⁻¹) | $10^6$ | $10^6$ | $10^6$ | $10^6$ | $10^6$ | $10^6$ |
| Activation energy, $\Delta F/\mu_0 b^3$ | 0.5 | 0.5 | 0.5 | 0.5 | 0.5 | 0.5 |

$$* \; \mu = \mu_0\left(1 + \frac{(T-300)}{T_M}\right)\frac{T_M}{\mu_0}\frac{d\mu}{dT}.$$

$$\dagger \; D_v = D_{0v}\exp-\frac{Q_v}{RT}; \; \delta D_b = \delta D_{0b}\exp-\frac{Q_b}{RT}; \; a_c D_c = a_c D_{0c}\exp-\frac{Q_c}{RT}.$$

‡ This value of $A$ refers to tensile stress and strain-rate. The maps relate shear stress and strain-rate. In constructing them we have used $A_s = (\sqrt{3})^{n+1}A$.

(a) Alers et al. (1960)
(b) Monma (1965).
(c) Wazzan (1965).
(d) Cannon and Stark (1969).
(e) See text.
(f) Chang and Himmel (1966); Overton and Gaffney (1955).
(g) Kuper et al. (1954), (1956).
(h) Estimated by setting $\delta D_{0B} = b.D_{0v}$ and $Q_B = 0.6Q_v$ or $Q_c$ (if known) $a_c D_{0c} = 4b^2 D_{0v}$ and $Q_c = 0.6Q_v$ or $Q_B$ (if known) unless otherwise stated in the text.
(i) Derived from comparison with experiment as described in the text.
(j) Chang and Himmel (1966).
(k) Neighbours and Alers (1958).
(l) Tomizuka and Sonder (1956).
(m) Hoffman and Turnbull (1951).
(n) Turnbull and Hoffman (1954).
(o) Sutton (1953); Lazarus (1959).
(p) Lundy and Murdock (1962).
(q) Balluffi (1970).
(r) Huntington (1958).
(s) Nachtrieb et al. (1959).
(t) Okkerse (1954).
(u) Data for γ-Fe are described in full in Chapter 8. They are listed here to permit comparison.

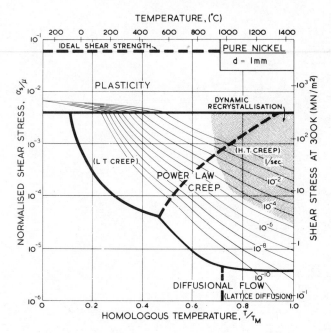

Fig. 4.3. Pure nickel of grain size 1 mm, work-hardened, as Fig. 4.1.

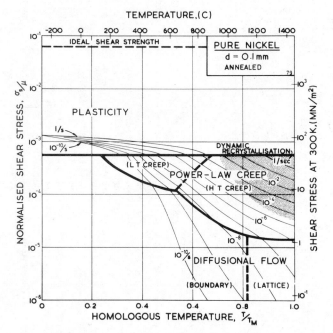

Fig. 4.4. Pure nickel of grain size 0·1 mm, annealed. The obstacle spacing, $l$, is taken as $2 \times 10^{-7}$ m.

although it is still well below the highest known densities and is not therefore a saturation or steady-state density. Annealing lowers the yield strength to about $\sigma_s/\mu = 10^{-3}$ for typical commercial purity f.c.c. metals (corresponding to $\ell \approx 2 \times 10^{-7}$ m); a map for annealed nickel is shown in Fig. 4.4. The activation energy for forest cutting is taken as $\Delta F = 0\cdot5\ \mu b^3$ (Table 2.1) leading to a flow strength which depends only weakly on temperature. This value is

slightly larger than the best estimates (Hirth and Lothe, 1968) but the effect of this difference is undetectable on the maps shown here.

Above about $0\cdot3\ T_M$, the f.c.c. metals start to creep. Diffusion (which is thought to control creep in these metals) is slower in the f.c.c. structure than in the more-open b.c.c. structure; this is reflected in lower creep-rates at the same values of $\sigma_s/\mu$ and $T/T_M$. The creep field is subdivided (Chapter 2, eqn. (2.20)) into a region of low-temperature, core-diffusion controlled creep in which the stress exponent is about $n + 2$, and a region of high-temperature, lattice-diffusion controlled creep in which the stress exponent is $n$. The power-law breaks down for all five metals near $\sigma_s/\mu = 10^{-3}$, corresponding to a value of $\alpha'$ of about $10^3$. Sufficient data are available for two of them (Cu and Al) to include this on the maps with real confidence.

Diffusional flow appears at high temperature and low stress. The field is subdivided (Chapter 2, eqn. (2.30)) into a region in which boundary diffusion controls the creep-rate and one in which lattice diffusion is controlling. When the grain size is large this field may be replaced by one of Harper–Dorn creep (Chapter 2, eqn. (2.23)); examples are given below in the sections on Al and Pb.

The effect of grain size on the maps is illustrated by Figs. 4.1, 4.2, 1.2 and 4.3. Taken in this order, they form a sequence between each of which the grain size increases by a factor of 10. In doing so, the power-law creep field expands at the expense of diffusional flow; lattice-diffusion becomes dominant, displacing boundary diffusion in the control of diffusional flow, and (as mentioned above) Harper–Dorn creep may appear, though there is no direct evidence for it in nickel.

The moduli of all five metals were calculated from single-crystal data using

$$\mu = \left( \frac{c_{44}}{2}(c_{11} - c_{12}) \right)^{\frac{1}{2}}.$$

It is this value which enters the anisotropic calculation of the elastic energy of a $1/2\langle110\rangle$ screw dislocation, and therefore appears in estimates of the force required to cause screws to intersect, or in calculations of the force required to bow an edge dislocation between strong obstacles. We have incorporated a linear temperature dependence—a reasonably accurate approximation for these metals. Plasticity, creep and diffusion data are discussed below.

### References for Section 4.1

Hirth, J. P. and Lothe, J. (1968) *Theory of Dislocations.* McGraw Hill.

## 4.2 ORIGINS OF THE DATA FOR THE f.c.c. METALS

### Nickel (Figs. 1.2 to 1.6 and 4.1 to 4.6)

Nickel (and its alloys, Chapter 7) is covered in more detail than any other material in this book. Figs. 4.1, 1.2, 4.2 and 4.3, show, in sequence, maps for grain sizes between 1 $\mu$m and 1 mm in the work-hardened state. Figs. 1.3 to 1.6 show other ways of plotting the same information. Finally, Fig. 4.4 shows a map describing nickel in the annealed state. All are based on the data plotted in Figs. 4.5 and 4.6, and summarized in Table 4.1.

The high-temperature creep parameters are based on the measurements of Weertman and Shahinian (1956) of the creep of 99·75% Ni with $n = 4·6$ in the high-temperature region. It has been shown, however (Dennison et al., 1966), that small amounts of impurities ($\sim 0·1\%$) may lower the creep rate of nickel by up to an order of magnitude. We have therefore used a value of $A$ of $3·0 \times 10^6$, almost an order of magnitude greater than would be derived from Weertman and Shahinian. This value provides a much closer correspondence to the peak flow stress in torsion data of Luton and Sellars (1969), and other various data near the H.T. creep–L.T. creep boundary of Fig. 4.5.

For low-temperature creep we have used the core-diffusion coefficient given by Canon and Stark (1969) for edge dislocations in a symmetric tilt boundary. The activation energy for this ($Q_c = 170$ kJ/mole) matches that found by Norman and Duran (1970) for creep in the L.T. creep regime. In addition, Norman and Duran find $n \simeq 7·0$, which supports the existence of the low-temperature creep field. Their strain-rates accurately match the numerical predication (using $A = 3·0 \times 10^6$). Weertman and Shahinian also find a low-temperature increase in the stress exponent, although their strain-rates are lower. We have not included power-law breakdown for nickel, though (by analogy with copper and aluminium) eqn. (2.26) with $\alpha' = 1000$ should give an approximate description.

The shaded field of dynamic recrystallization is based on the observations reported and reviewed by Ashby et al. (1979).

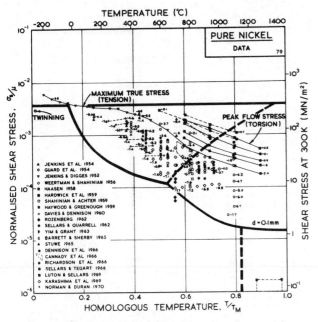

Fig. 4.5. Data for nickel, divided into blocks. Each block is fitted to a rate-equation. The numbers are $\log_{10} (\dot{\gamma})$.

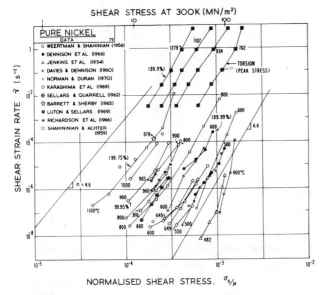

Fig. 4.6. Creep data for nickel. Data are labelled with the temperature in °C.

### References for nickel

Alers, G. A., Neighbours, J. R. and Sato, H. (1960) J. Phys. Chem. Solids **13**, 40.

Ashby, M. F., Gandhi, C. and Taplin, D. M. R. (1979) Acta Met. **27**, 699.

Barrett, C. R. and Sherby, O. D. (1965) Trans. AIME **233**, 1116.

Cannaday, J. E., Austin, R. J. and Saxer, R. K. (1966) Trans. AIME **236**, 595.

Canon, R. F. and Stark, J. P. (1969) J. Appl. Phys. **40**, 4361; **40**, 4366.

Davies, P. W. and Dennison, J. P. (1959–60) J. Inst. Met. **88**, 471.

Dennison, J. P., Llewellyn, R. J. and Wilshire, B. (1966) J. Inst. Met. **94**, 130.

Guard, R. W., Keeler, J .H. and Reiter, S. F. (1954) Trans. AIME **200**, 226.

Haasen, P. (1958) Phil. Mag. **3**, 384.

Hardwick, D., Sellars, C. M. and McG Tegart, W. J. (1961–62) *J. Inst. Met.* **90**, 21.

Hayward, E. R. and Greenough, A. P. (1959–60) *J. Inst. Met.* **88**, 217.

Jenkins, W. D. and Digges, T. G. (1952) *J. Res. N.B.S.* **48**, 313.

Jenkins, W. D., Digges, T. G. and Johnson, L. R. (1954) *J. Res. N.B.S.* **53**, 329.

Karashima, S., Oikawa, H. and Motomiya, T. (1969) *Trans. Jap. Inst. Met.* **10**, 205.

Luton, M. J. and Sellars, C. M. (1969) *Acta Met.* **17**, 1033.

Monma, K., Suto, H. and Oikawa, H. (1964) *J. Jap. Inst. Met.* **28**, 188.

Norman, E. C. and Duran, S. A. (1970) *Acta Met.* **18**, 723.

Richardson, G. J., Sellars, C. M. and McG Tegart, W. J. (1966) *Acta Met.* **14**, 1225.

Rozenberg, V. M. (1962) *Fiz. Metal. Metalloved.* **14**(1), 114.

Sellars, C. M. and Quarrell, A. G. (1961–62) *J. Inst. Met.* **90**, 329.

Sellars, C. M. and McG Tegart, W. J. (1966) *Mem. Sci. Rev. Met.* **63**, 731.

Shahinian, P. and Achter, M. R. (1959) *Trans. AIME* **215**, 37.

Stüwe, H. P. (1965) *Acta Met.* **13**, 1337.

Wazzan, A. R. (1965) *J. Appl. Phys.* **36**, 3596.

Weertman, J. and Shahinian, P. (1956) *Trans. AIME* **206**, 1223.

Yim, W. M. and Grant, N. J. (1963) *Trans. AIME* **227**, 868.

## Copper (Figs. 4.7 to 4.9)

A map for copper with a (typical) grain size of 100$\mu$m, is shown in Fig. 4.7; it includes the broad transition from power-law creep to plasticity which we call power-law breakdown (Chapter 2, eqn. (2.26)). It is based on data shown in Figs. 4.8 and

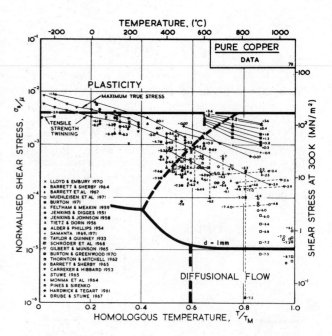

Fig. 4.8. Data for copper, divided into blocks. The numbers are $\log_{10}(\dot{\gamma})$.

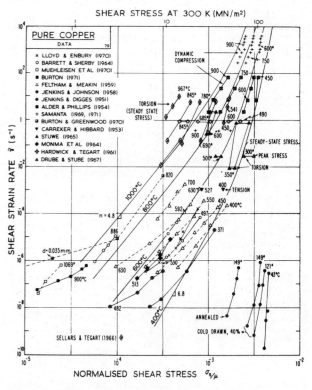

Fig. 4.9. Creep data for copper. Data are labelled with the temperature in °C.

4.9, and summarized in Table 4.1. Changes with grain size resemble those for nickel.

The primary references for the high-temperature creep of copper are Feltham and Meakin (1959) and Barrett and Sherby (1964). For the low-

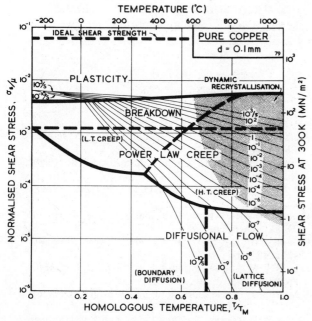

Fig. 4.7. Pure copper with a grain size of 0·1 mm, including power-law breakdown.

temperature creep field we have used a core diffusion activation energy of 117 kJ/mole, chosen to match the activation energy for low-temperature creep found by Barrett and Sherby (1964); there does not appear to be any experimental determination of core or boundary diffusion coefficients for copper. Power-law breakdown is calculated using eqn. (2.26) with $\alpha' = 7.94 \times 10^2$, chosen to match the dynamic compression data of Alder and Phillips (1954) and Samanta (1969, 1971).

The field of dynamic recrystallization is based on observations of Fleck *et al.* (1970) and of Ruckweid (1972).

## References for copper

Alder, J. F. and Phillips, V. A. (1954) *J. Inst. Met.* **83**, 80.

Barrett, C. R. and Shearby, O. D. (1964) *Trans. AIME* **230**, 1322.

Barrett, C. R. and Sherby, O. D. (1965) *Trans. AIME* **233**, 1116.

Barrett, C. R., Lytton, J. L. and Sherby, O. D. (1967) *Trans. AIME* **239**, 170.

Burton, B. and Greenwood, G. W. (1970) *Acta Met.* **18**, 1237.

Burton, B. (1971) *Met. Sci. J.* **5**, 11.

Carreker, R. P., Jr. and Hibbard, W. R., Jr. (1953) *Acta Met.* **1**, 654.

Chang, Y. A. and Himmel, L. (1966) *J. Appl. Phys.* **37**, 3567.

Drube, B. and Stuwe, H. P. (1967) *Z. Metallkde.* **58**, 799.

Feltham, P. and Meakin, J. D. (1959) *Acta Met.* **7**, 614.

Fleck, R. G., Cocks, G. J. and Taplin, D. M. R. (1970) *Met. Trans.* **1**, 3415.

Gilbert, E. R. and Munson, D. E. (1965) *Trans. AIME* **233**, 429.

Hardwick, D. and McG. Tegart, W. J. (1961–62) *J. Inst. Met.* **90**, 17.

Jenkins, W. D. and Digges, T. G. (1951) *J. Res. N.B.S.* **47**, 272.

Jenkins, W. D. and Johnson, C. R. (1958) *J. Res. N.B.S.* **60**, 173.

Kuper, A., Letlaw, H., Slifkin, L., Sonder, E. and Tomizuka, C. T. (1954) *Phys. Rev.* **96**, 1224; (1956) *ibid.*, errata **98**, 1870.

Lloyd, D. J. and Embury, J. D. (1970) *Met. Sci. J.* **4**, 6.

Monma, K., Suto, H. and Oikawa, H. (1964) *J. Jap. Inst. Met.* **28**, 253.

Muchleisen, E. C., Barrett, C. R. and Nix, W. D. (1970) *Scripta Met.* **4**, 995.

Overton, W. C. and Gaffney, J. (1955) *Phys. Rev.* **98**, 969.

Pines, B. Y. and Sirenko, A. F. (1963) *Fiz. Metal. Metalloved* **15**(4), 584.

Ruckweid, A. (1972) *Met. Trans.* **3**, 2999 and 3009.

Samanta, S. K. (1969) *Int. J. of Mech. Sci.* **11**, 433; (1971) *J. Mech. Phys. Solids* **19**, 117.

Schröder, K., Giannuzzi, A. J. and Gorscha, G. (1968) *Acta Met.* **16**, 469.

Stuwe, H. P. (1965) *Acta Met.* **13**, 1337.

Taylor, G. I. and Quinney, H. (1933) *Proc. R. Soc.* **A143**, 307.

Tietz, T. E. and Dorn, J. E. (1956) *Trans. AIME* **206**, 156.

Thornton, P. R. and Mitchell, T. E. (1962) *Phil. Mag.* **7**, 361.

## Silver (Figs. 4.10 to 4.12)

Fig. 4.10 shows a map for silver with a large grain size: 1 mm. It closely resembles those for copper and for nickel (which showed how grain size influences

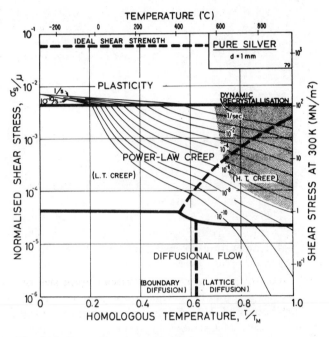

Fig. 4.10. Pure silver in grain size 1 mm.

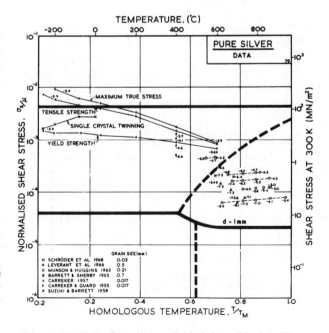

Fig. 4.11. Data for silver, divided into blocks. The numbers are $\log_{10}(\dot{\gamma})$.

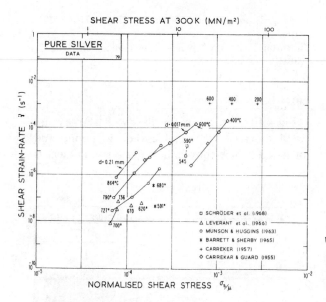

Fig. 4.12. Creep data for silver. Data are labelled with the temperature in °C.

the maps). It is based on data plotted in Figs. 4.11 and 4.12, and summarized in Table 4.1.

There has been less work on the creep of silver than on the other common f.c.c. metals. The high-temperature creep parameters are based on Leverant *et al.* (1966). For the low-temperature creep field, the average dislocation core diffusion coefficient given by Turnbull and Hoffman (1954) has been used. This gives good agreement with the creep data of Carreker and Guard (1955) and of Schröder *et al.* (1968).

The field of dynamic recrystallization is based on observations of Ashby *et al.* (1979).

### References for Silver

Ashby, M. F., Gandhi, C. and Taplin, D. M. R. (1979) *Acta Met.* **27**, 699.
Barrett, C. R. and Sherby, O. D. (1965) *Trans. AIME* **233**, 1116.
Carreker, R. P., Jr. (1957) *Trans. AIME* **209**, 112.
Carreker, R. P., Jr. and Guard, R. W. (1955) Report No. 55–R1–1414, General Electric Co., Schenectady, N.Y.; data taken from: Amin, K. E., Mukherjee, A. K. and Dorn, J. E. (1970) *J. Mech. Phys. Solids* **18**, 413.
Chang, Y. A. and Himmel, L. (1966) *J. Appl. Phys.* **37**, 3567.
Hoffman, R. E. and Turnbull, D. (1951) *J. Appl. Phys.* **22**, 634.
Leverant, G. R., Lenel, F. V. and Ansell, G. S. (1966) *Trans. ASM* **59**, 890.
Munson, D. E. and Huggins, R. A. (1963) DMS Report No. 63–4, Stanford University.
Neighbours, J. R. and Alers, G. A. (1958) *Phys. Rev.* **111**, 707.
Schröder, K., Giannuzzi, A. J. and Gorsha, G. (1968) *Acta Met.* **16**, 469.
Suzuki, H. and Barrett, C. S. (1958) *Acta Met.* **6**, 156.
Tomizuka, C. T. and Sonder, E. (1956) *Phys. Rev.* **103**, 11.
Turnbull, D. and Hoffman, R. E. (1954) *Acta Met.* **2**, 419.

### Aluminium (Figs. 4.13 to 4.16)

Two maps, Figs. 4.13 and 4.14, are given for aluminium to show how Harper–Dorn (eqn. (2.23))

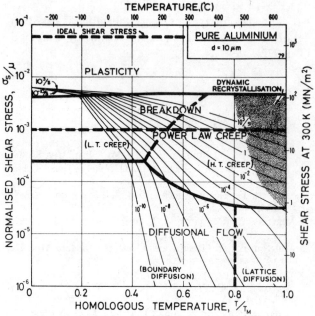

Fig. 4.13. Pure aluminium of grain size 10 μm, including power-law breakdown.

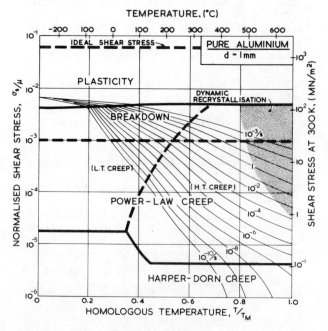

Fig. 4.14. Pure aluminium of grain size 1 mm. Harper–Dorn creep has displaced diffusional flow.

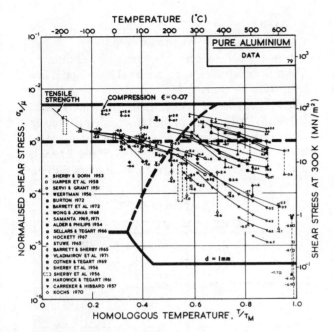

Fig. 4.15. Data for aluminium, divided into blocks. The numbers are $\log_{10}(\dot{\gamma})$.

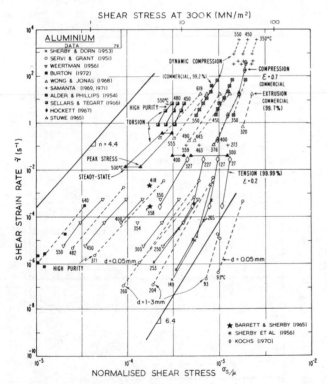

Fig. 4.16. Creep data for aluminium. Data are labelled with the temperature in °C.

creep displaces diffusional flow at large grain sizes. Both include power-law breakdown (eqn. (2.26)). The maps are based on the data plotted in Figs. 4.15 and 4.16, and summarized in Table 4.1.

The high-temperature creep parameters for aluminium are based on Weertman (1956) and Servi

and Grant (1951). At high temperatures, these studies show an activation energy in agreement with the diffusion coefficient of Lundy and Murdock (1962): $Q_v = 142$ kJ/mole. For low-temperature creep we have used the dislocation core diffusion coefficient cited by Balluffi (1970): $Q_c = 82$ kJ/mole, which provides good agreement with low-temperature creep experiments. The Harper–Dorn creep field is based on data of Harper et al. (1958). It appears on Fig. 4.14 for $d = 1$ mm, but is suppressed by diffusional flow at $d = 10$ μm, as shown in Fig. 4.13.

The power law breakdown region of aluminium has been extensively studied; experimental data have been fitted to a hyperbolic sine equation (eqn. (2.25)) by Wong and Jonas (1968) and Sellars and Tegart (1966). Data for commercial purity aluminium, obtained from dynamic compression tests (Alder and Phillips, 1954; Samanta, 1969, 1971; and Hockett, 1967) and extrusion tests (Wong and Jonas, 1968) show much lower strain-rates than data from high-purity aluminium tested in torsion (Sellars and Tegart, 1966) and from the extrapolation of creep data (Servi and Grant, 1951). Both sets of data cannot be matched by the same power-law breakdown parameter. The value used for both maps ($\alpha' = 1 \times 10^3$) provides an intermediate approximation. The field appears here as a general demonstration, not as an exact representation of experiment.

Above $0.8\, T_M$, rapid grain growth occurs in aluminium during creep (Servi and Grant, 1951) giving unstable creep. The shaded region refers to this rather than true dynamic recrystallization, which is rarely observed.

### References for aluminium

Alder, J. F. and Phillips, V. A. (1954) *J. Inst. Met.* **83**, 80.

Balluffi, R. W. (1970) *Phys. Stat. Sol.* **42**, 11.

Barrett, C. R. and Sherby, O. D. (1965) *Trans. AIME* **233**, 1116.

Barrett, C. R., Muehleisen, E. C. and Nix, W. D. (1972) *Mat. Sci. Eng.* **10**, 33.

Burton, B. (1972) *Phil. Mag.* **27**, 645.

Carreker, R. P., Jr. and Hibbard, W. J., Jr. (1957) *Trans. AIME* **209**, 1157.

Cotner, J. R. and McG Tegart, W. J. (1969) *J. Inst. Met.* **97**, 73.

Hardwick, D. and McG Tegart, W. J. (1961–62) *J. Inst. Met.* **90**, 17.

Harper, J. G., Shepard, L. A. and Dorr, J. E. (1958) *Acta Met.* **6**, 509.

Hockett, J. E. (1967) *Trans. AIME* **239**, 969.

Kocks, U. F. (1970) *Met. Trans.* **1**, 1121.

Lazarus, D. (1959) *Phys. Rev.* **76**, 545.

Lundy, T. S. and Murdock, J. F. (1962) *J. Appl. Phys.* **33**, 1671.

Samanta, S. K. (1969) *Int. J. Mech. Sci.* **11**, 433 (1971); *J. Mech. Phys. Solids* **19**, 117.

Sellars, C. M. and McG Tegart, W. J. (1966) *Mem. Sci. Rev. Met.* **63**, 731.

Servi, I. S. and Grant, N. J. (1951) *Trans. AIME* **191**, 909.

Sherby, O. D. and Dorn, J. E. (1953) *Trans. AIME* **197**, 324.

Sherby, O. D., Trozera, T. A. and Dorn, J. E. (1956) *Proc. ASTM* **56**, 789.

Sherby, O. D., Lytton, J. L. and Dorn, J. E. (1957) *Acta Met.* **5**, 219.

Stuwe, H. P. (1965) *Acta Met.* **13**, 1337.

Sutton, P. M. (1953) *Phys. Rev.* **91**, 816.

Weertman, J. (1956) *J. Mech. Phys. Solids* **4**, 230.

Wong, W. A. and Jonas, J. J. (1968) *Trans. AIME* **242**, 2271.

Vladimirova, G. V., Likhachev, V. A., Myshlyayev, M. M. and Olevskiy, S. S. (1971) *Fiz. Metal. Metalloved* **31(1)**, 177.

### Lead (Figs. 4.17 to 4.20)

Lead, like aluminium, exhibits Harper–Dorn creep when the grain size is large. This is illustrated by the two maps (Figs. 4.17 and 4.18). They are based on data plotted in Figs. 4.19 and 4.20 and summarized in Table 4.1.

Our high-temperature creep parameters are based on Mohamed *et al.* (1973): $n = 5.0$, $A = 2.5 \times 10^8$. These differ from those that could be derived from the single-crystal creep experiments of Weertman (1960), which show a slightly lower $n$ at high temperatures. We have used a diffusion activation energy of 109 kJ/mole (Resing and Nachtrieb, 1961) which is higher than the value found

by Mohamed *et al.* (1973), and lower than the value derived from Weertman (1960).

Low-temperature creep behaviour is indicated by a number of studies. Weertman's (1960) data at low temperatures show a higher stress exponent and a lower activation energy than at high. Data of Feltham (1956), again in the low-temperature creep regime, show an apparent activation energy of 92 kJ/mole with $n \geqslant 7$ (compared with 109 kJ/mole and 5 at high). Room temperature data of Gifkins

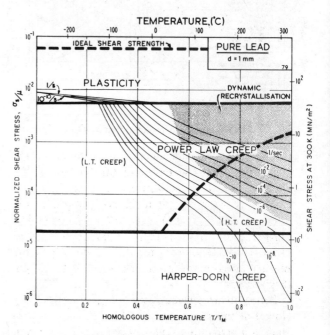

Fig. 4.18. Pure lead of grain size 1 mm. Harper–Dorn creep has displaced diffusional flow.

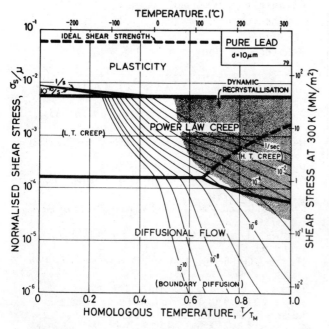

Fig. 4.17. Pure lead of grain size 10 $\mu$m.

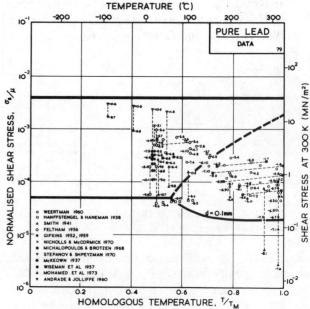

Fig. 4.19. Data for lead, divided into blocks. The numbers are $\log_{10}(\dot{\gamma})$.

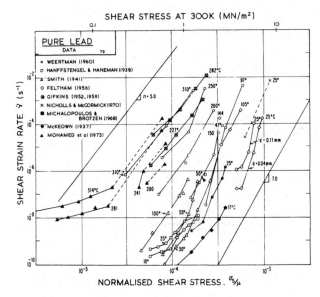

Fig. 4.20. Creep data for lead. Data are labelled with the temperature in °C.

(1952, 1959) and Nichols and McCormick (1970) show $n \approx 8$. The dislocation core diffusion coefficient is taken to match the boundary diffusion data of Okkerse (1954): $Q_c = 65$ kJ/mole. The Harper–Dorn creep field is based on Mohamed et al. (1973). It appears for $d = 1$ mm (Fig. 4.18), but is masked by diffusional flow when $d = 10\ \mu$m (Fig. 4.17).

Diffusion in lead, at a given $T/T_M$, is exceptionally slow. For this reason the rate of diffusional flow of lead, at a given homologous temperature, is slower than that of other f.c.c. metals.

Dynamic recrystallization is observed in lead at above $0{\cdot}7\ T_M$. The field is based on the observations of Hanson and Wheeler (1931), Greenwood and Worner (1939) and Gohn et al. (1946).

## References for lead

Andrade, E. N. da C. and Jolliffe, K. H. (1960) *Proc. R. Soc.* **A254**, 291.

Feltham, P. (1956) *Proc. Phys. Soc.* **B69**, 1173.

Gifkins, R. C. (1952) *J. Inst. Met.* **81**, 417; (1959) *Trans. AIME* **215**, 1015.

Gohn, G. R., Arnold, S. M. and Bouton, G. M. (1946) *Proc. ASTM* **46**, 990.

Greenwood, J. N. and Worner, H. K. (1939) *J. Inst. Met.* **64**, 135.

Hanffstengel, K. and Hanemann, H. (1938) *Z. Metallk.* **30**, 41.

Hanson, D. and Wheeler, M. A. (1931) *J. Inst. Met.* **45**, 229.

Huntington, H. B. (1958) *Solid State Physics* **7**, 213.

McKeown, J. (1937) *J. Inst. Met.* **60**, 201.

Michalopoulos, C. D. and Brotzen, F. R. (1968) *J. Inst. Met.* **96**, 156.

Mohamed, F. A., Murty, K. L. and Morris, J. W., Jr. (1973) *Met. Trans.* **4**, 935.

Nachtrieb, N. H., Resing, H. A. and Rice, S. A. (1959) *J. Chem. Phys.* **31**, 135.

Nicholls, J. H. and McCormick, P. G. (1970) *Met. Trans.* **1**, 3469.

Okkerse, B. (1954) *Acta Met.* **2**, 551.

Resing, H. A. and Nachtrieb, N. H. (1961) *J. Phys. Chem. Solids* **21**, 40.

Smith, A. A. (1941) *Trans. AIME* **143**, 165.

Stepanov, V. A. and Shpeysman, V. V. (1970) *Fiz. Metal. Metalloved.* **29(2)**, 375.

Weertman, J. (1960) *Trans. AIME* **218**, 207.

Wiseman, C. D., Sherby, O. D. and Dorn, J. E. (1957) *Trans. AIME* **209**, 57.

## Austenite: $\gamma$-iron (Figs. 8.1 to 8.3)

Parameters for austenite are included in Table 4.1 for comparative purposes. The map for iron, and the origins of the data, are described in detail in Chapter 8.

# CHAPTER 5

# THE B.C.C. TRANSITION METALS: W, V, Cr, Nb, Mo, Ta AND α-Fe

THE REFRACTORY b.c.c. metals have high melting points and moduli. Many of their applications are specialized ones which exploit these properties: tungsten lamp filaments (Chapter 19) operate at up to 2800°C, and molybdenum furnace windings to 2000°C, for example. They are extensively used as alloying elements in steels and in superalloys, raising not only the yield and creep strengths, but the moduli too. But above all, iron is the basis of all steels and cast irons, and it is this which has generated the enormous scientific interest in the b.c.c. transition metals.

Maps for six pure b.c.c. metals (W, V, Cr, Nb, Mo and Ta) are shown in the Figures of this Chapter. Those for β-Ti, α-iron and ferrous alloys are discussed separately in Chapters 6 and 8. The maps are based on data plotted on the figures and the parameters listed in Table 5.1.

## 5.1 GENERAL FEATURES OF THE MECHANICAL BEHAVIOUR OF B.C.C. METALS

Like the f.c.c. metals, maps for the b.c.c. metals show three principal fields: dislocation glide, power-law creep, and diffusional flow. The principal difference appears at low temperatures (below about $0.15$ $T_M$) where the b.c.c. metals exhibit a yield stress which rises rapidly with decreasing temperature, because of a lattice resistance (or Peierls' resistance). As a result, the flow stress of pure b.c.c. metals extrapolates, at 0 K, to a value close to $10^{-2}$ $\mu$, independent of obstacle content for all but extreme states of work-hardening. On the other hand, the normalized strength of the b.c.c. metals at high temperatures ($>0.5$ $T_M$) because diffusion in the more open b.c.c. lattice is faster than in the close-packed f.c.c. lattice.

It is important to distinguish the transition metals from both the b.c.c. alkali metals (Li, K, Na, Cs) and from the b.c.c. rare earths and transuranic metals (γ-La, δ-Ce, γ-Yb, γ-U, ε-Pu, etc.). Each forms an isomechanical group (Chapter 18); members of

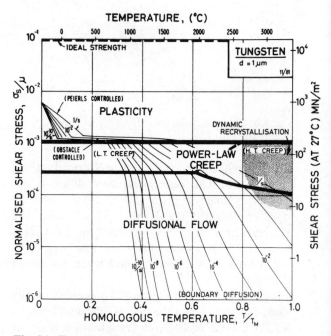

Fig. 5.1. Tungsten of grain size 1 $\mu$m. The obstacle spacing is $l = 2 \times 10^{-7}$ m.

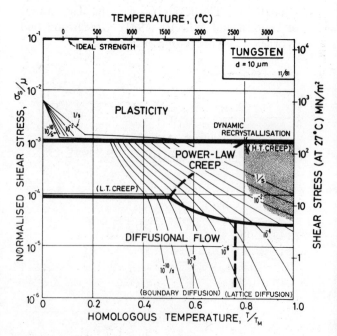

Fig. 5.2. Tungsten of grain size 10 $\mu$m.

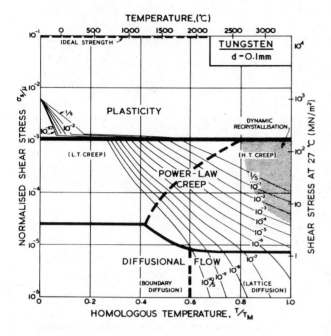

Fig. 5.3. Tungsten of grain size 100 μm.

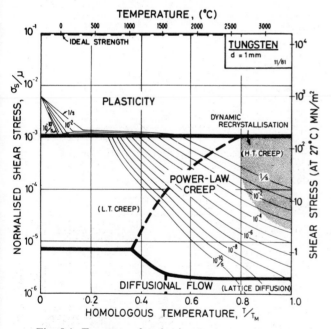

Fig. 5.4. Tungsten of grain size 1 mm.

each group have similar mechanical properties, but the groups differ significantly. The data and maps presented in this Chapter are for the b.c.c. transition metals only, and give no information about the other two groups.

One map is presented for most of the metals, computed for a "typical" grain size of 100 μm. The influence of grain size is illustrated in detail for tungsten by maps for grain sizes of 1, 10, $10^2$ and $10^3$ μm (Figs. 5.1 to 5.4). The other metals follow the same pattern.

Some of the b.c.c. transition metals show large elastic anistropy; then the various averages of the single crystal moduli differ significantly. We have taken the shear modulus at 300 K ($\mu_0$) as that appropriate for the anisotropic calculation of the energy of a $\frac{1}{2} \langle 111 \rangle$ screw dislocation (Hirth and Lothe, 1968), and have calculated a linear temperature dependence either from single-crystal data or, when there was none, from polycrystal data. This is adequate for all but niobium which shows anomalous behaviour (Armstrong et al., 1966) and iron (for which see Chapter 8).

Lattice diffusion has been studied in all these metals and is shown for each as a data plot (below). For some there is evidence that the activation energy decreases at lower temperatures. We have used a dual expression for the lattice diffusion coefficient to describe this "anomalous diffusion" in vanadium; parameters are given in Table 5.1. Similar behaviour has been demonstrated for tantalum (Pawel and Lundy, 1965) but can be well approximated by one simple Arrhenius relationship.

Complete data for grain boundary and dislocation core self-diffusion are available only for tungsten. For chromium, an activation energy for core diffusion has been reported. All other core and boundary diffusion coefficients have been estimated using the approximation $Q_c \approx Q_B \approx (2Q_v/3)$ (see Brown and Ashby, 1980, for an analysis of such correlations).

The parameters describing lattice-resistance controlled glide were obtained, when possible, by fitting experimental data to eqn. (2.12). This rate equation gives a good match to experiment; but it must be realized that the values of the parameters $\Delta F_p$ and $\hat{\tau}_p$ depend on the form chosen for $\Delta G$: a different choice (one with the form of that in eqn. (2.9), for instance) gives different values. Further, the set of values for $\Delta F_p$, $\hat{\tau}_p$ and $\dot{\gamma}_p$ obtained by fitting eqn. (2.12) to a given batch of data are not, in practice, unique; changing any one slightly still leads to an acceptable fit provided the others are changed to compensate. The set listed here gives a good fit to the data shown in the data plots.

The quantity $\hat{\tau}_p$ is determined approximately by extrapolating flow stress data to 0 K. It is more difficult to determine $\dot{\gamma}_p$ experimentally unless the flow stress is known over a wide range of strain rates. The data of Briggs and Campbell (1972) for molybdenum, and those of Raffo (1969) for tungsten, when fitted to eqn. (2.12), are well described by the value $\dot{\gamma}_p = 10^{11}$/s, and we have therefore used this for all the b.c.c. transition metals. Once $\dot{\gamma}_p$ and $\hat{\tau}_p$ are known, $\Delta F_p$ is found from the temperature-dependence of the flow stress. Finally, the rate-equation (eqn. (2.12)) is evaluated, and adjustments

**TABLE 5.1  The b.c.c. metals**

| Material | Tungsten | Vanadium | Chromium | Niobium | Molybdenum | Tantalum | α-Iron (r) |
|---|---|---|---|---|---|---|---|
| *Crystallographic and thermal data* | | | | | | | |
| Atomic volume, $\Omega$ (m³) | $1\cdot59 \times 10^{-29}$ | $1\cdot40 \times 10^{-29}$ | $1\cdot20 \times 10^{-29}$ | $1\cdot80 \times 10^{-29}$ | $1\cdot53 \times 10^{-29}$ | $1\cdot80 \times 10^{-29}$ | $1\cdot18 \times 10^{-29}$ |
| Burgers vector, $b$ (m) | $2\cdot74 \times 10^{-10}$ | $2\cdot63 \times 10^{-10}$ | $2\cdot50 \times 10^{-10}$ | $2\cdot86 \times 10^{-10}$ | $2\cdot73 \times 10^{-10}$ | $2\cdot86 \times 10^{-10}$ | $2\cdot48 \times 10^{-10}$ |
| Melting temperature, $T_M$ (K) | 3683 | 2173 | 2163 | 2741 | 2883 | 3271 | 1810 |
| *Modulus** | | | | | | | |
| Shear modulus at 300 K, $\mu_0$ (MN/m²) | $1\cdot60 \times 10^{5}$ (a) | $5\cdot01 \times 10^{4}$ (g) | $1\cdot26 \times 10^{5}$ (l) | $4\cdot43 \times 10^{4}$ (g) | $1\cdot34 \times 10^{5}$ (o) | $6\cdot12 \times 10^{4}$ (g) | $6\cdot4 \times 10^{4}$ (r) |
| Temperature dependence of modulus, $\dfrac{T_M}{\mu_0}\dfrac{d\mu}{dT}$ | $-0\cdot38$ (b) | $-0\cdot38$ (h) | $-0\cdot50$ (h) | $0\cdot0$ (b) | $-0\cdot42$ (o) | $-0\cdot42$ (b) | $-0\cdot81$ |
| *Lattice diffusion (normal)†* | | | | | | | |
| Pre-exponential, $D_{0v}$ (m²/s) | $5\cdot6 \times 10^{-4}$ (c) | $3\cdot6 \times 10^{-5}$ (i) | $2\cdot8 \times 10^{-5}$ (m) | $1\cdot1 \times 10^{-4}$ (a) | $5\cdot0 \times 10^{-5}$ (p) | $1\cdot2 \times 10^{-5}$ (q) | $2\cdot0 \times 10^{-4}$ |
| Activation energy, $Q_v$ (kJ/mole) | 585 (c) | 308 | 306 (m) | 401 (a) | 405 (p) | 413 (q) | 251 |
| *Lattice diffusion (anomalous)†* | | | | | | | |
| Pre-exponential $D_{0v}$ (m²/s) | — | $2\cdot14 \times 10^{-2}$ (j) | — | — | — | — | — |
| Activation energy $Q_v$ (kJ/mole) | — | 394 | — | — | — | — | — |
| *Boundary diffusion†* | | | | | | | |
| Pre-exponential, $D_{0b}$ (m³/s) | $3\cdot3 \times 10^{-13}$ (d) | $5\cdot0 \times 10^{-14}$ (k) | $5\cdot0 \times 10^{-15}$ (k) | $5\cdot0 \times 10^{-14}$ (k) | $5\cdot5 \times 10^{-14}$ (k) | $5\cdot7 \times 10^{-14}$ (k) | $1\cdot1 \times 10^{-12}$ (k) |
| Activation energy, $Q_b$ (kJ/mole) | 385 (d) | 209 (k) | 192 (k) | 263 (k) | 263 (k) | 280 (k) | 174 (k) |
| *Core diffusion†* | | | | | | | |
| Pre-exponential, $a_c D_{0c}$ (m⁴/s) | $7\cdot9 \times 10^{-22}$ (d) | $1\cdot0 \times 10^{-22}$ (d) | $1\cdot0 \times 10^{-23}$ (n) | $1\cdot3 \times 10^{-23}$ (k) | $1\cdot0 \times 10^{-22}$ (k) | $1\cdot0 \times 10^{-23}$ (k) | $1\cdot0 \times 10^{-23}$ (k) |
| Activation energy, $Q_c$ (kJ/mole) | 378 (d) | 209 (d) | 192 (n) | 263 (k) | 263 (k) | 280 (k) | 174 (k) |
| *Power-law creep* | | | | | | | |
| Exponent, $n$ | 4·7 (e) | 5·0 (e) | 4·3 (e) | 4·4 (e) | 4·85 (e) | 4·2 (e) | 6·9 (e) |
| Dorn constant‡, $A$ | $1\cdot1 \times 10^{8}$ (e) | $1\cdot0 \times 10^{8}$ (e) | $1\cdot3 \times 10^{6}$ (e) | $4\cdot0 \times 10^{7}$ (e) | $1\cdot0 \times 10^{8}$ (e) | $7\cdot5 \times 10^{5}$ (e) | $7\cdot0 \times 10^{13}$ (e) |
| *Obstacle-controlled glide* | | | | | | | |
| 0 K flow stress, $\hat{\tau}/\mu_0$ | $1\cdot4 \times 10^{-3}$ (f) | $1\cdot3 \times 10^{-3}$ (f) | $1\cdot3 \times 10^{-3}$ (f) | $1\cdot4 \times 10^{-3}$ (f) | $1\cdot0 \times 10^{-3}$ (f) | $1\cdot4 \times 10^{-3}$ (f) | $1\cdot7 \times 10^{-3}$ (f) |
| Pre-exponential, $\dot{\gamma}_0$(s⁻¹) | $10^{6}$ | $10^{6}$ | $10^{6}$ | $10^{6}$ | $10^{6}$ | $10^{6}$ | $10^{6}$ |
| Activation energy, $\Delta F/\mu_0 b^3$ | 0·5 | 0·5 | 0·5 | 0·5 | 0·5 | 0·5 | 0·5 |
| *Lattice-resistance-controlled glide* | | | | | | | |
| 0 K flow stress, $\hat{\tau}_p/\mu_0$ | $6\cdot5 \times 10^{-3}$ (f) | $1\cdot7 \times 10^{-2}$ (f) | $9\cdot8 \times 10^{-3}$ (f) | $1\cdot7 \times 10^{-2}$ (f) | $6\cdot6 \times 10^{-3}$ (f) | $1\cdot3 \times 10^{-2}$ (f) | $1\cdot0 \times 10^{-2}$ (f) |
| Pre-exponential, $\dot{\gamma}_p$(s⁻¹) | $10^{11}$ | $10^{11}$ | $10^{11}$ | $10^{11}$ | $10^{11}$ | $10^{11}$ | $10^{11}$ |
| Activation energy, $\Delta F_p/\mu_0 b^3$ | $5\cdot2 \times 10^{-2}$ (f) | $8\cdot8 \times 10^{-2}$ (f) | $5\cdot9 \times 10^{-2}$ (f) | $8\cdot9 \times 10^{-2}$ (f) | $5\cdot7 \times 10^{-2}$ (f) | $6\cdot6 \times 10^{-2}$ (f) | $1\cdot0 \times 10^{-1}$ |

32

$$* \quad \mu = \mu_0\left(1 + \frac{(T-300)}{T_M}\frac{T_M}{\mu_0}\frac{d\mu}{dT}\right)$$

$$\dagger \quad D_v = D_{0v}\exp-\left(\frac{Q_v}{RT}\right); \quad \delta D_b = \delta D_{0b}\exp-\left(\frac{Q_b}{RT}\right);$$

$$a_c D_c = a_c D_{0c}\exp-\left(\frac{Q_c}{RT}\right)$$

‡ This value of $A$ refers to tensile stress and strain-rate. The maps relate shear stress and strain rate. In constructing them we have used $A_s = (\sqrt{3})^{n+1} A$.

(a) Lundy et al. (1965).
(b) Köster (1948).
(c) Robinson and Sherby (1969).
(d) Kreider and Bruggeman (1967).
(e) Derived by fitting creep data to the diffusion-controlled creep equation, as described in the text.
(f) Derived by fitting low-temperature yield data to the glide equations for obstacle and lattice resistance controlled glide, as described in the text.
(g) Bolef (1961).
(h) Armstrong and Brown (1964).
(i) Peart (1965).

(j) See text for vanadium.
(k) Estimated by setting $D_{0B} = b \cdot D_{0v}$ and $Q_B = 0.66\, Q_v$ or $Q_c$ (if known), $a_c D_{0c} = 4b^2 D_{0v}$ and $Q_c = 0.66\, Q_v$ or $Q_B$ (if known), unless otherwise stated in text.
(l) Bolef and de Klerk (1963).
(m) Hagel (1962).
(n) Gleiter and Chalmers (1972).
(o) Bolef and de Klerk (1962).
(p) Askill and Tomlin (1963).
(q) Pawel and Lundy (1965).
(r) Data for $\alpha$-iron are described in full in Chapter 8. They are listed here to permit comparison.

made to $\hat{\tau}_p$ and $\Delta F_p$ to give the best fit to the data.

The yield stress of some b.c.c. metals decreases as the metal is made purer. There is debate as to whether the Peierls' stress $\hat{\tau}_p$ results from an intrinsic lattice resistance or from small concentrations of interstitial impurities. The question need not concern us here, except that it must be recognized that the yield parameters refer to a particular level of purity.

The critical resolved shear stress of b.c.c. single crystals is related to the polycrystalline shear strength by the Taylor factor, $M_s = 1.67$ (Chapter 2, Section 2.2). The data plots of this chapter record both; it can be seen that polycrystal data and single-crystal critical resolved shear stress data (for comparable purities) differ by about this factor, at all temperatures. Table 5.1 records the polycrystal shear strength. As discussed in Chapter 3 (Section 3.2), the lattice resistance-controlled and obstacle-controlled glide are treated as alternative mechanisms: that leading to the slowest strain rate is controlling. This results in a sharp corner in the strain rate contours. A more complete model of the mechanism interaction would smooth the transition, as the data plots suggest. We have arbitrarily chosen an obstacle spacing of $\ell = 2 \times 10^{-7}$ m (or a dislocation density of $\rho = 2.5 \times 10^{13}/\text{m}^2$). This value is lower than that used for f.c.c. maps, and describes a lower state of work-hardening. It is not necessarily that of the samples recorded in the data plots, although it gives a fair description of most of the data.

### References for Section 5.1

Armstrong, P. E., Dickinson, J. M. and Brown, H. L. (1966) *Trans. AIME* **236**, 1404.

Briggs, T. L. and Campbell, J. D. (1972) *Acta Met.* **20**, 711.

Brown, A. M. and Ashby, M. F. (1980) *Acta Met.* **28**, 1085.

Hirth, J. P. and Lothe, J. (1968) *Theory of Dislocations.* McGraw-Hill, p. 435.

Pawel, R. E. and Lundy, T. S. (1965) *J. Phys. Chem. Solids* **26**, 937.

Raffo, P. L. (1969) *J. Less Common Metals* **17**, 133.

## 5.2 ORIGINS OF THE DATA FOR THE B.C.C. METALS

### Tungsten (Figs. 5.1 to 5.7)

Figs. 5.1 to 5.4 show maps for tungsten of grain size 1 $\mu$m, 10 $\mu$m, 100 $\mu$m, and 1 mm, showing the influence of grain size. They are based on the data plotted in Figs. 5.5 to 5.7, and on the parameters listed in Table 5.1.

Lattice diffusion data for tungsten are plotted in Fig. 5.7. There is good agreement between the data plotted as the lines labelled, 3, 5 and 6 (Andelin

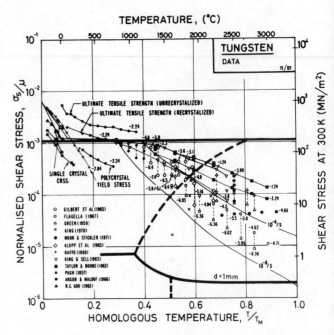

Fig. 5.5. Data for tungsten, divided into blocks. Each block is fitted to a rate-equation. The numbers are $\log_{10}(\dot{\gamma})$.

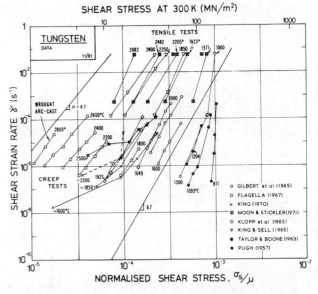

Fig. 5.6. Creep data for tungsten. Data are labelled with the temperature in °C.

*et al.*, 1965; Robinson and Sherby, 1969; Kreider and Bruggeman, 1967). We have taken the co-efficients cited by Robinson and Sherby (1969) because they typify these data. Boundary and core diffusion co-efficients are from Kreider and Bruggeman (1967).

The high-temperature creep of tungsten has been reviewed by Robinson and Sherby (1969), who demonstrated that most of the available data can be divided into high-temperature creep above 2000°C and low-temperature creep below. The high-temperature data are those of Flagella (1967)

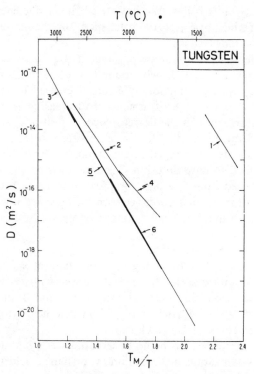

Fig. 5.7. Lattice diffusion data for tungsten. The data are from (1) Vasilev and Chernomorchenko (1956); (2) Danneberg (1961); (3) Andelin *et al.* (1965); (4) Neumann and Hirschwald (1966); (5) Robinson and Sherby (1969); and (6) Kreider and Bruggeman (1967).

Brodrick, R. F. and Fritch, D. J. (1964) *Proc. ASTM* **64**, 505.

Danneberg, W. (1961) *Metall.* **15**, 977.

Flagella, P. N. (1967) GE-NMPO, GEMP-543, Aug. 31, presented at Third International Symposium—High Temperature Technology, Asilomar, Calif., September.

Glasier, L. F., Allen, R. D. and Saldinger, I. L. (1959) "Mechanical and Physical Properties of the Refractory Metals", Aerojet—General Corp., Report No. M1826.

Green, W. V. (1959) *Trans. AIME* **215**, 1057.

King, G. W. (1970) Westinghouse Report BLR 90284-2.

King, G. W. and Sell, H. G. (1965) *Trans. AIME* **233**, 1104.

Klopp. W. D., Witzke, W. R. and Raffo, P. L. (1965) *Trans AIME* **233**, 1860.

Koo, R. C. (1963) *Acta Met.* **11**, 1083.

Köster, W. (1948) *Z. Metallk.* **39**, 1.

Kreider, K. G. and Bruggeman, G. (1967) *Trans. AIME* **239**, 1222.

Lundy, T. S., Winslow, F. R., Pawel, R. E. and McHargue, C. J. (1965) *Trans. AIME* **233**, 1533.

Moon, D. M. and Stickler, R. (1971) *Phil. Mag.* **24**, 1087.

Neumann, G. M. and Hirschwald, W. (1966) *Z. Naturforsch.* **21a**, 812.

Pugh, J. W. (1957) *Proc. ASTM* **57**, 906.

Raffo, P. L. (1969) *J. Less-Common Metals* **17**, 133.

Robinson, S. L. and Sherby, O. D. (1969). *Acta Met.* **17**, 109 (estimated value).

Taylor, J. L. and Boone, D. H. (1963) *Trans. ASM* **56**, 643.

Vasilev, V. P. and Chernomorchenko, S. G. (1956) *Zavodsk. Lab.* **22**, 688, AEC-tr-4726.

—for wrought arc-cast tungsten—and of King and Sell (1965). These data show faster creep-rates than those of Green (1959) and Flagella (1967) for powder-metallurgy tungsten. The low-temperature creep region is represented by the data of Gilbert *et al.* (1965) which show $n \cong 7$ between 1300°C and 1900°C with an apparent activation energy of about 376 kJ/mole (although they are nearly an order of magnitude slower than the data of Flagella (1967) in the overlapping temperature range). This general behaviour is also indicated by other papers. The low-temperature yield parameters for tungsten are based on the polycrystalline yield data of Raffo (1969) (see also Chapter 2 and Fig. 2.3). These data are in general agreement with single-crystal critical resolved shear stress data of Koo (1963) and Argon and Maloof (1966).

The field of dynamic recrystallization is based on the observations of Glasier *et al.* (1959) and of Brodrick and Fritch (1964) who tested tungsten up to 0·998 $T_M$.

### References for tungsten

Andelin, R. L., Knight, J. D. and Kahn, M. (1965) *Trans. AIME* **233**, 19.

Argon, A. S. and Maloof, S. R. (1966) *Acta Met.* **14**, 1449.

### Vanadium (Figs. 5.8 to 5.10)

Fig. 5.8 shows a map for vanadium with a grain size of 0·1 mm. It is based on data shown in Fig. 5.9 and parameters listed in Table 5.1. There are so

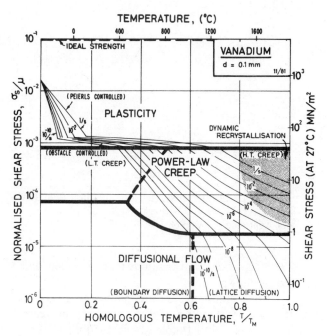

Fig. 5.8. Vanadium of grain size 100 μm.

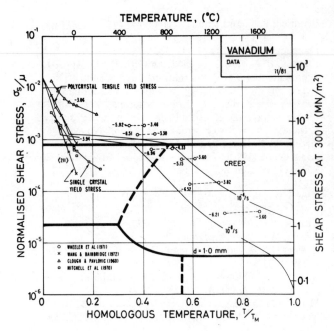

Fig. 5.9. Data for vanadium, divided into blocks. The numbers are $\log_{10}(\dot{\gamma})$.

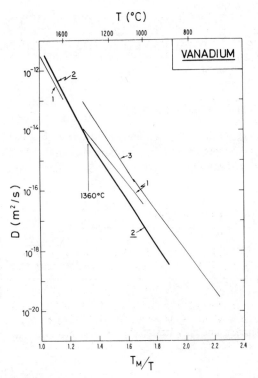

Fig. 5.10. Lattice diffusion data for vanadium. The data are from (1) Lundy and Mehargue (1965); (2) Peart (1965); and Agarwala et al. (1968).

few creep data for vanadium that no creep data plot is shown.

Lattice diffusion data are plotted in Fig. 5.10. It is found that the activation energy decreases from about 390 kJ/mole above 1350°C to about

310 kJ/mole below. We have used the coefficients determined by Peart (1965), labelled 2 on the figure, as the most reliable.

The dislocation creep parameters are based on Wheeler et al. (1971), who found that the activation for creep, like that for diffusion, decreased (from 472 to 393 to 318 kJ/mole) with decreasing temperature, though the decrease occurred at a lower temperature. At their lowest temperatures the activation energy dropped to 226 kJ/mole which they ascribed to core diffusion. The stress exponent, $n$, increased from 5 at high temperatures to 8 at low temperatures in accordance with the expected low-temperature creep behaviour (Chapter 2, Section 2.4).

The lattice-resistance parameters are derived from the high-purity polycrystalline data of Wang and Bainbridge (1972). They agree well with those derived from single-crystal data (Wang and Brainbridge, 1972; Mitchell et al., 1970) when the Taylor factor conversion is included.

There appear to be no observations of dynamic recrystallization in vanadium. The shaded field is that typical of the other b.c.c. metals.

### References for vanadium

Agarwala, R. P., Murarka, S. P. and Anand, M. S. (1968) *Acta Met.* **16**, 61.

Armstrong, P. E. and Brown, H. L. (1964) *Trans. AIME* **230**, 962.

Bolef, D. I. (1961) *J. Appl. Phys.* **32**, 100.

Clough, W. R. and Pavlovic, A. S. (1960) *Trans. ASM* **52**, 948.

Lundy, T. S. and Mehargue, C. J. (1965) *Trans. AIME* **233**, 243.

Peart, R. F. (1965) *J. Phys. Chem. Solids* **26**, 1853.

Mitchell, T. E., Fields, R. J. and Smialek, R. L. (1970) *J. Less-Common Metals* **20**, 167.

Wang, C. T. and Bainbridge, D. W. (1972) *Met Trans.* **3**, 3161.

Wheeler, K. R., Gilbert, E. R., Yaggee, F. L. and Duran, S. A. (1971) *Acta Met.* **19**, 21.

### Chromium (Figs. 5.11 to 5.13)

Fig. 5.11 shows a map for chromium of grain size 0·1 mm. It is based on the data shown in Fig. 5.12 and the parameters listed in Table 5.1. There are so few measurements of creep of chromium that no creep plot is given.

Lattice diffusion data are plotted in Fig. 5.13. We have chosen the coefficients of Hagel (1962) (labelled 6) which come closest to describing all the available data.

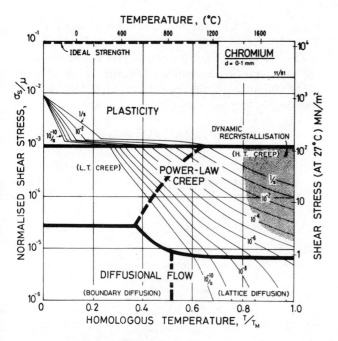

Fig. 5.11. Chromium with a grain size of 100 μm.

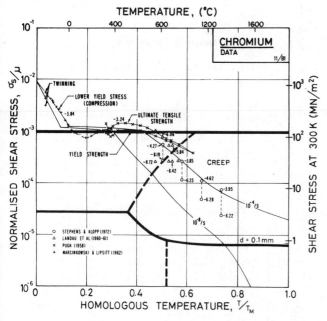

Fig. 5.12. Data for chromium, divided into blocks. The numbers are $\log_{10}(\dot{\gamma})$.

The dislocation creep parameters are derived from the data of Stephens and Klopp (1972) who tested high-purity iodide–chromium. Their data at 1316°C to 1149°C show a stress exponent of $n \cong 4.5$; data at 816°C and 982°C show $n \cong 6.5$. The lower-temperature data, however, show no tendency toward a lower activation energy. There is, therefore, no conclusive evidence for (or against) a low-temperature creep field in chromium. The parameters describing the lattice resistance are based on the data of Marcinkowski and Lipsitt (1962) for

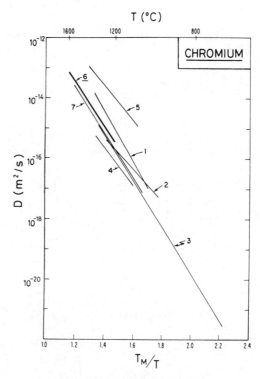

Fig. 5.13. Lattice diffusion data for chromium. The data are from (1) Gruzin *et al.* (1959); (2) Paxton and Gondolf (1959); (3) Bokshtein *et al.* (1959) and (1961); (4) Bogdanov (1960); (5) Ivanov *et al.* (1962); (6) Hagel (1962); and (7) Askill and Tomlin (1965).

polycrystals. Twinning is observed in chromium at 88 K.

There are no reports of dynamic recrystallization in chromium. The shaded field is that typical of other b.c.c. metals.

### References for chromium

Askill, J. and Tomlin, D. H. (1965) *Phil. Mag.* **11**, 467.

Bogdanov, N. A. (1960) *Russ. Met. Fuels* (English translation), **3**, 95.

Bokshtein, S. Z., Kishkin, S. T. and Moroz, L. M. (1959) Investigation of the structure of Metals by Radioactive Isotope Methods, Moscow; (1961) AEC-tr-4505.

Bolef, D. I. and de Klerk, J. (1963) *Phys. Rev.* **129**, 1063.

Gleiter, H. and Chalmers, B. (1972) *Prog. Mat. Sci.* **16**, 93.

Gruzin, P. L., Pavlinov, L. V. and Tyutyunnik, A. D. (1959) *Izv. Akad. Nauk., SSSR Ser. Fiz.* **5**, 155.

Hagel, W. C. (1962) *Trans. AIME* **224**, 430.

Ivanov, L. I., Matveeva, M. P., Morozov, V. A. and Prokoshkin, D. A. (1962) *Russ. Met. Fuels* (English translation), **2**, 63, V.

Landau, C. S., Greenaway, H. T. and Edwards, A. R. (1960–61) *J. Inst. Met.* **89**, 97.

Marcinkowski, M. J. and Lipsitt, P. A. (1962) *Acta Met.* **10**, 95.

Paxton, H. W. and Gondolf, E. G. (1959) *Arch. Eisenhüttenw.* **30**, 55.

Pugh, J. W. (1958) *Trans. ASM* **50**, 1072.
Stephens, J. R. and Klopp, W. D. (1972) *J. Less-Common
Metals* **27**, 87.

### Niobium (columbium) (Figs. 5.14 to 5.17)

A map for niobium with a grain size of 100 μm, is shown in Fig. 5.14. It is based on data shown in Figs. 5.15 and 5.16 and on the parameters listed in Table 5.1.

The elastic constants of niobium have an anomalous temperature dependence (Armstrong *et al.*, 1966). Because of this, we have neglected the temperature-dependence of the modulus.

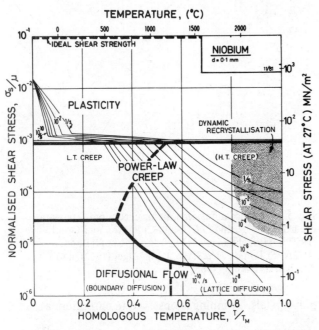

Fig. 5.14. Niobium (columbium) with a grain size of 100 μm.

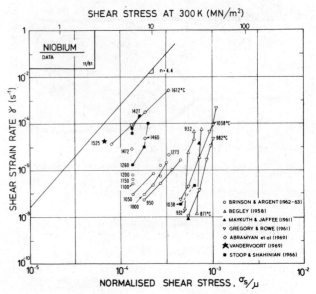

Fig. 5.16. Creep data for niobium. Data are labelled with the temperature in °C.

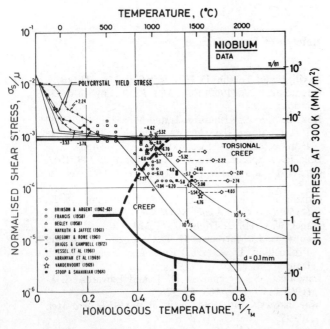

Fig. 5.15. Data for niobium, divided into blocks. The numbers are $\log_{10} (\dot{\gamma})$.

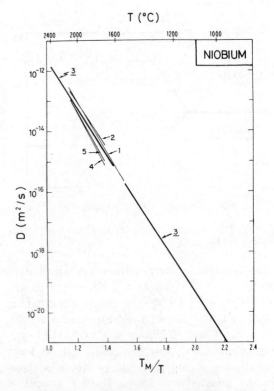

Fig. 5.17. Lattice diffusion data for niobium. The data are from (1) Resnick and Castleman (1960); (2) Peart *et al.* (1962); (3) Lundy *et al.* (1965); (4) Lyubinov *et al.* (1964); single crystals; and (5) Lyubinov *et al.* (1964), polycrystals.

Lattice diffusion data are plotted in Fig. 5.17. We have used coefficients derived from the measurements of Lundy *et al.* (1965), labelled *3*, which lie centrally through the remaining data, and cover an exceptionally wide range of temperature.

Creep data for niobium are somewhat limited. The parameters listed in Table 5.1 are based on the measurements of Brunson and Argent (1962); these agree well with the data of Stoop and Shahinian (1966) but not with those of Abramyan *et al.* (1969), which show slower strain-rates.

The low-temperature yield behaviour of niobium has been extensively studied for both polycrystals and single crystals. We have derived yield parameters from the data of Briggs and Campbell (1972), which agree well with earlier studies.

There appear to be no studies above 1400°C. The dynamic recrystallization field is arrived at by analogy with other b.c.c. metals.

### References for niobium

Abramyan, E. A., Ivanov, L. I. and Yanushkevich, V. A. (1969) *Fiz. Metal. Metalloved* **28(3)**, 496.

Begley, R. T. (1958) Westinghouse Report WADC TR-57-344, Part II.

Bolef, D. I. (1961) *J. Appl. Phys.* **32**, 100.

Briggs, T. L. and Campbell, J. D. (1972) *Acta Met.* **20**, 711.

Brunson, G. and Argent, B. B. (1962) *J. Inst. Met.* **91**, 293.

Francis, E. L. (1958) UKAEA Report IGR R/R 304, Risley, Library and Information Department, Warrington, Lancashire.

Gregory, D. P. and Rowe, G. H. (1961) *Columbium Metallurgy*, Proc. of AIME Symposium, Boulton Landing. Interscience, N.Y., p. 309.

Lundy, T. S., Winslow, F. R., Pawel, R. E. and McHargue, C. J. (1965) *Trans. AIME* **233**, 1533.

Lyubinov, V. D., Geld, P. V. and Shveykin, G. P. (1964) *Izv. Akad. Nauk. S.S.S.R. Met. i Gorn Delo* **5**, 137; *Russ. Met. Mining* **5**, 100.

Maykuth, D. J. and Jaffee, R. I. (1961) *Columbium Metallurgy*, Proc. of AIME Symposium, Boulton Landing. Interscience, N.Y., p. 223.

Peart, R. F., Graham, D. and Tomlin, D. H. (1962) *Acta Met.* **10**, 519.

Resnick, R. and Castleman, L. S. (1960) *Trans. AIME* **218**, 307.

Stoop, J. and Shahinian, P. (1966) *High Temperature Refractory metals*, Part 2, AIME Symposium. Gordon & Breach Science Publishers, N.Y., p. 407.

Vandervoort, R. R. (1969) *Trans. AIME* **245**, 2269.

Wessel, E. T., France, L. L. and Begley, R. T. (1961) *Columbium Metallurgy*, Proc. of AIME Symposium, Boulton Landing. Interscience, N.Y., p. 459.

### Molybdenum (Figs. 5.18 to 5.21)

Fig. 5.18 shows a map for molybdenum with a grain size of 100 μm. It is based on the data shown

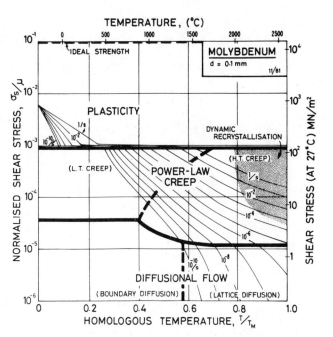

Fig. 5.18. Molybdenum with a grain size of 100 μm.

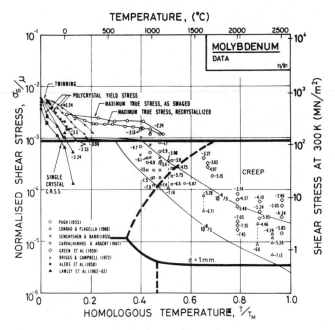

Fig. 5.19. Data for molybdenum, divided into blocks. The numbers are $\log_{10}(\dot{\gamma})$.

in Figs. 5.19 and 5.20 and on the parameters listed in Table 5.1.

Lattice diffusion data for molybdenum are plotted in Fig. 5.21. There is substantial agreement between all investigators. We have used the parameters derived by Askill and Tomlin (1963) from studies of polycrystals (labelled *6*).

Creep of molybdenum has been well studied. The high-temperature creep parameters are derived

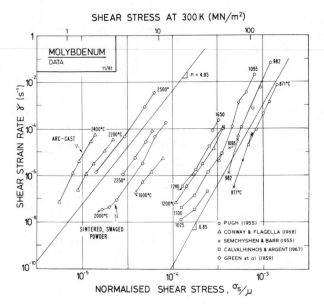

Fig. 5.20. Creep data for molybdenum. Data are labelled with the temperature in °C.

existence of a low-temperature creep field. Carvalhinos and Argent (1967), Pugh (1955) and Semchyshen and Barr (1955) all found $n = 6$–$8$ for $T = 0.4$–$0.53$ $T_M$, and an activation energy lower than that for volume diffusion.

The low-temperature yield parameters are derived from data of Briggs and Campbell (1972). These are not the lowest known yield strengths, and therefore do not describe molybdenum of the highest purity. Lawley *et al.* (1962), found that the polycrystalline yield stress can be further lowered by nearly a factor of 2 by repeated zone refining, although the change in purity cannot be detected.

Above 2000°C ($0.8$ $T_M$) molybdenum of commercial purity shows dynamic recrystallization (Glasier *et al.*, 1959; Hall and Sikora, 1959); pure molybdenum does so at a slightly lower temperature. The positioning of the dynamic recrystallization field is based on an average of these observations.

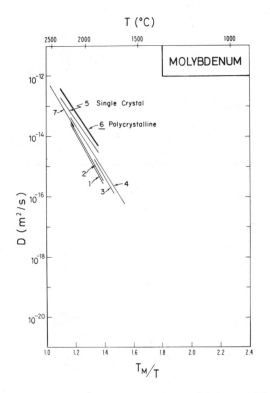

Fig. 5.21. Lattice diffusion data for molybdenum. The data are from (1) Borisov *et al.* (1959); (2) Gruzin *et al.* (1959); (3) Brofin *et al.* (1960); (4) Danneburg and Krautz (1961); (5) Askill and Tomlin (1963); single crystals; (6) Askill and Tomlin (1963), polycrystals; and (7) Pavlinov and Bikov (1964).

from the data of Conway and Flagella (1968) (which are more than an order of magnitude faster in creep rate than those of Green *et al.*, 1959), giving an exponent of $n = 4.85$. Several studies suggest the

### References for molybdenum

Alers, G. A., Armstrong, R. W. and Bechtold, J. H. (1958) *Trans. AIME* **212**, 523.

Askill, J. and Tomlin, D. H. (1963) *Phil. Mag.* **8(90)**, 997.

Bolef, D. I. and de Klerk, J. (1962) *J. Appl. Phys.* **33**, 2311.

Borisov, Y. V., Gruzin, P. L. and Pavlinov, L. V. (1959) *Met. i. Metalloved. Chistykh Metal.* **1**, 213.

Briggs, T. L. and Campbell, J. D. (1972) *Acta Met.* **20**, 711.

Brofin, M. B., Bokshtein, S. Z. and Zhukhovitskii, A. A. (1960) *Zavodsk. Lab.* **26(7)**, 828.

Carvalhinhos, H. and Argent, B. B. (1967) *J. Inst. Met.* **95**, 364.

Conway, J. B. and Flagella, P. N. (1968) *Seventh Annual Report—AEC Fuels and Materials Development Program*, GE–NMPO, GEMP–1004; also (1971) *Creep Rupture Data for the Refractory Metals to High Temperatures*. Gordon & Breach, p. 610.

Danneberg, W. and Krautz, E. (1961) *Naturforsch.* **16(a)**, 854.

Glasier, L. F., Allen, R. D. and Saldinger, I. L. (1959) "Mechanical and Physical Properties of Refractory Metals", Aerojet General Corp Report No. M1826.

Green, W. V., Smith, M. C. and Olson, D. M. (1959) *Trans. AIME* **215**, 1061.

Gruzin, P. L., Pavlinov, L. V. and Tyutyunnik, A. D. (1959) *Izv. Akad. Nauk S.S.S.R. Ser. Fiz.* **5**, 155.

Hall, R. W. and Sikora, P. F. (1959) "Tensile Properties of Molybdenum and Tungsten from 2500 to 3700 F", NADA Memo 3-9E.

Lawley, A., Van den Sype, J. and Maddin, R. (1962) *J. Inst. Met.* **91**, 23.

Pavlinov, L. V. and Bikov, V. N. (1964) *Fiz. Met. i Metalloved* **18**, 459.

Pugh, J. W. (1955) *Trans. ASM* **47**, 984.

Semchyshen, M. and Barr, R. Q. (1955) Summary Report NR 039-002 to Office of Naval Research, Climax Molybdenum Co.

## Tantalum (Figs. 5.22 to 5.25)

Fig. 5.22 shows a map for tantalum with a grain size of 100 μm. It is based on data plotted in Figs. 5.23 and 5.24 and on the parameters listed in Table 5.1.

Lattice diffusion data are plotted in Fig. 5.25. We have used the parameters of Pawel and Lundy

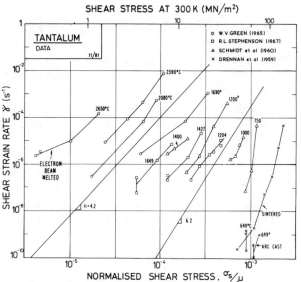

Fig. 5.24. Creep data for tantalum. Data are labelled with temperature in °C.

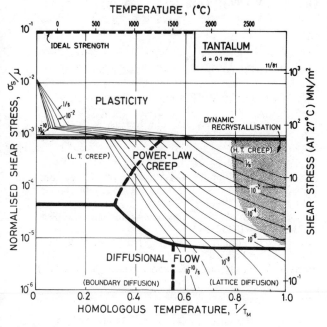

Fig. 5.22. Tantalum with a grain size of 100 μm.

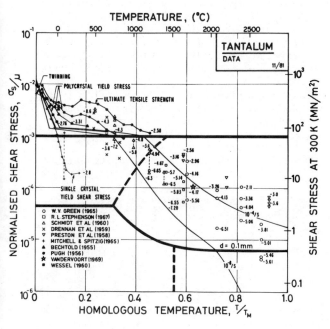

Fig. 5.23. Data for tantalum, divided into blocks. The numbers are $\log_{10}(\dot\gamma)$.

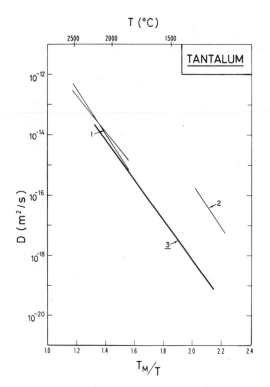

Fig. 5.25. Lattice diffusion data for tantalum. The data are from (1) Eager and Langmuir (1953); (2) Gruzin and Meshkov (1955); (3) Pawel and Lundy (1965).

(1965), whose data are in good agreement with measurements of Eager and Langmuir (1953), and which cover an exceptionally wide range of temperature.

The high-temperature creep parameters for tantalum are taken from Green (1965). His steady-state data show an activation energy which increases with temperature (as pointed out by Flinn and Gilbert, 1966) but this may be explained by the fact that the highest temperature tests lie in the field of dynamic recrystallization. The stress exponent is $n \cong 4 \cdot 2$. There is some indication of low-temperature creep behaviour: the data of Schmidt *et al.* (1960) at 1000°C and 1200°C show $n \cong 6$. The yield parameters for tantalum are based on data of Wessel as cited by Bechtold *et al.* (1961), and are in good agreement with the single-crystal data of Mitchell and Spitzig (1965), adjusted by the appropriate Taylor factor (Chapter 2, Section 2.2).

The existence of a field of dynamic recrystallization can be inferred from the data reported by Green (1965), Preston *et al.* (1958) and Glasier *et al.* (1959). Its position was determined partly from these observations, and partly by analogy with other b.c.c. metals.

### References for tantalum

Bechtold, J. H. (1955) *Acta Met.* **3**, 249.
Bechtold, J. H., Wessel, E. T. and France, L. L. (1961) *Refractory Metals and Alloys* (edited by M. Semchyshen and J. J. Harwood), AIME, p. 49.
Bolef, D. I. (1961) *J. Appl. Phys.* **32**, 100.
Drennan, D. C., Lanston, M. E., Slunder, C. J. and Dunleavy, J. G. (1959) Battelle Memorial Inst., EMI Report 1326.
Eager, R. L. and Langmuir, D. B. (1953) *Phys. Rev.* **89**, 911, errata, p. 890; Langmuir, D. B. (1952) *Phys. Rev.* **86**, 642.
Glasier, L. F., Allen, R. D. and Saldinger, I. L. (1959) "Mechanical and Physical Properties of the Refractory Metals", Aerojet-General Corp. Report No. M1826.
Green, W. V. (1965). *Trans. AIME* **233**, 1818.
Gruzin, P. L. and Meshkov, V. I. (1955) *Vopr. Fiz. Met. i Metalloved.*, Sb. Nauchn. Rabot. Inst. Metallofiz. Akad. Nauk. Ukr., S.S.R. 570, AEC-tr-2926.
Köster, W. (1948) *Z. Metallk.* **39**, 1.
Mitchell, T. E. and Spitzig, W. A. (1965) *Acta Met.* **13**, 1169.
Pawel, R. E. and Lundy, T. S. (1965) *J. Phys. Chem. Solids* **26**, 927.
Preston, J. B., Roo, W. P. and Kattus, J. R. (1958) Tech. Report, Southern Research Inst., WADC TR 57-049, Part I.
Pugh, J. W. (1966) *Trans. ASM* **48**, 677.
Schmidt, F. F., Klopp, W. D., Albrecht, W. M., Holdon, F. L., Ogden, H. R. and Jaffee, R. I. (1960) Batelle Memorial Inst. Report. WADD TR 59-13.
Stephenson, R. L. (1967) Oak Ridge Technical Report ORNL-TM-1944.
Vandervoort, R. R. (1969) *Trans. AIME* **245**, 2269.

# CHAPTER 6

# THE HEXAGONAL METALS: Zn, Cd, Mg AND Ti

THE HEXAGONAL metals form the basis of a range of industrial alloys of which those based on magnesium, on zinc and on titanium are the most familiar. These metals and alloys resemble the f.c.c. metals in many regards, but their structure, though close-packed, is of lower symmetry. This necessitates new modes of slip, some of which are opposed by a considerable lattice-resistance, making them more prone to twinning and to cleavage fracture. Maps and data for four of them (Zn, Cd, Mg and Ti) are shown in Figs. 6.1 to 6.13. The parameters used to compute the maps are listed in Table 6.1.

## 6.1 GENERAL FEATURES OF THE MECHANICAL BEHAVIOUR OF HEXAGONAL METALS

The maps we show here for the hexagonal metals resemble somewhat those for the f.c.c. metals. The fields of plasticity, power-law creep and diffusional flow are of the same general size, though they are displaced to slightly higher stresses. The h.c.p. metals differ from the f.c.c. in two regards. First, twinning is much more prevalent, becoming an important mode of deformation at low temperatures. And second, the low symmetry means that at least two different classes of slip system must operate for polycrystal plasticity. Three of the metals considered here (Zn, Cd and Mg) have c/a ratios which are greater than 1·6, and slip most easily on the basal plane (0001), with no measurable lattice resistance; but the concomitant prismatic or pyramidal slip, or twinning, all require that at low temperatures a lattice resistance of general magnitude $5 \times 10^{-3} \mu$ must be overcome. Ti, like Zr and Hf, has a c/a ratio of less than 1·6, and slips most easily on the prism planes (10$\bar{1}$0); the lattice resistance on the basal and pyramidal planes is larger. In all h.c.p. metals this lattice resistance falls with increasing temperature (like that of the b.c.c. metals) until it is obscured by obstacle-strengthening associated with work-hardening. A satisfactory theoretical treatment and rate-equation for macroscopic plasticity involving coupled slip on dissimilar

systems is not at present available. We have avoided the difficulty by presenting maps for heavily worked polycrystals, such that work-hardening determines the flow strength at all temperatures—that is why they resemble those for f.c.c. metals.

At very high temperatures ($<0·8$ $T_M$) there is evidence, for Zn, Cd and Mg, for a creep field of anomalously high activation energy. It is generally explained as a result of a rapid drop in the resistance to shear on non-basal systems, though it may simply reflect the onset of dynamic recrystallization. The mechanism is not yet clear; but since it results in important changes in strength, we have included it in an empirical way, as described in the text (Section 6.2 and in note (f) of Table 6.1).

The maps for Ti (Figs. 6.10, 6.11 and 17.4) are divided at the phase boundary (1115 K; 0·6 $T_M$) into two parts. There is a discontinuity in strain-rate at the phase boundary; at high stresses the creep-rate increases (by a factor of about 10) on transforming from $\alpha$- to $\beta$-Ti; at low stresses it decreases. Because of this, the strain-rate/stress map (Fig. 6.11) has an inaccessible region (heavily shaded) and an overlapping region (broken field boundaries) separating the two phases.

Ti resembles Hf and Zr in having low shear moduli in certain planes, an unexpectedly low diffusion coefficient in the h.c.p. $\alpha$-phase, and anomalous diffusion in the b.c.c. $\beta$-phase, suggesting an activation energy which increases with temperature. Power-law creep in both $\alpha$-Ti and $\alpha$-Zr has an activation energy which is larger, by a factor of about 2, than that for lattice diffusion. There is no satisfactory explanation for this at present; it may be related to the requirement of slip on two (or more) dissimilar slip systems for polycrystal plasticity. The activation energy for creep in $\beta$-Ti is close to that for self-diffusion.

One consequence of the anisotropic lattice resistance in these metals is that texture has an unusually profound effect on strength. Textured, extruded zinc, for example, creeps (under a given load) at rates which differ by a factor of 100 in the longitudinal and transverse direction (Edwards *et al.*, 1974)—a much larger factor than that found for

**TABLE 6.1: The h.c.p. metals**

| Materials | Zinc | Cadmium | Magnesium | $\alpha$-Ti | $\beta$-Ti |
|---|---|---|---|---|---|
| *Crystallographic and thermal data* | | | | | |
| Atomic volume, $\Omega$ (m$^3$) | $1.52 \times 10^{-28}$ | $2.16 \times 10^{-28}$ | $2.33 \times 10^{-28}$ | $1.76 \times 10^{-29}$ (p) | $1.81 \times 10^{-29}$ (p) |
| Burger's vector, $b$ (m) | $2.67 \times 10^{-10}$ | $2.93 \times 10^{-10}$ | $3.21 \times 10^{-10}$ | $2.95 \times 10^{-10}$ (p) | $2.86 \times 10^{-10}$ (p) |
| Melting temperature, $T_m$ (K) | 693 | 594 | 924 | 1933 | 1933 |
| *Modulus\** | | | | | |
| Shear modulus at 300 K, $\mu_0$ (MN/m$^2$) | $4.93 \times 10^4$ (a) | $2.78 \times 10^4$ (i) | $1.66 \times 10^4$ (l) | $4.36 \times 10^4$ (q) | $2.05 \times 10^4$ (u) |
| Temperature dependence of modulus, $\dfrac{T_m}{\mu_0}\dfrac{d\mu}{dT}$ | $-0.50$ (a) | $-0.59$ (i) | $-0.49$ (l) | $-1.2$ (q) | $-0.5$ (u) |
| *Lattice diffusion†* | | | | | |
| Pre-exponential, $D_{0v}$ (m$^2$/s) | $1.3 \times 10^{-5}$ (b) | $5.0 \times 10^{-6}$ (j) | $1.0 \times 10^{-4}$ (l) | $8.6 \times 10^{-10}$ (r) | $1.9 \times 10^{-7}$ (v) |
| Activation energy, $Q_v$ (kJ/mole) | 91.7 (b) | 76.2 (j) | 135 (l) | 150 (r) | 153 (v) |
| *Boundary diffusion†* | | | | | |
| Pre-exponential, $\delta D_{0b}$ (m$^3$/s) | $1.3 \times 10^{-14}$ (c) | $5.0 \times 10^{-14}$ (j) | $5.0 \times 10^{-12}$ (m) | $3.6 \times 10^{-16}$ (s) | $5.4 \times 10^{-17}$ |
| Activation energy, $Q_b$ (kJ/mole) | 60.5 (c) | 54.4 (j) | 92 (m) | 97 (s) | 153 |
| *Core diffusion†* | | | | | |
| Pre-exponential, $a_c D_{0c}$ (m$^4$/s) | $2.6 \times 10^{-24}$ (d) | $1.0 \times 10^{-22}$ (d) | $3.0 \times 10^{-23}$ (n) | $7.8 \times 10^{-29}$ (d) | $1.6 \times 10^{-26}$ |
| Activation energy, $Q_c$ (kJ/mole) | 60.5 (d) | 54.4 (d) | 92 (n) | 97 (d) | 153 |
| *Power-law creep* | | | | | |
| Exponent, $n$ | 4.5 (e) | 4.0 (k) | 5.0 (o) | 4.3 (f,s) | 4.3 (w) |
| Dorn constant, $A$ | $4 \times 10^4$ (e) | $6.8 \times 10^3$ (k) | $1.2 \times 10^6$ (o) | $7.7 \times 10^4$ (f,s) | $1 \times 10^5$ (w) |
| Activation energy, $Q_{cr}$ (kJ/mole) | 152 (f) | — | 230 (f) | 242 (f,s) | — |
| Pre-exponential for creep, $D_{0cr}$ (m$^2$/s) | 8.1 (f) | — | $8.9 \times 10^2$ (f) | $1.3 \times 10^{-2}$ (f,s) | — |
| *Obstacle-controlled glide* | | | | | |
| 0 K flow stress, $\hat{\tau}/\mu_0$ | $6.7 \times 10^{-3}$ (g) | $7.5 \times 10^{-3}$ (g) | $8.0 \times 10^{-3}$ (g) | $1.5 \times 10^{-2}$ (g) | $5 \times 10^{-3}$ (t) |
| Pre-exponential, $\dot{\gamma}_0$ (s$^{-1}$) | $10^6$ | $10^6$ | $10^6$ | $10^6$ | $10^6$ |
| Activation energy, $\Delta F/\mu_0 b^3$ | 0.5 (h) | 0.5 (h) | 0.5 (h) | 0.14 (h) | 0.14 |

44

$$* \quad \mu_0 = (\tfrac{1}{3}c_{44}(c_{11} - c_{12}))^{\frac{1}{3}}; \quad \mu = \mu_0 \left(1 + \left(\frac{T - 300}{T_M}\right)\left(\frac{T_M}{\mu_0}\frac{d\mu}{dT}\right)\right)$$

$$\dagger \quad D_v = D_{0v} \exp - \left(\frac{Q_v}{RT}\right); \quad \delta D_b = \delta D_{0b} \exp - \left(\frac{Q_b}{RT}\right); \quad a_c D_c = a_c D_{0c} \exp - \left(\frac{Q_c}{RT}\right).$$

‡ This value of $A$ refers to tensile stress and strain-rate. The maps refer to shear stress and strain-rate. In computing them we have used $A_s = (\sqrt{3})^{n+1}A$.

(a) Huntington (1958).
(b) Peterson and Rothman (1967).
(c) Wadja (1954).
(d) $Q_c$ is assumed equal to that for boundary diffusion. The pre-exponentials $a_c D_{0c}$ are chosen to be consistent with creep data in the low-temperature creep regime.
(e) Selected as the best fit to the data plots for zinc: see text.
(f) For zinc and magnesium we show two regions of high-temperature creep, one with an activation energy equal to that for lattice diffusion, the other with a higher activation energy listed as $Q_{cr}$. The maps are computed using, for high temperature creep:

$$\dot{\varepsilon} = \frac{A\mu b}{kT}\{\max(D_v, D_{cr})\}\left(\frac{\sigma}{\mu}\right)^n$$

where $\quad D_{cr} = D_{0cr} \exp - \left(\frac{Q_{cr}}{RT}\right)$.

For $\alpha$-Ti, the map is computed using $D_{cr}$ alone. See text for further discussion.

(g) This corresponds to a heavily worked state, but is below the ultimate strength.
(h) Typical values for obstacle-controlled glide, when forest dislocations are principal obstacles.
(i) Garland and Silverman (1960).
(j) Wadja et al. (1955).
(k) See text.
(l) Shewmon (1956).
(m) Chosen to fit Coble-creep data. See text.
(n) Chosen to fit creep data as described in text.
(o) Chosen to fit creep data of Fig. 6.9; see text.
(p) Barrett and Massalski (1966).
(q) Mean values of bounds calculated by Simmons and Wang (1971) from data of Fisher and Renken (1964); see Fig. 6.12.
(r) Dyment and Libanati (1968), tracer diffusion.
(s) Derived from the diffusional flow data of Malakondaiah (1980) and consistent with the power-law creep data of Doner and Conrad.
(t) Interstitial impurities appear to determine the flow strength down to 4·2 K in all published data. Twinning is observed at low temperatures (Conrad et al., 1973).
(u) Calculated from the data of Fisher and Dever (1968) as described in the text, assuming $T_M/\mu_0 \, d\mu/dT_m = -0.5$.
(v) de Reca and Libanati (1968); see Fig. 6.13 and text.
(w) Fitted to data of Bühler and Wagener (1965).

45

f.c.c. or b.c.c. metals. Some of the scatter in the data may derive from texture differences between samples. Strictly, a given map describes only a particular texture, tested in a particular direction.

The lattice diffusion coefficients of the h.c.p. metals depend only slightly on direction. We have used that for diffusion parallel to the $c$-axis. The parameters are sufficiently well established that diffusion plots like those shown for the b.c.c. metals are unnecessary for all but Ti.

Further data and maps for Ti can be found in Chapter 17, Section 17.2. The field of drag-controlled plasticity, visible on Fig. 6.10, is explained there.

### References for Section 6.1

Edwards, G. R., McNelly, T. R. and Sherby, O. D. (1974) *Scripta Met.* **8**, 475.

Hirth, J. P. and Lothe, J. (1968) *Theory of Dislocations.* McGraw-Hill, p. 429.

## 6.2  ORIGINS OF THE DATA FOR HEXAGONAL METALS

### Zinc (Fig. 6.1 to 6.3)

Fig. 6.1 shows a map for zinc with a grain size of 0·1 mm. It is based on the data plotted in Figs. 6.2 and 6.3, and summarized in Table 6.1.

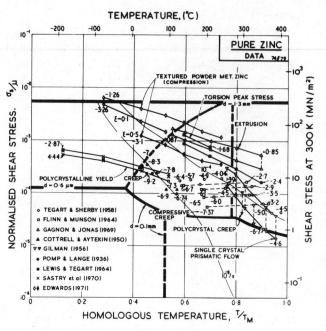

Fig. 6.2. Data for zinc, divided into blocks. The numbers are $\log_{10}(\dot{\gamma})$.

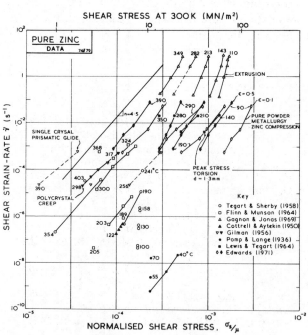

Fig. 6.3. Creep data for pure zinc, labelled with the temperature in °C.

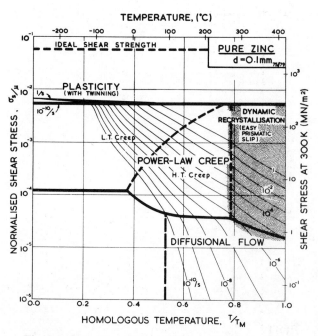

Fig. 6.1.  Pure zinc with a grain size of 0·1 mm.

The creep field is based primarily on the work of Tegart and Sherby (1958) and of Flinn and Munson (1964). Both groups found unexpectedly high activation energies at temperatures above $0.8\,T_M$: Flinn and Munson report that, above 270°C, the activation energy for creep, $Q_{cr}$, is about 152 kJ/mole; above 350°C Tegart and Sherby found 159 kJ/mole (compared with about 92 kJ/mole for lattice diffusion). Both groups report a slightly lower value of $n$ for the high-temperature regime. We have used two power-laws (Table 6.1, note f) and have employed the value $Q_{cr} = 152$ kJ/mole for the upper part, and 92 kJ/mole for the lower part of the high-temperature creep field, and used $n = 4.5$ for both. This combination leads to a map which fits well the plots of the experimental data shown as Figs. 6.2 and 6.3. The higher activation matches the value, 159 kJ/mole, found by Gilman (1956) for prismatic glide in single crystals of zinc, and for that reason this high-temperature region is often attributed to easy prismatic slip. We think it is more likely that it is a consequence of dynamic recrystallization (Section 2.4), which usually raises the apparent activation energy for creep.

Figs. 6.2 and 6.3 include the data of Edwards (1971) for powder-metallurgy zinc containing 1.2 vol.% of ZnO. Edwards reported an activation energy of 54–59 kJ/mole at low temperatures (0.5 to 0.6 $T_M$) and a larger energy of 113 kJ/mole at higher temperatures (0.8 to 0.95 $T_M$). We have assumed that the lower activation energy reflects a contribution of core diffusion; there are, however, no data for pure zinc which clearly lie in the low-temperature regime.

**References for zinc**

Cottrell, A. H. and Aytekin, V. (1950) *J. Inst. Met.* **77**, 389.

Edwards, G. R. (1971) "The Mechanical Behaviour of Powder Metallurgy Zinc and Zinc–Tungsten Particulate Composites", Ph.D. thesis, Stanford University.

Flinn, J. E. and Munson, D. E. (1964) *Phil. Mag.* **10**, 861.

Gagnon, G. and Jones, J. J. (1969) *Trains. AIME* **245**, 2581.

Gilman, J. J. (1956) *Trans. AIME* **206**, 1326.

Huntington, H. B. (1958) *Solid State Physics* **7**, 213.

Lewis, G. P. and Tegart, W. J. McG. (1963–64) *J. Inst. Met.* **92**, 249.

Peterson, N. L. and Rothman, S. J. (1967) *Phys. Rev.* **163**, 645.

Pomp, A. and Lange, W. (1936) *Mitt. Kaiser-Wilhelm Institut Eisenforschung zu Dusseldorf* **18**, 51.

Sastry, D. H., Prasad, Y. V. R. K. and Vasu, K. I. (1970) *Met. Trans.* **1**, 1827.

Tegart, W. J. McG. and Sherby, O. D. (1958) *Phil. Mag.* **3**, 1287.

Wadja, E. S. (1954) *Acta Met.* **2**, 184.

**Cadmium (Fig. 6.4 to 6.6)**

Fig. 6.4 shows a map for cadmium with a grain size of 0.1 mm. It is based on the data plotted in Figs. 6.5 and 6.6, and summarized in Table 6.1.

We have derived the low-temperature plasticity

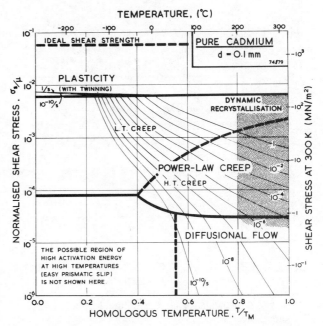

Fig. 6.4. Pure cadmium with a grain size of 0.1 mm.

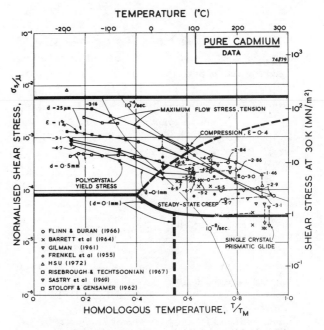

Fig. 6.5. Data for cadmium divided into blocks. The numbers are $\log_{10} (\dot{\gamma})$.

and the creep parameters from the more recent compression tests of Hsu (1972) (Figs. 6.5 and 6.6). His tests on cast cadmium showed strain softening which appears on our Figures as a difference between the peak stress and that at a tensile strain, $\varepsilon$, of 0.4. We have adjusted the parameters of the rate-equations to match the peak stress. At high temperatures the activation energy observed by Hsu, 75–84 kJ/mole, matches that for lattice self-

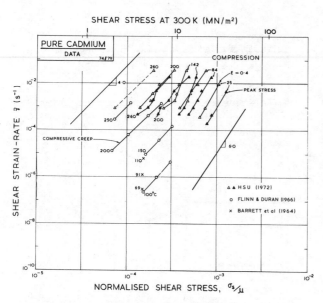

Fig. 6.6. Creep data for pure cadmium, labelled with the temperature in °C.

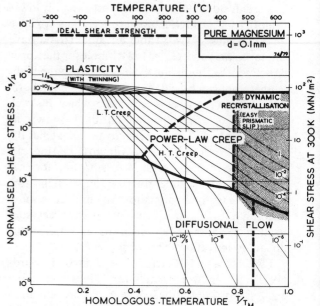

Fig. 6.7. Pure magnesium with a grain size of 0·1 mm.

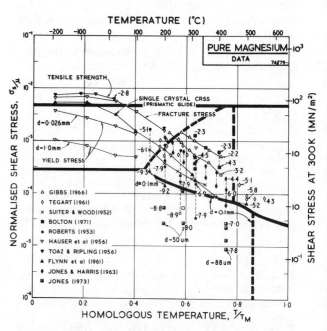

Fig. 6.8. Data for magnesium, divided into blocks. The numbers are $\log_{10}(\dot{\gamma})$.

diffusion (76 kJ/mole). At lower temperatures he found 40–60 kJ/mole, suggesting core diffusion (54 kJ/mole).

There is some suggestion of a high-temperature sub-field, controlled by prismatic glide, or reflecting dynamic recrystallization, like that shown on the map for zinc. Flinn and Duran (1966) for instance, report a regime above 150°C in which $Q_{cr}$ is about 127 kJ/mole, and a regime at a somewhat lower temperature in which it is about 87 kJ/mole. Hsu's work, however, does not explicitly support such a subdivision.

## References for cadmium

Barrett, C. R., Ardell, A. J. and Sherby, O. D. (1964) *Trans. AIME* **230**, 200.

Frenkel, R. E., Sherby, O. D. and Dorn, J. E. (1955) *Acta Met.* **3**, 470.

Flinn, J. E. and Duran, S. A. (1966) *Trans. AIME* **236**, 1056.

Garland, C. W. and Silverman, J. (1960) *Phys. Rev.* **119**, 1218.

Gilman, J. J. (1961) *Trans. AIME* **221**, 456.

Hsu, Shu-en (1972) "The Mechanical Behaviour of Cadmium Base Composites", Ph.D thesis, Stanford University.

Risebrough, N. R. and Teghtsoonian, E. (1967) *Can. J. Phys.* **45**, 591.

Sastry, D. H., Prasad, Y. V. R. K. and Vasu, K. I. (1969) *Acta Met.* **17**, 1453.

Stoloff, N. S. and Gensamer, M. (1963) *Trans AIME* **227**, 70.

Wadja, E. S., Shirn, G. A. and Huntington, H. B. (1955) *Acta Met.* **3**, 39.

## Magnesium (Fig. 6.7 to 6.9)

Fig. 6.7 shows a map for magnesium with a grain size of 0·1 mm. It is based on data shown in Figs. 6.8 and 6.9 and summarized in Table 6.1.

The parameters describing low-temperature plasticity are based on the polycrystal yield data of Toaz and Ripling (1956) and Hauser et al. (1956), and on the single-crystal data of Flynn et al. (1961).

In reviewing the creep data for magnesium,

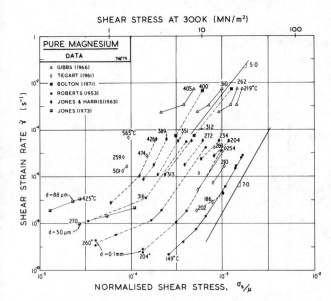

SHEAR STRESS AT 300K (MN/m²)

Fig. 6.9. Creep data for magnesium, labelled with the temperature in °C.

Crossland and Jones (1972) found reasonable consistency between published data if that derived from impure magnesium, containing substantial amounts of MgO, was excluded. They concluded that the activation energy for creep was 92 kJ/mole between 0·5 and 0·67 $T_M$, rising to 230 kJ/mole at higher temperatures. We have found, however, that the modulus-corrected data could be equally well described, from 0·5 to 0·78 $T_M$, by a single activation energy: 134 kJ/mole, close to that for self-diffusion. Above 0·78 $T_M$ we have used their value of 230 kJ/mole, which is consistent also with data reported by Tegart (1961); it is generally assumed that this reflects the climb-plus-glide motion of prismatic dislocations (as in zinc), though (as with zinc) it may reflect dynamic recrystallization. We have treated the lower activation energy (92 kJ/mole) as that of core diffusion which appears in the "low-temperature creep" region of the map.

Crossland and Jones (1972) report stress exponents of $n = 5$ at low temperatures and $n = 10$ at high. The larger value appears to be based on data of Jones and Harris (1963) at 0·76 $T_M$; data at the same temperature but lower stresses shows $n \simeq 4$. Tegart (1961) reported $n = 5·5$ at low temperatures and $n = 4$ at high. Lacking any sound theoretical basis, we have resolved this conflict by using a single value, $n = 5$, for both regions (that of lattice-diffusion control, and that of prismatic glide or dynamic-recrystallization control). At lower temperatures the rate-equation for creep (eqn. (2.21)) predicts an increase in the stress exponent which is consistent with the data of Roberts (1953). The pre-exponential for core diffusion was

(in this instance) chosen so that the equation matches Roberts' data at 0·46 $T_M$.

Within the Coble-creep regime, we have chosen values of $\delta D_{0b}$ and $Q_b$ which lead to an approximate fit to published data. Jones (1973) reported an activation energy of 80 kJ/mole, while Crossland and Jones (1972) report 105 kJ/mole. We have used the intermediate value of 92 kJ/mole.

The field of dynamic recrystallization is based on the observations of Gandhi and Ashby (1979).

## References for Magnesium

Bolton, C. J. (1971) Unpublished work, quoted by Crossland, I. G. and Jones, R. B. (1972) *Met. Sci. J.* **6**, 162.

Crossland, I. G. and Jones, R. B. (1972) *Met. Sci. J.* **6**, 162.

Flynn, P. W., Mote, J. and Dorn, J. E. (1961) *Trans. AIME* **221**, 1148.

Gandhi, C. and Ashby, M. F. (1979) *Acta Met.* **27**, 1565.

Gibbs, G. B. (1966) *Phil. Mag.* **13**, 317.

Hauser, F. E., Landon, P. R. and Dorn, J. E. (1956) *Trans. AIME* **206**, 589.

Huntington, H. B. (1958) *Solid State Physics* **7**, 213.

Jones, R. B. (1973) *J. Sheffield Univ. Metallurgical Soc.* **12**.

Jones, R. B. and Harris, J. E. (1963) from *Proc. of the Joint International Conf. on Creep*, vol. I, p. 1. London: Inst. Mech. Eng.

Roberts, C. S. (1953) *Trans. AIME* **197**, 1021.

Shewmon, P. G. (1956) *Trans. AIME* **206**, 918.

Suiter, J. W. and Wood, W. A. (1952–53) *J. Inst. Met.* **81**, 181.

Tegart, W. J. McG. (1961) *Acta Met.* **9**, 614.

Toaz, M. W. and Ripling, E. J. (1956) *Trans. AIME* **206**, 936.

## Titanium (Figs. 6.10 to 6.13 and 17.4)

Figs. 6.10, 6.11 and 17.4 show maps for Ti with a grain size of 0·1 mm. They are based on data plotted on the figures and summarized in Table 6.1.

The shear modulus of α-Ti is taken as a mean of the bounds calculated by Simmons and Wang (1971) from the single-crystal data of Fisher and Renken (1964), as shown in Fig. 6.12. We have normalized throughout by the melting point of the β-phase (1934 K); doing so leads to the large normalized temperature dependence of −1·24. (If, instead, Ardell's (1963) estimate for $T_M$ for the α-phase, 1957 ± 30 K, based on thermodynamic reasoning, is used, the temperature dependence, −1·12, is still large.) Modulus data for the β-phase are much more limited. Fisher and Dever (1968) give single-crystal constants at 1273 K, from which:

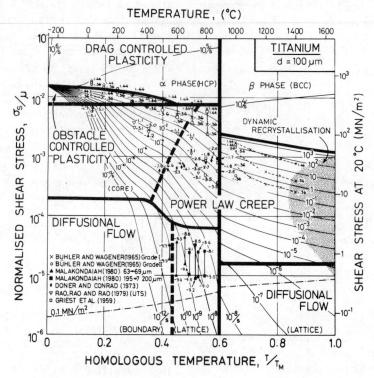

Fig. 6.10. A stress/temperature map for commercially pure titanium with a grain size of 0·1 mm, showing data.

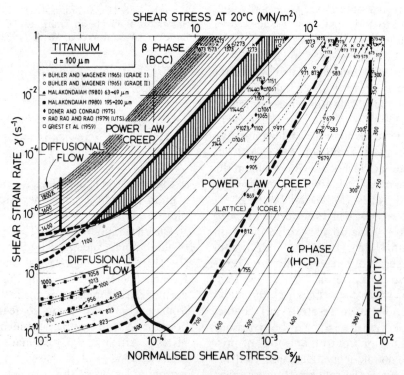

Fig. 6.11. A strain-rate/stress map for commercially pure titanium with a grain size of 0·1 mm, showing data.

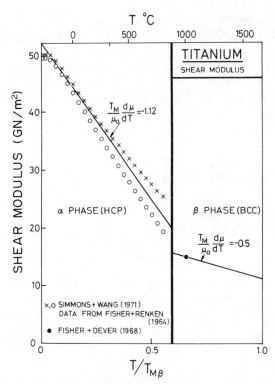

Fig. 6.12. Shear moduli for titanium.

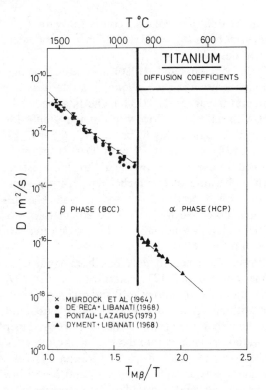

Fig. 6.13. Diffusion data for titanium.

$$\mu^{\beta}_{1273 \text{ K}} = (\tfrac{1}{2} c_{44} (c_{11} - c_{12}))^{\frac{1}{2}}$$

is found to be 15·3 GN/m². Assuming a normalized temperature dependence of $-0·5$, leads to the value of $\mu_0$ given in Table 6.1. The anomalously low moduli of both phases reflects their low stiffness with respect to the shear which causes the $\alpha \to \beta$ phase transformation at 1155 K.

Lattice diffusion in $\alpha$-Ti has an anomalously low activation energy. The normalized value $(Q_v/RT_M)$ typical of h.c.p. metals is $17·3 \pm 0·5$ (Brown and Ashby, 1980). The best data for $\alpha$-Ti are probably those of Dyment and Libanati (1968), shown in Fig. 6.13 who report $Q_v = 150·1$ kJ/mol, or (using Ardell's estimate of the melting point), $Q_v/RT_M = 10·3$. A low pre-exponential ($D_0 = 8·6 \times 10^{-10}$ m²/s) partly compensates for this, but the diffusion coefficient at $\tfrac{1}{2}T_M$, for example, is still much larger than in most close-packed metals. These anomalous values may be because the $\alpha$-phase is only just stable with respect to the more open $\beta$- and $\omega$-phases; then the more open structure of the activated state associated with a diffusive jump has a lower energy than in normal close-packed metals. The low $Q_v$ and $D_{0v}$ cannot be reconciled with data for either power-law creep or for diffusional flow (reviewed below).

The b.c.c. $\beta$-phases of Ti, Zr, and Hf all have non-linear diffusion plots with an activation energy which increases with temperature (see, for example, Murdock *et al.*, 1964). Sanchez and de Fontaine

(1978) explain this as caused by metastable $\omega$-phase embryos which form as the $\beta \to \alpha$ phase boundary is approached; the embryos have a structure like that of the activated state associated with a diffusive jump. The "typical" normalized activation energy $(Q_v/RT_M)$ for the b.c.c. transition metals is $17·8 \pm 2$ (Brown and Ashby, 1980) whereas Murdock *et al.* report 15·6 at high temperatures falling to 8·1 at the phase boundary (Fig. 6·13). To a tolerable approximation, all the data can be approximated by a single straight line corresponding to an activation energy of 152 kJ/mol (de Reca and Libanati, 1968). This is the value we have used.

There are no direct measurements of boundary diffusion in titanium. For $\alpha$-Ti, coefficients have been inferred from diffusional creep data for fine-grained wires (see below). The anomalously rapid diffusion in $\beta$-Ti near the phase boundary is a consequence of its marginal stability with respect to other competing structures ($\alpha$ and $\omega$), so that the grain boundary, although more open, may not be more permeable. Lacking data, we set $Q^{\beta}_b$ equal to the mean activation energy for lattice diffusion in the $\beta$-phase, and $D^{\beta}_{0b}$ equal to $2b$ times the pre-exponential for this diffusion.

In analysing the creep data for $\alpha$-Ti, it is perhaps best to start with the extensive study of Mala-kondaiah (1980), who tested helical springs of commercial-purity titanium (0·3% oxygen equivalent) in the regime of diffusional flow. Large grained (200–400 $\mu$m) samples showed the characteristics of lattice-diffusion control, consistent with $Q_v = 242$ kJ/mol and $D_{0v} = 1.3 \times 10^{-2}$ m$^2$/s (parameters which describe power-law creep well also, see below). Malakondaiah (1980) also tested springs with a fine grain size (63–69 $\mu$m) in a relatively low temperature range (832–873 K). The data are consistent with diffusional flow with boundary diffusion control. Assuming this to be so gives $Q_b = 97$ kJ/mol and $D_{0b} = 3.6 \times 10^{-16}$ m$^3$/s. Data from small-grained specimens above 900 K are inconsistent, perhaps because, at this higher temperature, some grain growth occurred during the test. The diffusional flow field for $\alpha$-Ti is based on these coefficients. Malakondaiah's data are plotted on Figs. 6.10 and 6.11

Power-law creep in $\alpha$-Ti has been studied by Doner and Conrad (1973) and Griest et al. (1959). Their data, shown in Figs. 6.10 and 6.11, are consistent with an activation energy of 242 kJ/mole, the same as that found for lattice-diffusional creep by Malakondaiah, whose data show an upturn at higher stresses which helps to position the field-boundary between diffusional flow and power-law creep. Doner and Conrad's (1973) estimate of power-law breakdown (for which $\alpha' = 300$) has been incorporated into the maps, which were computed using eqn. (2.26). Further data of Rao et al. (1979) and of Bühler and Wagener (1965) are consistent with this picture, and give, in addition, an indication of a transition to core-diffusion-controlled power-law creep. This field is based on their data, using the same activation energy as that for boundary diffusional creep, 97 kJ/mol.

Low-temperature plasticity in $\alpha$-Ti is extremely sensitive to the concentration of the interstitial impurities, H, C, N and O; the flow strength of "commercially pure" $\alpha$-Ti varies by a factor of 3.5, depending on this concentration (Bühler and Wagener, 1965). Fig. 16.10 is fitted to data for Bühler and Wagener's (1965) Grade 1 (0.25 wt.% oxygen equivalent) titanium, which is softer than their less-pure Grade 2, but harder than Rao et al.'s (1979) 99.8 wt.% titanium.

Creep data for $\beta$-Ti are more limited. Those of Bühler and Wagener (1965) are shown on Figs. 16.10 and 16.11; they lie in the regime of power-law breakdown. Assuming $\alpha' = 10^3$ (eqn. (2.26)) the data can be fitted by a power of 4.3, and an activation energy equal to the average value obtained from tracer-diffusion studies: 152 kJ/mol. There are no data for diffusional flow; the field is based on the diffusion coefficient of de Reca and Libanati (1968).

## References for titanium

Ardell, A. J. (1963) *Acta Met.* **11**, 591.

Barrett, C. and Massalski, T. B. (1966) *The Structure of Metals*, 3rd ed. McGraw-Hill, New York.

Brown, A. M. and Ashby, M. F. (1980) *Acta Met.* **28**, 1085.

Bühler, H. and Wagener, H. W. (1965) *Bänder Bläche Rohre* **6**, 677.

Conrad, H., Doner, M. and de Meester, B. (1973) in *Titanium Science and Technology* (eds. Jaffee, R. I. and Burke, H. M.). Plenum Press and TMS–AIME, p. 969.

de Reca, N. E. W. and Libanati, C. M. (1968) *Acta Met.* **16**, 1297.

Doner, M. and Conrad, H. (1973) *Met. Trans.* **4**, 2809.

Dyment, F. and Libanati, C. M. (1968) *J. Mat. Sci.* **3**, 349.

Dyment, F. (1980) in *Titanium '80* (eds. Kimura, H. and Izumi, O.). TMS–AIME, p. 519.

Fisher, E. S. and Dever, D. (1968) in *The Science, Technology and Applications of Titanium* (eds. Jaffee, R. I. and Promisel, N. E.). Pergamon.

Fisher, E. S. and Renken, C. J. (1964) *Phys. Rev.* **135**, A482.

Griest, A. J., Sabroff, A. M. and Frost, P. D. (1959) *Trans. ASM* **51**, 935.

Malakondaiah, G. (1980) Ph.D. thesis, Banaras Hindu University, Varanasi.

Murdock, J. E., Lundy, T. S. and Stansbury, E. E. (1964) *Acta Met.* **12**, 1033.

Pontau, A. E. and Lazarus, D. (1979) *Phys. Rev.* **B19**, 4027.

Rao, Y. K. M., Rao, V. K. and Rao, P. R. (1979) *Scripta Met.* **13**, 851.

de Reca, N. E. W. and Libanati, C. M. (1968) *Acta Met.* **16**, 1297.

Sanchez, J. N. and de Fontaine, D. (1978) *Acta Met.* **26**, 1083.

Santos, E. and Dyment, F. (1975) *Phil. Mag.* **31**, 309.

Simmons, G. and Wang, H. (1971) in *Single Crystal Elastic Constants and Calculated Aggregate Properties*, 2nd edn. MIT Press.

## CHAPTER 7

# NON-FERROUS ALLOYS:
# NICHROMES, T-D NICKELS AND NIMONICS

UP TO this point we have considered only pure metals. Although such maps are instructive, they are rarely useful for engineering design: even the simplest alloys are more complicated. In the first place, important physical constants such as the diffusion coefficients and moduli are changed by alloying. And alloying introduces new microscopic processes which oppose dislocation motion: solution strengthening, precipitation hardening and dispersion hardening. Their influence on the rate equations was discussed in Chapter 2, and can be seen in the maps presented here.

It is impractical to give a complete treatment of the deformation of alloys. Instead we shall consider six alloys based on nickel: two nichromes, deriving their strength from a solid solution of chromium: a Ni–ThO₂ alloy exemplifying dispersion hardening; two Ni–Cr–ThO₂ alloys which combine solution and dispersion hardening; and MAR–M200, a nickel-based superalloy which combines solution, dispersion and precipitation hardening.

We have chosen these because they illustrate many of the effects of alloying, and are sufficiently well characterized that maps for them can be constructed. They are shown in Figs. 7.1, 7.2 and 7.4 to 7.9. The parameters used to compute them are listed in Table 7.1; those for pure nickel are included for comparison. Ferrous alloys are described in Chapter 8.

## 7.1 GENERAL FEATURES OF THE DEFORMATION OF NICKEL-BASED ALLOYS

The various influences of alloying on deformation maps are best seen by comparing the maps of this chapter with those for pure nickel, of the same grain size, shown in Figs. 1.2 and 4.4.

A solid solution, illustrated by the nichromes (Figs. 7.1 and 7.2) raises the yield strength. The effect is masked by work-hardening in Fig. 1.2, but

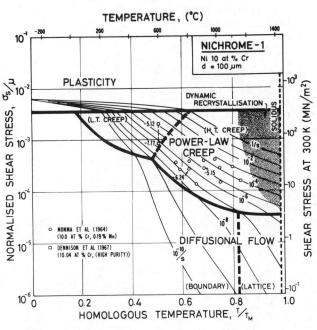

Fig. 7.1. Nickel–10 at.% chromium of grain size 100 μm, showing data. The temperature is normalized by the melting point of pure nickel (1726 K).

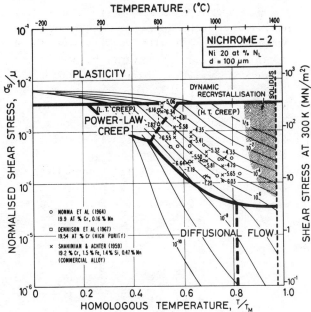

Fig. 7.2. Nickel–20 at.% chromium of grain size 100 μm, showing data. The temperature is normalized by the melting point of pure nickel (1726 K).

53

**TABLE 7.1  Nickel alloys**

| Material | Nickel (a) | Ni-10 at. % Cr | Ni-20 at. % Cr | Ni-1 vol. % ThO₂ | Ni-13·5 wt. % Cr -1 vol. % ThO₂ | Ni-22·6 wt. % Cr -1 vol. % ThO₂ | MAR-M200 |
|---|---|---|---|---|---|---|---|
| *Crystallographic and thermal data* | | | | | | | |
| Atomic volume, $\Omega$ (m³) | $1\cdot09 \times 10^{-29}$ (a) | $1\cdot1 \times 10^{-29}$ | $1\cdot1 \times 10^{-29}$ | $1\cdot09 \times 10^{-29}$ (g) | $1\cdot09 \times 10^{-29}$ (j) | $1\cdot1 \times 10^{-29}$ (j) | $1\cdot1 \times 10^{-29}$ (j) |
| Burger's vector, $b$ (m) | $2\cdot49 \times 10^{-10}$ | $2\cdot5 \times 10^{-10}$ | $2\cdot5 \times 10^{-10}$ | $2\cdot49 \times 10^{-10}$ | $2\cdot5 \times 10^{-10}$ | $2\cdot5 \times 10^{-10}$ | $2\cdot5 \times 10^{-10}$ |
| Melting temperature, $T_M$ (K) | 1726 | 1700 | 1673 | 1726 | 1688 | 1660 | 1600 |
| *Modulus\** | | | | | | | |
| Shear modulus at 300 K, $\mu_0$ (MN/m²) | $7\cdot89 \times 10^4$ | $8\cdot24 \times 10^4$ (b) | $8\cdot31 \times 10^4$ (b) | $7\cdot89 \times 10^4$ | $8\cdot24 \times 10^4$ | $8\cdot31 \times 10^4$ | $8\cdot0 \times 10^4$ |
| Temperature dependence of modulus, $\dfrac{T_M}{\mu_0}\dfrac{d\mu}{dT}$ | $-0\cdot64$ | $-0\cdot5$ (b) | $-0\cdot5$ (b) | $-0\cdot64$ | $-0\cdot5$ | $-0\cdot5$ | $-0\cdot5$ |
| *Lattice diffusion† for nickel* | | | | | | | |
| Pre-exponential, $D_{0v}$ (m²/s) | $1\cdot9 \times 10^{-4}$ | $3\cdot3 \times 10^{-4}$ (c) | $1\cdot6 \times 10^{-4}$ (c) | $1\cdot9 \times 10^{-4}$ | $3\cdot3 \times 10^{-4}$ | $1\cdot6 \times 10^{-4}$ | $1\cdot6 \times 10^{-4}$ |
| Activation energy, $Q_v$ (kJ/mole) | 284 | 293 | 285 | 284 | 293 | 285 | 285 |
| *Lattice diffusion† for chromium* | | | | | | | |
| Pre-exponential, $D_{0v}$ (m²/s) | — | $1\cdot4 \times 10^{-4}$ (c) | $1\cdot9 \times 10^{-4}$ (c) | — | $1\cdot4 \times 10^{-4}$ | $1\cdot9 \times 10^{-4}$ | — |
| Activation energy, $Q_b$ (kJ/mole) | — | 278 (c) | 283 (c) | — | 278 | 283 | — |
| *Boundary diffusion†* | | | | | | | |
| Pre-exponential, $\delta D_{0b}$ (m³/s) | $3\cdot5 \times 10^{-15}$ | $2\cdot8 \times 10^{-15}$ (d) | $2\cdot8 \times 10^{-15}$ (d) | $3\cdot5 \times 10^{-15}$ | $2\cdot8 \times 10^{-15}$ | $2\cdot8 \times 10^{-15}$ | $2\cdot8 \times 10^{-15}$ |
| Activation energy, $Q_b$ (kJ/mole) | 115 | 115 (d) | 115 (d) | 115 | 115 | 115 | 115 |
| *Core diffusion†* | | | | | | | |
| Pre-exponential, $a_c D_{0c}$ (m⁴/s) | $3\cdot1 \times 10^{-23}$ | $2 \times 10^{-25}$ (d) | $1 \times 10^{-25}$ (d) | $3\cdot1 \times 10^{-23}$ | $2\cdot0 \times 10^{-25}$ | $1\cdot0 \times 10^{-25}$ | $10^{-26}$ |
| Activation energy, $Q_c$ (kJ/mole) | 170 | 170 (d) | 170 (d) | 170 | 170 | 170 | 170 |
| *Power-law creep* | | | | | | | |
| Exponent, $n$ | 4·6 | 5·2 (e) | 4·6 (e) | 8·0 (h) | 6·3 (h) | 7·2 (h) | 7·7 (k) |
| Dorn constant‡, $A$ | $3\cdot0 \times 10^6$ | $3\cdot8 \times 10^7$ (e) | $1\cdot22 \times 10^5$ (e) | $5\cdot0 \times 10^{15}$ (h) | $1\cdot2 \times 10^9$ (h) | $1\cdot5 \times 10^{11}$ (h) | (k) |
| Activation energy, $Q_{cr}$ (kJ/mole) | — | — | — | — | — | — | 556 (k) |
| Pre-exp for creep, $A'$ (s⁻¹) | — | — | — | — | — | — | $5\cdot3 \times 10^{34}$ (k) |
| *Obstacle-controlled glide* | | | | | | | |
| 0 K flow stress, $\hat{\tau}/\mu_0$ | $6\cdot2 \times 10^{-3}$ | $6\cdot3 \times 10^{-3}$ | $6\cdot3 \times 10^{-3}$ | $1\cdot0 \times 10^{-3}$ (i) | $1\cdot0 \times 10^{-3}$ (i) | $1\cdot0 \times 10^{-3}$ (i) | $8\cdot3 \times 10^{-3}$ (l) |
| Pre-exponential, $\dot{\gamma}_0$ (s⁻¹) | $10^6$ | $10^6$ | $10^6$ | $10^6$ | $10^6$ | $10^6$ | $10^6$ (l) |
| Activation energy, $\Delta F/\mu_0 \, b^3$ | 0·5 | 0·5 | 0·5 | 2·0 | 2·0 | 2·0 | 2·0 (l) |
| *Stacking-fault energy, $\gamma_{SF}$ (J/m²)* | 0·24 (f) | 0·18 (f) | 0·10 (f) | 0·24 (f) | 0·14 (f) | 0·80 (f) | — (f) |

54

$$*D_v = D_{0v} \exp\left(\frac{Q_v}{RT}\right); \quad D_b = \delta D_{0b} \exp -\left(\frac{Q_b}{RT}\right); \quad a_c D_c = a_c D_{0c} \exp -\left(\frac{Q_c}{RT}\right).$$

$$\dagger \, \mu = \mu_0 \left(1 + \frac{(T-300)}{T_M} \frac{T_M}{\mu_0} \frac{d\mu}{dT}\right).$$

$\ddagger$ This value of $A$ refers to tensile stress and strain rate. The maps relate shear stress and strain-rate. In computing them we have used $A_s = (\sqrt{3})^{n+1} A$.

(a) The data for pure nickel are documented in Chapter 4.
(b) Calculated from polycrystalline Young's moduli measured by Benieva and Polotskii (1961) using $\mu = E/2(1+v)$ and $v = \frac{1}{3}$.
(c) Monma et al. (1964a).
(d) See text.
(e) Based on the measurements of Monma et al. (1964b).
(f) Estimated by Wilcox and Clauer (1969).
(g) The lattice parameter, moduli and diffusion coefficients are assumed to be identical with those of pure nickel.
(h) Based on the work of Wilcox and Clauer (1969).
(i) For recrystallized Ni–ThO$_2$; see text.
(j) The lattice parameters, moduli and diffusion coefficients are based on the data for Ni–Cr alloys.
(k) The activation energy for creep of MAR–M200 differs from that for diffusion. The data have been fitted to the equation:

$$\dot{\varepsilon} = A'\left(\frac{\sigma}{\mu}\right)^n \exp -\left\{\frac{Q_{cr}}{RT}\right\}$$

or, equivalently:

$$\dot{\gamma} = (\sqrt{3})^{n+1} A' \left(\frac{\sigma_s}{\mu}\right)^n \exp -\left(\frac{Q_{cr}}{RT}\right)$$

The constants $A'$, $Q_{cr}$ and $n$ were adjusted to fit the data of Webster and Piearcey (1967) and of Kear and Piearcey (1967).
(l) The yield strength and its temperature dependence were chosen to fit the data of Piearcey et al. (1967).

becomes clearer if Fig. 7.1 is compared with Fig. 4.4. The solute slows power-law creep throughout the field: the 10% nichrome creeps roughly 10 times more slowly than pure nickel, at a given stress and temperature; and the 20% nichrome creeps almost

3 times slower still. There is an influence on diffusional flow, too: the diffusion coefficients are changed (Fig. 7.3) and the solute can exert a drag on the grain boundary dislocations which are the sinks and sources of matter during diffusional flow. (This last effect is discussed further in Chapter 17, Section 17.3, but is not included here.) The relatively larger retardation of power-law creep causes the

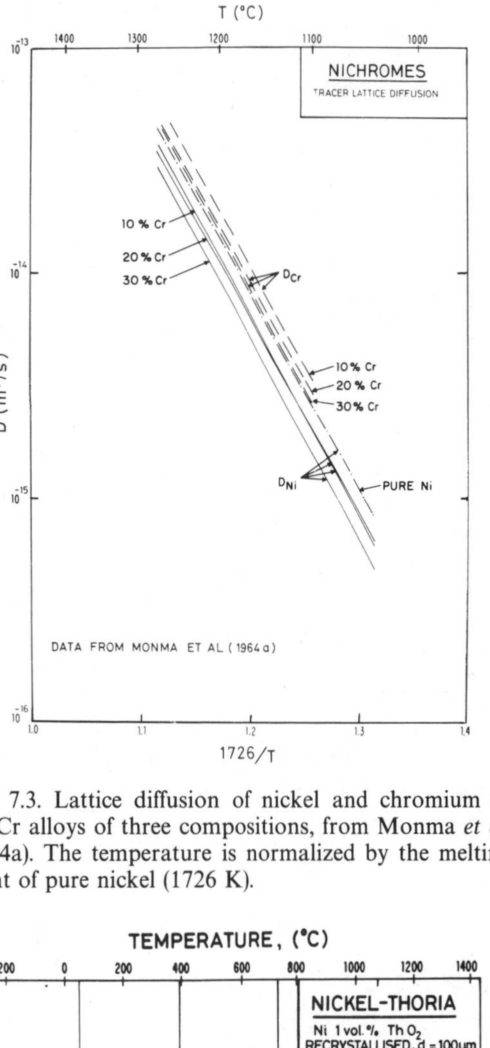

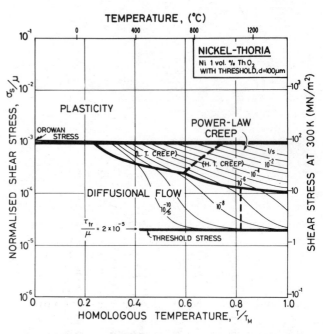

Fig. 7.3. Lattice diffusion of nickel and chromium in Ni–Cr alloys of three compositions, from Monma *et al.* (1964a). The temperature is normalized by the melting point of pure nickel (1726 K).

Fig. 7.5. Nickel 1 vol.% thoria of grain size 100 $\mu$m, showing the way the map is changed if the dispersion pins grain boundary dislocations giving a threshold stress $\hat{\tau}^{TR} = 2 \times 10^{-5} \mu$.

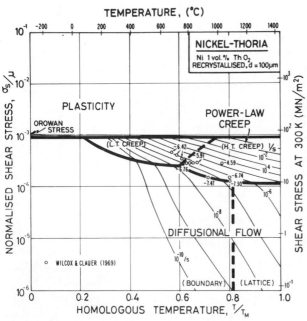

Fig. 7.4. Nickel 1 vol.% thoria of grain size 100 $\mu$m.

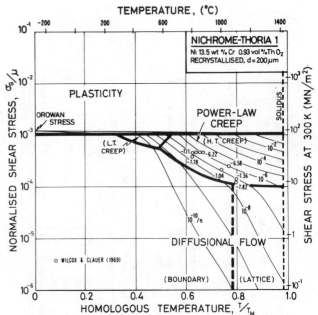

Fig. 7.6. Nickel 13·5 wt.% Cr–0·93 vol.% ThO$_2$ of grain size 200 $\mu$m, recrystallized.

power-law creep field to shrink and that of diffusional flow to expand relative to pure nickel.

A dispersion also raises the yield strength. The maps for thoriated nickel and nichromes (Figs. 7.4, 7.5, 7.6 and 7.7) are for recrystallized material, and are properly compared with Fig. 4.4 for pure nickel. The yield strength is found to relate directly to the particle spacing $\ell$ through the Orowan formula (Wilcox and Clauer, 1969):

$$\sigma_y \approx \frac{\mu b}{\ell}$$

The most striking difference, however, is the retardation of power-law creep, which is $10^{-2}$ to $10^{-3}$ times slower than in pure nickel at the same stress and temperature. Both the dispersion and the solid solution influence diffusional flow also (Chapter 17, Section 17.3) but to a lesser extent, so that the field of power-law creep shrinks dramatically.

Diffusional flow, too, is influenced by alloying (Chapter 17, Section 17.3). A solid solution reduces the mobility of the grain-boundary dislocations which are the sinks and sources of matter during diffusional flow, slowing the rate of creep and changing the stress dependence, particularly when the grain size is small. A fine dispersion of particles pins grain-boundary dislocations, creating a threshold stress $\tau_{th}$ below which they cannot move, and no creep is possible. The expected change in deformation map is illustrated by Fig. 7.5 for nickel–thoria, which is based on the same data as Fig. 7.4 but includes a threshold stress, $\tau_{tr}$ of $2 \times 10^{-5}$ $\mu$ (Chapter 17, eqn. (17.35)). Such a threshold may

exist for all dispersion-strengthened alloys, but it is generally small and data for it are meagre.

The combined effects of a solid solution and a dispersion can be seen, particularly in the power-law creep field, by a careful comparison of the maps for nichrome (Figs. 7.1 and 7.2), thoriated nickel (Fig. 7.4) and thoriated nichromes (Figs. 7.6 and 7.7). The two strengthening methods, acting together, slow creep more effectively than either one act-

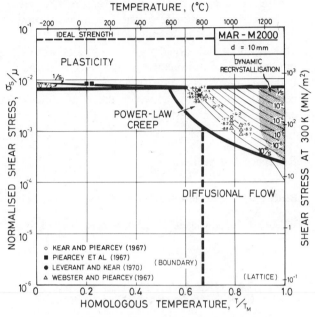

Fig. 7.8. MAR–M200 with a grain size of 10 mm, as cast, showing data from single-crystal samples.

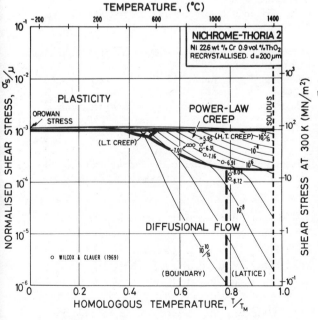

Fig. 7.7. Nickel 22·6 wt.% Cr–0·9 vol.% ThO$_2$ of grain size 200 $\mu$m, recrystallized.

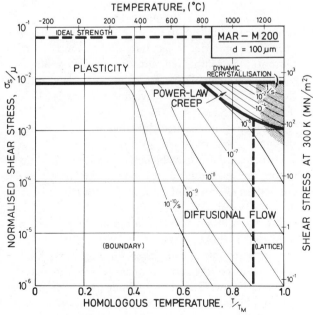

Fig. 7.9. MAR–M200 with a grain size of 100 $\mu$m, as cast.

ing alone. In addition, the dispersion suppresses dynamic recrystallization.

The nickel-based superalloy (Betteridge and Heslop, 1974) combine a concentrated solid solution (typically of Cr and Co) with a large volume fraction of the precipitate $Ni_3(Al,Ti)$ and a lesser volume fraction of carbides, much of it in grain boundaries. The maps shown here (Figs. 7.8 and 7.9) are largely based on data for MAR–M200, but their purpose is to illustrate the major features of the superalloys generally. The yield strength is raised to roughly the ultimate strength of nickel, and power-law creep is drastically slowed. Very little is known about diffusional flow in the superalloys, but its field of dominance is likely to be large. Dynamic recrystallization is anticipated at temperatures above those at which the $\gamma'$ precipitate and the $M_{23}C_6$ and $M_7C_3$ carbides dissolve.

### References for Section 7.1

Betteridge, W. and Heslop, J. (1974) *Nimonic Alloys*, 2nd edn. Arnold.
Wilcox, B. A. and Clauer, A. H. (1969) *Met. Sci. J.* **3**, 26.

## 7.2 ORIGINS OF THE DATA FOR NICKEL-BASED ALLOYS

### Solution-strengthened nickel: the nichromes (Figs. 7.1 and 7.2)

The maps for both Ni–10% Cr (Fig. 7.1) and Ni–20% Cr (Fig. 7.2) are plotted with the temperature scale normalized to the melting temperature of pure nickel in order to facilitate direct comparison. On both, the solidus temperature is marked. Shear moduli were calculated from the polycrystal data for Young's moduli of Benieva and Polotskii (1961), using $\mu_0 = \frac{3}{8}E_0$.

Alloying slows lattice diffusion in these alloys (Monma *et al.*, 1964a) as shown in Fig. 7.3. Table 7.1 lists the diffusion coefficients for both nickel and chromium; in evaluating the rate equations we calculated an effective diffusion coefficient from these, as described in Chapter 2, eqn. (2.27).

We have found no data for boundary or core diffusion in these alloys, and have therefore used the same activation energies as for pure nickel, but have adjusted the pre-exponentials in the ways outlined below. Gleiter and Chalmers (1972) point out that boundary diffusion may be either enhanced or reduced by the addition of a solute; the same should apply to dislocation core diffusion. Because the presence of chromium reduces the rate of lattice diffusion of nickel, we have assumed that boundary diffusion is similarly reduced, and have taken $D_{0B}$ to be 80% of the value for pure nickel.

For core diffusion there is direct evidence for such a reduction. Dennison *et al.* (1967) find an activation energy for creep at 0·5 $T_M$ of 284 kJ/mole (10% Cr) and 309 kJ/mole (20% Cr), which match the activation energies for lattice diffusion. The location of their data on the map also requires a lower core-diffusion coefficient. Accordingly, we have used $a_c D_{0c} = 2 \times 10^{-25}$ m$^4$/s for 10% Cr and $1 \times 10^{-25}$ m$^4$/s for 20 Cr. This reduces the extent of the low temperature creep field compared with pure nickel, although it still leaves the data of Dennison *et al.* within that field. A similar reduction is found in the stainless steels and is discussed in Chapter 8.

The parameters for power-law creep are derived from Monma *et al.* (1964b) whose data are plotted on the maps. Their data agree well with the creep data for commercial Ni–20% Cr of Shahinian and Achter (1959).

Dynamic recrystallization is observed in monels (Ni–Cu solid solutions) at above 1050°C and strain rates of $10^{-5}$/s (Gandhi and Ashby, unpublished work). Similar behaviour is expected in the nichromes and is indicated as a shaded field.

### References for nickel–chromium alloys

Benieva, T. Ya and Polotskii, I. G. (1961) *Fiz. Metalloved.* **12**, 584.
Dennison, J. P., Llewellyn, R. J. and Wilshire, B. (1967) *J. Inst. Met.* **95**, 115.
Gleiter, H. and Chalmers, B. (1972) *Prog. Mat. Sci.* **16**, 1.
Monma, K., Suto, H. and Oikawa, H. (1964a) *J. Jap. Inst. Met.* **28**, 188.
Monma, K., Suto, H. and Oikawa, H. (1964b) *J. Jap. Inst. Met.* **28**, 253.
Shahinian, P. and Achter, M. R. (1959) *Trans. ASM* **51**, 244.
Wilcox, B. A. and Clauer, A. H. (1969) *Met. Sci. J.* **3**, 26.

### Dispersion-strengthened nickel: nickel–thoria and nickel–chromium–thoria alloys (Figs. 7.4 to 7.7)

The maps are based on creep data reported by Wilcox and Clauer (1969). Figs. 7.4 and 7.5 show Ni–1 vol.% ThO$_2$; Fig. 7.6 is for Ni–13 wt% Cr–1 vol.% ThO$_2$; and Fig. 7.7 is for Ni–22·6 wt.% Cr–1 vol.% ThO$_2$, all three in the recrystallized state. The creep data of Wilcox and Clauer are plotted on each. Diffusion coefficients can reasonably be assumed to be the same as those for pure nickel and the Ni–Cr alloys. Moduli, too, have been set equal to those for nickel and nichrome, although this is less reasonable because, in the worked state, the dispersion-strengthened alloys have a marked texture which lowers the modulus in the rolling direction.

The grain size for the Ni–1% $ThO_2$ alloy ($d = 0.1$ mm) is an estimate made from published micrographs. A slightly larger grain size ($d = 0.2$ mm) was used for the Ni–Cr–$ThO_2$ alloys because the smaller value incorrectly places much of the creep data in the diffusional flow field (the high observed stress exponent of between 6 and 8 indicates power-law creep). There are no reports of diffusional flow in thoriated nickel and nichromes. The field was calculated using the diffusion coefficients for pure nickel and for the nichromes described earlier. The threshold stress shown in Fig. 7.5 is based on creep data for Cu–$Al_2O_3$ and Au–$Al_2O_3$ alloys, and on theory, presented in Chapter 17, Section 17.3.

The dislocation glide field is based on the initial yield stress for the fully recrystallized alloy (well approximated by the Orowan stress, $\mu_0 b/\ell$, where $\ell$ is the spacing of thoria particles). The plasticity fields extend to lower stress than do those for the Ni–Cr alloys because the latter describe the work-hardened state. The thoriated alloys show a higher rate of work-hardening, and are normally used in a work-hardened state, when their low-temperature strength is higher than of the solid solution alloys.

The thoria particles in these alloys are stable up to the melting point and completely suppress dynamic recrystallization.

### Reference for nickel–thoria alloys

Wilcox, B. A. and Clauer, A. H. (1969) *Met. Sci. J.* **3**, 26.

### Nickel-based superalloys: MAR–M200 (Figs. 7.8 and 7.9)

MAR–M200 is a nickel-based superalloy strengthened by a solid solution of W, Co and Cr, and by precipitates of $Ni_3(Ti,Al)$ and carbides of the types $M_{23}C_6$ and $M_7C_3$. A typical composition is given in Table 7.2. Fig. 7.8 shows a map corresponding to grain size of 10 mm. On it are plotted data for single crystal MAR–M200.

**TABLE 7.2  Nominal composition of MAR–M200 in wt.%**

| Al | Ti | W | Cr | Nb | Co | C | B | Zr | Ni |
|----|----|------|-----|-----|------|------|-------|------|-----|
| 5.0 | 2.0 | 12.5 | 9.0 | 1.0 | 10.0 | 0.15 | 0.015 | 0.05 | Bal |

Alloying lowers the melting point substantially (to 1600 K), and raises the shear modulus slightly (to 80 $GN/m^2$) compared with pure nickel. Lacking more complete data, we have taken the coefficient for lattice and boundary diffusion to be the same as those for high-alloy nichrome. The dispersion of $Ni_3(Al,Ti)$, superimposed on the heavy solid-solution strengthening of the W and Cr, gives MAR–M200 (and alloys like it) a yield strength comparable with the ultimate strength of pure nickel, although it is less dependent on temperature (Ver Snyder and Piearcey, 1966). The alloying also reduces greatly the rate of power-law creep, the field of which is based on the data of Webster and Piearcey (1967), Kear and Piearcey (1967) and Leverant and Kear (1970).

These differences can be seen by comparing Figs. 7.9, for fine-grained MAR–M200 with that for pure nickel of the same grain size (Fig. 1.2). Precipitation strengthening and solution hardening have raised the yield line, and have reduced drastically the size of the power-law creep field. They also change the rate of diffusional flow (Whittenberger, 1977, 1981) though, since there are no experimental data for MAR–M200, we have made the assumption that it occurs at the same rate as it would in Ni–20% Cr alloy (but see Chapter 17, Section 17.3, and Fig. 7.5).

We are not aware of observations of dynamic recrystallization in MAR–M200, but above 1000°C the $\gamma'$ phase dissolves, and at a slightly higher temperature the grain boundary carbides do so also ($M_{23}C_6$ at 1040 to 1095°C; $M_7C_3$ at 1095 to 1150°C; Betteridge and Heslop, 1974). This means that above 0.9 $T_M$ the alloy is a solid solution, and if the data cited earlier for solid solutions can be used as a guide, we would expect dynamic recrystallization. The shaded field is based on this reasoning.

### References for MAR–M200

Betteridge, W. and Heslop, J. (1974) *Nimonic Alloys*, 2nd edn. Arnold.

Kear, B. H. and Piearcey, B. J. (1967) *Trans. Met. Soc. AIME* **239**, 1209.

Leverant, G. R. and Kear, B. H. (1970) *Met. Trans.* **1**, 491.

Piearcey, B. J., Kear, B. H. and Smashey, R. W. (1967) *Trans. ASM* **60**, 634.

Ver Snyder, F. L. and Piearcey, B. J. (1966) *SAE June* **74**, 36.

Webster, G. A. and Piearcey, B. J. (1967) *Met. Sci. J.* **1**, 97.

Whittenberger, J. D. (1977) *Met. Trans.* **8A**, 1155.

Whittenberger, J. D. (1981) *Met. Trans.* **12A**, 193.

# CHAPTER 8

# PURE IRON AND FERROUS ALLOYS

THIS chapter describes maps for pure iron, a low-alloy ferritic steel, and two stainless steels. Because it is the basis of all ferrous alloys, pure iron is treated in more detail than any other material in this book. Its behaviour is complicated by two crystallographic and one magnetic phase change, but a vast body of data exists, adequately covering the temperature range from $0.05\ T_M$ to $0.9\ T_M$. This allows the construction of a map (Fig. 8.1) which, although complicated, gives a reasonably accurate and complete summary of its mechanical behaviour.

The steels are treated at a less detailed level. It is impractical here to try to analyse a wide range of steels; not only are there a large number of them, but, when strength is derived from thermomechanical processing, one map properly describes only one state of heat treatment and mechanical history. Rather our aim is to illustrate the broad features that characterize two important classes of steels which are used at both low and high temperatures (a ferritic 1% Cr–Mo–V steel, and types 316 and 304 stainless steels). Many of these features are

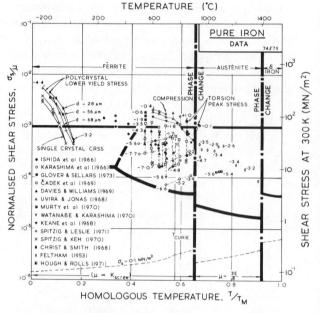

Fig. 8.2. Data for pure iron, divided into blocks and labelled with $\log_{10}(\dot{\gamma})$.

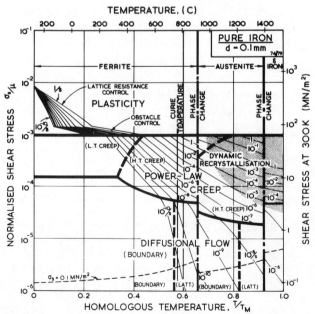

Fig. 8.1. Pure iron of grain size 100 $\mu$m.

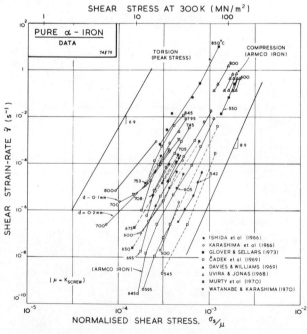

Fig. 8.3. Creep data for pure $\alpha$-iron, and labelled with the temperature in °C.

60

common to a range of similar steels—and the illustration of the method and results may help the reader who wishes to characterize a steel of particular interest to himself.

The low-alloy ferritic steels are used for applications where a high strength at or above room temperature is required. They are widely used for steam pipes and turbine casings, where they are required to carry load at temperatures up to 550°C. The stainless steels are used over an even wider temperature range—from 4·2 K (in cryogenic equipment) to over 650°C (in chemical engineering applications, and in nuclear power plant). The particular steels analysed here are typical of their class, after a typical thermomechanical treatment. The maps for them are based on data presented below, and on the parameters listed in Table 8.1.

## 8.1 GENERAL FEATURES OF THE MECHANICAL BEHAVIOUR OF IRON AND FERROUS ALLOYS

The map for *pure iron* (Fig. 8.1) is based on data shown in Figs. 8.2 to 8.6 (Frost and Ashby, 1977). It is complicated by two crystallographic and one magnetic phase change. It shows three sections corresponding to the three allotropic forms of iron: $\alpha$, $\gamma$ and $\delta$. At each crystallographic phase change the strength has a sharp discontinuity, and at the magnetic phase change its derivative with respect to temperature changes—largely because of similar changes in the moduli and the diffusion coefficients. But although it is complicated in this way, the data for iron are extensive, and the map is a good fit to a large body of mechanical measurements.

The $\alpha$-region of the map shows a field of dislocation-glide in much of which a lattice resistance limits the strain-rate; a power-law creep field, with core-diffusion important at the low temperature end; and a field of diffusional flow, with boundary diffusion important at the low temperature end.

The creep fields are truncated at 910°C by the phase-change to austenite, where the field boundaries and the strain-rate contours suffer a sharp discontinuity. This is because all the physical properties (lattice parameter, Burger's vector, modulus,

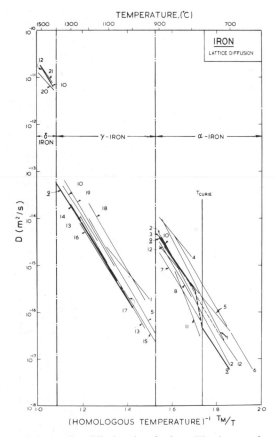

Fig. 8.5. Lattice diffusion data for iron. The data are from: (1) Birchenall and Mehl (1948); (2) Birchenall and Mehl (1950); (3) Buffington *et al.* (1951); (4) Gruzin (1952); (5) Zhukhovitskii and Geodakyan (1955); (6) Golikov and Borisov (1955); (7) Leymonie and Lacombe (1957, 1958, 1959, 1960); (8) Borg and Birchenall (1960); (9) Buffington *et al.* (1961); (10) Graham and Tomlin (1963); (11) Amonenko *et al.* (1964); (12) James and Leak (1966); (13) Gruzin (1953); (14) Mead and Birchenall (1956); (15) Bokshtein *et al.* (1957); (16) Gertsriken and Pryanishnikov (1958); (17) Bokshtein *et al.* (1959); (18) Bogdanov (1962); (19) Sparke *et al.* (1965); (20) Staffansson and Birchenall (1961); and (21) Borg *et al.* (1963).

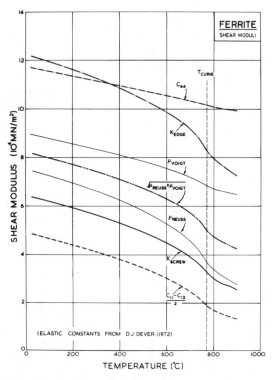

Fig. 8.4. Shear moduli for $\alpha$-iron.

**TABLE 8.1  Iron and ferrous alloys**

| Material | α-Iron (Ferro) | α-Iron (Para) | γ-Iron | δ-Iron | 1% Cr-Mo-V Steel (Ferro) | 1% Cr-Mo-V Steel (Para) | 304 Stainless | 316 Stainless |
|---|---|---|---|---|---|---|---|---|
| *Crystallographic and thermal data* | | | | | | | | |
| Atomic volume, $\Omega$ (m³) | $1{\cdot}18 \times 10^{-29}$ | | $1{\cdot}21 \times 10^{-29}$ | $1{\cdot}18 \times 10^{-29}$ | $1{\cdot}18 \times 10^{-29}$ | | $1{\cdot}21 \times 10^{-29}$ | $1{\cdot}21 \times 10^{-29}$ |
| Burger's vector, $b$ (m) | $2{\cdot}48 \times 10^{-10}$ | | $2{\cdot}58 \times 10^{-10}$ | $2{\cdot}48 \times 10^{-10}$ | $2{\cdot}48 \times 10^{-10}$ | | $2{\cdot}58 \times 10^{-10}$ | $2{\cdot}58 \times 10^{-10}$ |
| Temperature range (K) | 0–1184 | | 1184–1665 | 1665–1810 | 0–1753 | | 0–1680 | 0–1680 |
| *Modulus** | | | | | | | | |
| Shear modulus at 300 K, $\mu_0$ (MN/m²) | $6{\cdot}4 \times 10^4$ (a) | $6{\cdot}92 \times 10^4$ | $8{\cdot}1 \times 10^4$ (h) | $3{\cdot}9 \times 10^4$ (j) | $8{\cdot}1 \times 10^4$ (l) | | $8{\cdot}1 \times 10^4$ (p) | $8{\cdot}1 \times 10^4$ (p) |
| Temperature dependence of modulus, $\dfrac{T_M}{\mu_0}\dfrac{d\mu}{dT}$ | $-0{\cdot}81$ (b) | $-1{\cdot}31$ | $-0{\cdot}91$ (h) | $-0{\cdot}72$ (j) | $-1{\cdot}09$ (l) | | $-0{\cdot}85$ (p) | $-0{\cdot}85$ (p) |
| *Lattice diffusion†* | | | | | | | | |
| Pre-exponential, $D_{0v}$ (m²/s) | $2{\cdot}0 \times 10^{-4}$ (c) | $1{\cdot}9 \times 10^{-4}$ | $1{\cdot}8 \times 10^{-5}$ (c) | $1{\cdot}9 \times 10^{-4}$ (k) | $2{\cdot}0 \times 10^{-4}$ (k) | $1{\cdot}9 \times 10^{-4}$ (k) | $3{\cdot}7 \times 10^{-5}$ (q) | $3{\cdot}7 \times 10^{-5}$ (q) |
| Activation energy, $Q_v$ (kJ/mole) | 251 (c) | 239 | 270 (c) | 239 (k) | 251 (k) | 239 (k) | 280 (q) | 280 (q) |
| *Boundary diffusion†* | | | | | | | | |
| Pre-exponential, $D_{0b}$ (m³/s) | $1{\cdot}1 \times 10^{-12}$ (d) | | $7{\cdot}5 \times 10^{-14}$ (d) | $1{\cdot}1 \times 10^{-12}$ (k) | $1{\cdot}1 \times 10^{-12}$ (k) | | $2{\cdot}0 \times 10^{-13}$ (r) | $2{\cdot}0 \times 10^{-13}$ (r) |
| Activation energy, $Q_b$ (kJ/mole) | 174 (d) | | 159 (d) | 174 (k) | 174 (k) | | 167 (r) | 167 (r) |
| *Core diffusion†* | | | | | | | | |
| Pre-exponential, $a_c D_{0c}$ (m⁴/s) | $1{\cdot}0 \times 10^{-23}$ (e) | | $1{\cdot}0 \times 10^{-23}$ (e) | $1{\cdot}0 \times 10^{-23}$ (k) | $1{\cdot}0 \times 10^{-24}$ (m) | | — | — |
| Activation energy, $Q_c$ (kJ/mole) | 174 (e) | | 159 (e) | 174 (k) | 174 (k) | | — | — |
| *Power-law creep* | | | | | | | | |
| Exponent, $n$ | $6{\cdot}9$ (f) | | $4{\cdot}5$ (i) | $6{\cdot}9$ (k) | $6{\cdot}0$ (n) | | $7{\cdot}5$ (s) | $7{\cdot}9$ (t) |
| Dorn constant‡, $A$ | $7{\cdot}0 \times 10^{13}$ (f) | | $4{\cdot}3 \times 10^{5}$ (i) | $7{\cdot}0 \times 10^{13}$ (k) | $1{\cdot}1 \times 10^{4}$ (n) | | $1{\cdot}5 \times 10^{12}$ (s) | $1{\cdot}0 \times 10^{10}$ (t) |
| *Obstacle-controlled glide* | | | | | | | | |
| 0 K flow stress, $\hat{\tau}/\mu$ | $1{\cdot}7 \times 10^{-3}$ | | — | — | $6{\cdot}2 \times 10^{-3}$ | | $6{\cdot}5 \times 10^{-3}$ (s) | $6{\cdot}5 \times 10^{-3}$ (s) |
| Pre-exponential, $\dot{\gamma}_0$ (s⁻¹) | $10^6$ | | — | — | $10^6$ | | $10^6$ | $10^6$ |
| Activation energy, $\Delta F/\mu_0 b^3$ | $0{\cdot}5$ | | — | — | $2{\cdot}0$ (o) | | $0{\cdot}5$ | $0{\cdot}5$ |
| *Lattice-resistance-controlled glide* | | | | | | | | |
| 0 K flow stress, $\hat{\tau}_p/\mu_0$ | $1{\cdot}0 \times 10^{-2}$ (g) | | — | — | $1{\cdot}0 \times 10^{-2}$ (k) | | — | |
| Pre-exponential, $\dot{\gamma}_p$ (s⁻¹) | $10^{11}$ | | — | — | $10^{11}$ (k) | | — | |
| Activation energy, $\Delta F_p/\mu_0 b^3$ | $0{\cdot}1$ | | — | — | $0{\cdot}1$ (k) | | — | |

$^*$ $\mu = \mu_0\left(1 + \dfrac{(T-300)}{T_M}\dfrac{T_M}{\mu_0}\dfrac{d\mu}{dT}\right)$ except for ferrite: see note (a). In normalizing the temperature dependence of the modulus we have used $T_M = 1810$ K.

$\dagger$ $D_v = D_{0v}\exp\left(-\dfrac{Q_v}{RT}\right)$; $\delta D_b = \delta D_{0b}\exp\left(-\left(\dfrac{Q_b}{RT}\right)\right)$; $a_c D_c = a_c D_{0c}\exp\left(-\left(\dfrac{Q_c}{RT}\right)\right)$.

$\ddagger$ This value of $A$ refers to tensile stress and strain rate. The maps relate shear stress and strain rate. In constructing them, we have used $A_s = (\sqrt{3})^{n+1}\,A$.

(a) The modulus for $\alpha$-iron was set equal to $K_{\text{screw}}$ (see text and Fig. 8.4) and calculated from the single crystal moduli of Denver (1972).

(b) In normalizing the temperature dependence we have used $T_M = 1810$ K. In computing the ferromagnetic region ($T \leqslant 1043$ K) we used a non-linear temperature dependence.

$$\mu = \mu_0\left(1 - \dfrac{(T-300)}{T_M}\dfrac{T_M}{\mu_0}\dfrac{d\mu}{dT}\right.$$
$$\left. - K_1(T-573)^2 - K_2(T-923)^2\right)$$
$$\text{when } T > 573 \text{ K} \qquad \text{when } T > 923 \text{ K}$$
with $K_1 = 3.2 \times 10^{-2}$ MN/m$^2$ ($^\circ$K)$^2$
and $K_2 = 2.4 \times 10^{-2}$ MN/m$^2$ ($^\circ$K)$^2$

(c) Buffington et al. (1961). A smoothed transition is taken between paramagnetic and ferromagnetic $\alpha$-iron:
$$D_v = fD_{v-\text{PARA}} + (1-f)D_{v-\text{FERRO}}$$
$$f = \dfrac{1}{2} + \dfrac{0.5(1-1043)}{(|(T-1043)| + 20)}$$

(d) James and Leak (1965).

(e) Estimated, based on boundary diffusion coefficients.

(f) Derived by fitting creep data to the creep equations (eqn. (2.21)) as described in the text.

(g) Derived by fitting low-temperature yield data for a $\alpha$-iron to the equation for lattice-resistance controlled glide (eqn. (2.12)).

(h) Using $\mu = 3E/8$, from Köster (1948).

(i) Based on creep data of Feltham (1953)—see text.

(j) Extrapolated from $\alpha$-Fe $K_{\text{screw}}$, following the method of Fe-3.1% Si given by Lytton (1964) (see text).

(k) Assumed to be the same as for $\alpha$-iron.

(l) Based on Young's modulus values from the CEGB-Universities Collaborative Project on 1% Cr-Mo-V steel (1974).

(m) The estimate used for $\alpha$-iron was reduced by one order of magnitude to be consistent with creep data. This is reasonable in a solid solution alloy: see text.

(n) CEGB-Universities Collaborative Project on 1% Cr-Mo-V steel (1974).

(o) Because of the precipitation hardening in this alloy, the obstacle-controlled yield calculation used $\Delta F = 2\ \mu b^3$. For the other alloys, $\Delta F = 0.5\ \mu b^3$ was used.

(p) Based on analogy to $\alpha$-iron, using the temperature dependence given by Blackburn (1972).

(q) Perkins et al. (1974). Value for tracer diffusion of $^{59}$Fe in Fe-17 wt.% Cr-13 wt.% Ni.

(r) Perkins et al. (1974). These values are a roughly weighted average for the boundary diffusion of Fe, Cr, and Ni in Fe-17Cr-12Ni.

(s) Blackburn (1972).

(t) Based, as described in the text, on creep data of: Blackburn (1972); Garofalo et al. (1963); and Challenger and Moteff (1973).

63

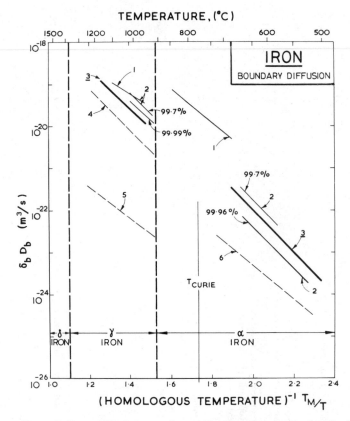

Fig. 8.6. Grain boundary diffusion of iron. The data are from: (1) Borisov *et al.* (1964); (2) Guiraldeng and Lacombe (1965); (3) James and Leak (1965); (4) Lacombe *et al.* (1963); (5) Bokshtein *et al.* (1959); and (6) Gertsriken and Pryanishnikov (1959).

diffusion coefficients, etc.) change sharply here, and again at the $\gamma$–$\delta$ phase transformation at 1381°C; the most important change is that of a factor of about 100 in the lattice diffusion coefficient.

Austenite shows a field of power-law creep, and a subdivided diffusional-flow field (for the grain size used here: 0·1 mm). The regime of $\delta$-iron shows one of each, neither subdivided. Flow in iron is further complicated by the loss of ferromagnetism at the Curie temperature: 770°C. This temperature is marked on the map; its effects are discussed in Section 8.2.

The problem of properties which depend on the thermomechanical treatment becomes important with the creep resistant *low-alloys ferritic steels* such as the 1% Cr–Mo–V steel described by Fig. 8.7. These alloys derive most of their strength from precipitation-hardening; small changes in thermomechanical history have large effects on the mechanical behaviour; and over-ageing in service is an important consideration (Honeycombe, 1980). It would be useful to know how the field boundaries of the maps move during ageing, but at present there is insufficient experimental data to do more than guess at this. In practice, however, the steel is

used in one of a few standard states and is limited in its use to temperatures below 600°C because of oxidation. Ageing in service becomes important only above 550°C, so the map has some general usefulness below this temperature.

The alloy starts to transform to austenite at about 700°C, and is never used structurally above this temperature, where its strength would be little better than that of pure austenite. For completeness we have shown the contours and field boundaries for pure austenite (though this ignores the small solution-hardening effect that the alloying elements would exert in the $\gamma$-phase). It emphasizes the high flow stress and creep resistance of the dispersion-strengthened ferrite. The alloying suppresses the b.c.c. $\delta$-phase, which therefore does not appear on the map.

The map is a tolerable fit to the available data (which are plotted on it), and is drawn for a grain size close to that of the alloys from which the data came. It shows that the diffusional-flow field lies just below the area covered by the data (one point may lie in this field), and suggests that diffusional flow could become the dominant mechanism in applications of the alloy at above 450°C if the pre-

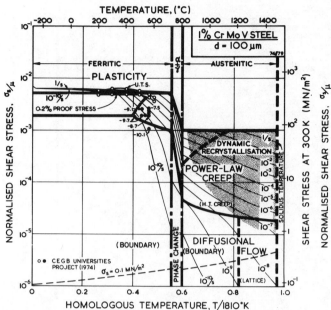

Fig. 8.7. A 1% Cr–Mo–V steel, of grain size 100 μm, showing data.

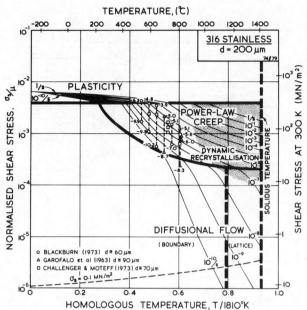

Fig. 8.9. A stress/temperature map for type 316 stainless steel of grain size 200 μm, showing data.

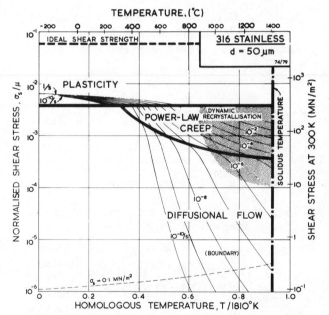

Fig. 8.8. A stress/temperature map for type 316 stainless steel of grain size 50 μm.

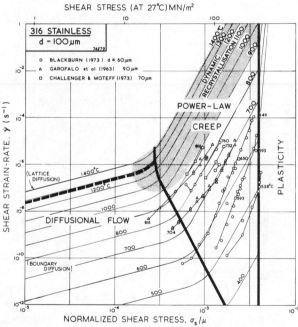

Fig. 8.10. A strain-rate/stress map for type 316 stainless steel of grain size 100 μm, showing data.

cipitates do not inhibit the grain boundary acting as sinks and sources for diffusion (Chapter 17, Section 17.3).

In some ways the f.c.c. *stainless steels* (Figs. 8.8 to 8.13) are simpler than the ferritic steels (Parr and Hanson, 1965). Both 304 and 316 stainless derive a large part of their strength from solid-solution hardening, a mechanism which is largely unaffected by thermal or mechanical processing. But, parti-

cularly in 316, part of the strength is due to a precipitate of carbides. Its contribution, too, depends on the previous history of the steel, and may change during the service life. For that reason, a map for 316 stainless describes the steel in one structural state only: bar-stock in the as-received condition, in the case of the maps shown here. Strictly, a new map should be constructed for each condition of

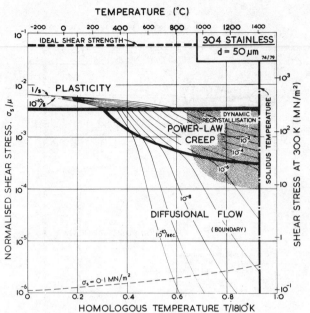

Fig. 8.11. A stress/temperature map for type 304 stainless steel of grain size 50 μm.

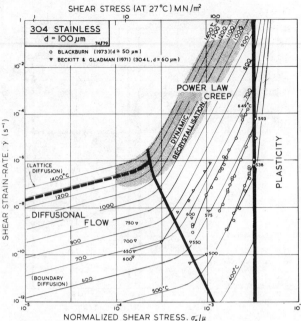

Fig. 8.13. A strain-rate/temperature map for type 304 stainless steel of grain size 100 μm, showing data.

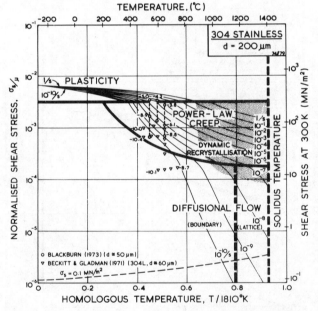

Fig. 8.12. A stress/temperature map for type 304 stainless steel of grain size 200 μm, showing data.

the steel, though in practice the changes may be slight, and the map shown below gives at least an approximate picture of its mechanical behaviour.

An alternative form of map is shown in Figs. 8.10 and 8.13: one with axes of strain-rate and stress—a more conventional way of presenting creep data. The data themselves are shown in these figures (which have been constructed for a grain size of 100 μm, close to that of the specimens from which the data were obtained). These plots give a good way

of checking the position of field boundaries against experiments: both show evidence for a transition to diffusional flow at low stresses and for power-law breakdown at high.

### References for Section 8.1

Frost, H. J. and Ashby, M. F. (1977) *Fundamental Aspects of Structural Alloy Design* (eds. Jaffee, R. I. and Wilcox, B. A.), Plenum, p. 27.

Honeycombe, R. W. K. (1980) *Steels, Microstructure and Properties*. Arnold.

Parr, J. G. and Hanson, A. (1965) *An Introduction to Stainless Steel*. Arnold.

## 8.2 ORIGINS OF THE DATA FOR PURE IRON AND FERROUS ALLOYS

### Pure iron (Fig. 8.1)

Fig. 8.1 shows a map for pure iron with a grain size of 100 μm. It is based on data plotted in Figs. 8.2 to 8.6, and summarized in Table 8.1.

The crystallographic data (Burger's vector, $b$; atomic volume, $\Omega$) for α- and γ-iron are from Taylor and Kagle (1963). Those for δ were extrapolated from data for α.

The modulus at 300 K and its temperature dependence present special problems for α-iron. Single crystals are elastically anisotropic, so the appropriate average of the single-crystal moduli must first be chosen. Fig. 8.4 shows various possible choices. The moduli $c_{44}$ and $\frac{1}{2}(c_{11} - c_{12})$, the extremes of the single-crystal shear moduli, were calculated from the measurements of Dever (1972). The Voigt and Reuss averages, though closer together, still differ considerably. The figure also shows the functions of the single-crystal moduli appropriate for the calculation of the energy of a $\frac{1}{2}\langle 111 \rangle$ edge ($K_{\text{edge}}$) and $\frac{1}{2}\langle 111 \rangle$ screw ($K_{\text{screw}}$) dislocation in $\alpha$-iron on a (110) plane, using the results of Hirth and Lothe (1968). In what follows, we have used $K_{\text{screw}}$.

This figure illustrates the second problem with $\alpha$-iron. Because of the magnetic transformation at 770°C, the modulus changes with temperature in a non-linear way (it also influences the diffusion coefficients—see below). Instead of the linear temperature correction to $\mu_0$, which is usually adequate, we have here used a power-law for $\mu$, giving a close approximation to the variation of $K_{\text{screw}}$ with temperature (see Table 8.1, note b). Because the map is plotted on normalized axes, the anomalous behaviour of $K_{\text{screw}}$ is not apparent. It becomes so when the true stress is plotted: see the line labelled $\sigma_s = 0.1$ MN/m$^2$ on Fig. 8.1.

The modulus of $\gamma$-iron, and its temperature dependence, are poorly characterized. We calculated both from Köster's (1948) measurements of Young's modulus, $E$, by setting $\mu = 3E/8$. The modulus of $\delta$-iron, too, presents a problem. A linear extrapolation from paramagnetic $\alpha$-iron goes to zero at a temperature below the melting point. We have therefore extrapolated the polycrystalline modulus of Fe–3·1% Si (as proposed by Lytton, 1964) and corrected this to give a screw-dislocation modulus by multiplying it by the ratio of $K_{\text{screw}}$ ($\alpha$-iron, 912°C) to $\mu$ (Fe–3·1% Si, 912°C).

Like the moduli, the diffusion coefficients for iron present special problems. Data for lattice diffusion are shown in Fig. 8.5 which illustrates the large, sharp decrease in $D_v$ when ferrite transforms to austenite, and the similar increase when this reverts to $\delta$-iron. A subtler influence is that of the magnetic transformation: diffusion in ferrite does not follow an Arrhenius law, but is about 3 times larger at 770°C than one would expect by extrapolation from lower temperatures (and thus has a temperature dependent activation energy: Buffington et al., 1961; Kučera et al., 1974). We have used the lattice diffusion coefficients of Buffington et al. (1961) labelled 9 on Fig. 8.3, for both $\alpha$- and $\gamma$-iron, and have assumed, as Fig. 8.5 suggests, that the coefficients for $\delta$-iron are the same as those for $\alpha$-iron.

The discontinuity in boundary diffusion rates (Fig. 8.6) at the phase boundaries is nothing like as large as that for lattice diffusion, presumably because the change in grain-boundary structure is not as profound as the change from a non-close-packed to a close-packed arrangement of atoms that occurs within the grains. We used the coefficients of James and Leak (1965), labelled 3 on Fig. 8.6, which lie broadly through the available data. The core diffusion parameters, too, are based on their measurements.

The power-law creep parameters for $\alpha$-iron ($n$ and $A$) and information about core diffusion ($a_c D_c$) were obtained by fitting the power-law creep equation to the data plotted in Figs. 8.2 and 8.3. For high-temperature creep we have used the value $n = 6·9$ of Ishida et al. (1966), which describes all the data reasonably well (Fig. 8.3). Except near the Curie temperature, the activation energy for high-temperature creep is close enough to that for lattice diffusion for the two to be equated; the anomaly is properly incorporated into the map through the non-linear temperature-dependence of the modulus and the diffusion anomaly described above.

Although there is little evidence for a low-temperature creep field for ferrite, experience in constructing maps for other b.c.c. metals (Chapter 5) indicates that it is usually present. The core-diffusion coefficients are based on the grain-boundary data for ferrite (James and Leak, 1965). Using these gives the small field shown in Fig. 8.1, and puts almost all the data of Fig. 8.3 into the high-temperature creep field—a result which is consistent with the data.

There are fewer data for creep of $\gamma$-iron than for $\alpha$-iron (Fig. 8.2). The bulk of the data is due to Feltham (1953) whose data we have used to derive the creep parameters $n$ and $A$. In using the map of Fig. 8.1 it should be remembered that the stress is normalized with respect to a different modulus above and below the transformation temperature:

$$\mu(\alpha\text{-Fe, } 912°\text{C}) = 2·5 \times 10^4 \text{ MN/m}^2$$

$$\mu(\gamma\text{-Fe, } 912°\text{C}) = 4·5 \times 10^4 \text{ MN/m}^2$$

This normalization conceals the true magnitude of the discontinuity of strain-rate at 912°C: it is larger than it appears from Fig. 8.1. The difference in normalizations is shown by the dashed line of constant shear stress $\sigma_s = 0.1$ MN/m$^2$. The strain-rate contours should be displaced accordingly to give the discontinuity in $\dot{\gamma}$ at constant $\sigma_s$.

The creep parameters of $\delta$-iron were set equal to those for $\alpha$-iron. We suspect that $n = 6·9$ may be a little high, but having no data to guide us, we have used it.

The glide parameters for $\alpha$-iron were determined

by fitting the equations for lattice-resistance and obstacle-controlled glide (eqns. 2.12 and 2.9) to the data shown in Fig. 8.2.

Pure iron shows dynamic recrystallization above $0.7\ T_M$ (1000°C). The positioning of the shaded field is based on data of Wray (1975 a, b; 1976) and Wray and Holmes (1975).

## References for pure iron

Amonenko, V. M., Blinkin, A. M. and Ivantsov, I. G. (1966) *Phys. Metals Metallog. U.S.S.R.* (English transl.) **17(1)**, 54.

Birchenall, C. E. and Mehl, R. F. (1948) *J. Appl. Phys.* **19**, 217.

Birchenall, C. E. and Mehl, R. F. (1950) *Trans. Met. Soc. AIME* **188**, 144.

Bogdanov, N. A. (1962) *Russ. Met. Fuels* (English transl.) **2**, 61.

Bokshtein, S. Z., Kishkin, S. T. and Moroz, L. M. (1957) *Metalloved. i Term. Obrabotka Metal.* **2**, 2.

Bokshtein, S. Z., Kishkin, S. T. and Moroz, L. M. (1959) *Radioactive Studies of Metal Surfaces*. Moscow, USSR.

Bokshtein, S. Z., Kishkin, S. T. and Moroz, L. M. (1961) "Investigation of the Structure of Metals and Radioactive Isotope Methods", State Publishing House of the Ministry of Defence Industry, Moscow (1959): AEC-tr-4505.

Borg, R. J. and Birchenall, C. E. (1960) *Trans. Met. Soc. AIME* **218**, 980.

Borg, R. J., Lai, D. Y. F. and Krikorian, O. (1963) *Acta Met.* **11**, 867.

Borisov, V. T., Golikov, V. M. and Scherbedinskiy, G. V. (1964) *Fiz. Metal. Metalloved.* **17(6)**, 881.

Buffington, F. S., Bakalar, I. D. and Cohen, M. (1951) *Physics of Powder Metallurgy*. McGraw-Hill, NY, p. 92.

Buffington, F. S., Hirano, K. and Cohen, M. (1961) *Acta Met.* **9**, 434.

Čadek, J., Pahutova, M., Čiha, K. and Hostinsky, T. (1969) *Acta Met.* **17**, 803.

Christ, B. W. and Smith, G. V. (1968) *Mem. Sci. Rev. Met.* **65**, 207.

Davies, P. W. and Williams, K. R. (1969) *Acta Met.* **17**, 897.

Dever, D. J. (1972) *J. Appl. Phys.* **43**, 3293.

Feltham, P. (1953) *Proc. Phys. Soc. Lond.* **B66**, 865.

Gertsriken, S. D. and Pryanishnikov, M. P. (1958) *Vopr. Fiz. Met. i Metalloved., Sb. Nauchn. Rabot. Inst. Metallofiz. Akad. Nauk Ukr. SSR* **9**, 147.

Gertsriken, S. D. and Pryanishnikov, M. P. (1959) *Issled. po Zharoprochn. Splavam* **4**, 123.

Glover, G. and Sellars, C. M. (1973) *Met. Trans.* **4A**, 765.

Golikov, V. M. and Borisov, V. T. (1958) "Problems of Metals and Physics of Metals", 4th Symposium, Consultants Bureau, New York, (1957); AEC-tr-2924.

Graham, D. and Tomlin, D. H. (1963) *Phil Mag.* **8**, 1581.

Gruzin, P. L. (1952) *Probl. Metalloved. i Fiz. Met.* **3**, 201.

Gruzin, P. L. (1953) *Izv. Akad. Nauk SSSR Otd. Tekhn. Nauk.* **3**, 383.

Guiraldeng, P. and Lacombe, P. (1965) *Acta Met.* **13**, 51.

Hirth, J. P. and Lothe, J. (1968) *Theory of Dislocations*. McGraw-Hill.

Hough, R. R. and Rolls, R. (1971) *Met. Sci. J.* **5**, 206.

Ishida, Y., Cheng, C.-Y. and Dorn, J. E. (1966) *Trans. AIME* **236**, 964.

James, D. W. and Leak, G. M. (1965) *Phil. Mag.* **12**, 491.

James, D. W. and Leak, G. M. (1966) *Phil. Mag.* **14**, 701.

Karashima, S., Oikawa, H. and Watanabe, T. (1966) *Acta Met.* **14**, 791.

Keane, D. M., Sellars, C. M. and McG. Tegart, W. J. (1968) *Deformation under Hot Working Conditions*. The Iron and Steel Institute, London, Special Report, **108**, p. 21.

Köster, W. (1948) *Z. Metallkde.* **9**, 1.

Kučera, J., Million, B., Ruzickova, J., Foldyna, V. and Jakobova, A. (1974) *Acta Met.* **22**, 135.

Lacombe, P., Guiraldeng, P. and Leymonie, C. (1963) *Radioisotopes in the Physical Science Industries*. IAEA, Vienna, p. 179.

Leymonie, C. and Lacombe, P. (1957) *Compt. Rendu* **245**, 1922; (1958) *Rev. Met.* (Paris) **55**, 524; (1959) *Metaux (Corrosion-Ind.)* **34**, 457; (1960) *Metaux (Corrosion-Ind.)* **35**, 45.

Lytton, J. L. (1964) *J. Appl. Phys.* **35**, 2397.

Mead, H. W. and Birchenall, C. E. (1956) *Trans. Met. Soc. AIME* **206**, 1336.

Murty, K. L., Gold, M. and Ruoff, A. L. (1970) *J. Appl. Phys.* **41**, 4917.

Sparke, B., James, D. W. and Leak, G. M. (1965) *J. Iron Steel Inst.* (*London*) **203(2)**, 152.

Spitzig, W. A. and Keh, A. S. (1970) *Acta Met.* **18**, 1021.

Spitzig, W. A. and Leslie, W. C. (1971) *Acta Met.* **19**, 1143.

Staffansson, L. I. and Birchenall, C. E. (1961) AFOSR-733.

Taylor, A. and Kagle, B. J. (1963) *Crystallographic Data on Metal and Alloy Structures*. Dover Publications, Inc.

Uvira, J. L. and Jonas, J. J. (1968) *Trans. AIME* **242**, 1619.

Watanabe, T. and Karashima, S. (1970) *Trans. Jap. Inst. Met.* **11**, 159.

Wray, P. J. (1975a) *Met. Trans.* **6A**, 1379; (1975b) *Met. Trans.* **6A**, 1197; (1976) Met. Trans. **7A**, 1621.

Wray, P. J. and Holmes, M. F. (1975) *Met. Trans.* **6A**, 1189

Zhukhovitskii, A. A. and Geodakyan, V. A. (1955) *Primenenie Radioaktivn. Izotroov v Metallurg. Sb.* **34**, 267; AEC-tr-3100, *Uses of Radioactive Isotopes in Metallurgy Symposium XXXIV*, Pt. 2, p. 52.

## 1% Cr–Mo–V steel (Fig. 8.7)

The map shown in Fig. 8.7 describes a particular batch of 1% Cr–Mo–V steel (CEGB–Universities Collaborative Project, 1974). Its composition is given in Table 8.2. All tests were carried out on forged material after a standard heat treatment designed to give a fine-grained, tempered bainite containing a carbide dispersion. (The heat-treat-

**TABLE 8.2    Composition of the steel, wt. %**

| C | Si | Mn | Ni | Cr | Mo | V | S | P | Fe |
|---|----|----|----|----|----|----|----|----|----|
| 0·24 | 0·29 | 0·64 | 0·21 | 1·02 | 0·57 | 0·29 | 0·10 | 0·16 | bal |

ment was as follows: soak at 1000°C; furnace cool to 690°C and hold for 70 hours; air cool; reheat and soak in salt bath at 975°C; quench into a second salt bath at 450°C; air cool; reheat to $700 \pm 3$°C for 20 hours.)

The modulus $\mu_0$ and its temperature were calculated from data for Young's modulus for the polycrystalline alloy (CEGB–Universities Project, 1974) taking $\mu_0 \simeq \frac{3}{8}E_0$. The elaborate treatment used to describe the modulus of pure ferrite is unnecessary here: the Curie temperature, and the elastic anomalies associated with it, lie above the $\alpha$–$\gamma$ phase transition.

The lattice and boundary diffusion coefficients are those for pure ferrite. Though the alloying elements will change the diffusion coefficient for steady-state mass transport (Chapter 2, Section 2.4) the effect is slight in a low-alloy steel, and can be ignored.

The data for dislocation glide obtained by the CEGB–Universities Project are well described by eqn. (2.9) with a large value of $\Delta F = 2\mu b^3$. This implies strong obstacles (presumably the carbides) and a flow stress which, when normalized, is insensitive to temperature or strain-rate (it is probably the Orowan stress for the carbide dispersion; see Chapter 2, Table 2.1).

The data for power-law creep are barely adequate for the construction of a map; they are derived from constant stress data for "Piece 2" of the CEGB–Universities Project (1974). "Piece 4" crept one-third as fast, emphasizing that heat-treatable alloys such as this one are sensitive to slight variations in composition or heat treatment.

The observed activation energy for creep (250–370 kJ/mole) lies close to that for self-diffusion. The observations suggest that core-diffusion is suppressed in this alloy (see also Chapter 7). We have suppressed it by reducing the pre-exponential by a factor of 10: all the data then lie in the high-temperature creep field.

There is no experimental verification of diffusional flow in this alloy. The simple treatment of Chapter 2 (eqn. 2.29)), on which the map is based, suggests that Coble-creep could become the dominant deformation mode at low strain rates above 450°C. But a fine dispersion may suppress diffusional flow (Chapter 17, Section 17.3). This field may, therefore, lie lower than we have shown it; only further experiments at low stresses can resolve this point.

## References for 1% Cr–Mo–V steel

CEGB–Universities Collaborative Project on 1% Cr–Mo–V Steel (1974) available from CERL, Leatherhead, Surrey.

## 316 and 304 Stainless Steels (Figs. 8.8 to 8.13)

Maps and data for 316 and 304 stainless steels are shown in Figs. 8.8 to 8.13. They are based on data which are plotted on the maps themselves, and on the parameters listed in Table 8.1. The specifications of the two alloys allows a certain latitude of composition. The maps are based on published data derived from three batches of 316 and two of 304; these compositions are listed in Table 8.3. (Both alloys are also available in a low-carbon specification designated 316L and 304L, with less than 0·03% carbon. Their properties differ significantly from those described by the maps shown here.)

The modulus and its temperature dependence are from Blackburn (1972). The diffusion coefficients of the components in these alloys are well documented. Perkins et al. (1973) list both lattice and boundary diffusion data for Fe, Cr and Ni in an alloy with a composition close to that of the two steels. Strictly, the proper diffusion coefficient for steady-state transport is one based on eqn. (2.27). While Perkins' data are sufficiently complete to allow this to be calculated, the three major components in the steel diffuse at so nearly equal rates that it is simplest to take the coefficient for the principal component, iron; this choice gives an adequate description of the creep data described below.

Boundary diffusion, too, should properly be described by a combination of diffusion coefficients. Again, we have taken a single coefficient (167 kJ/mole amongst the quoted values of $150.7 \pm 9.6$ for Cr, $177 \pm 17.2$ for Fe and $133.9 \pm 8.0$ for Ni), and a pre-exponential that places the coefficient in the centre of the scattered data.

The power-law creep parameters were derived from the blocks of data shown in Figs. 8.10 and 8.13. A simple power-law does not describe the data very well: much of it was obtained at high stresses, where the power-law starts to break down. For 316 (Fig. 8.10) we used a value $n = 7.0$, which lies between that for the high-temperature, low-stress data of Garofalo et al. (1963) and the low-temperature, high-stress data of Blackburn (1972): over the range they investigated, $n$ varies from 4 to 10. Garofalo himself fits his data to a $(\sinh \beta'\sigma_s)^{n'}$ law (eqn. 2.25), which is more flexible than a power-law but less satisfactory than eqn. (2.26), for reasons explained in Chapter 2, Section 2.4. Further, an activation energy for creep, derived by appropriately replotting the data of Fig. 8.10, is rather higher than that for diffusion. The pre-exponential constant, $A$, is chosen to match the data of Garofalo et al. at 732°C, and that of Blackburn at 650°C, but because of the discrepancy between the activation energies for

**TABLE 8.3** Composition of the stainless steels (wt.%)

| | Cr | Ni | Mo | C | Si | Mn | P | S | N | Ti | Al | B |
|---|---|---|---|---|---|---|---|---|---|---|---|---|
| *AISI 316* | | | | | | | | | | | | |
| Nominal 316, Parr and Hanson (1965) | 16–18 | 10·14 | 2·0–3·0 | 0·08 max | 1·0 max | 2·0 max | 0·045 max | 0·03 max | — | — | — | — |
| Blackburn (1972) | 17·8 | 13·6 | 2·4 | 0·05 | 0·4 | 1·7 | 0·01 | 0·02 | 0·04 | 0·003 | 0·026 | <0·005 |
| Garofalo et al. (1963) | 18 | 11·4 | 2·15 | 0·07 | 0·38 | 1·94 | 0·01 | 0·006 | 0·05 | — | 0·003 | — |
| Challenger and Moteff (1973) | 18·16 | 13·60 | 2·47 | 0·086 | 0·52 | 1·73 | 0·01 | 0·006 | 0·05 | <0·005 | <0·005 | <0·0005 |
| *AISI 304* | | | | | | | | | | | | |
| Nominal 304, Parr and Hanson (1965) | 18–20 | 8·0–12·0 | | 0·08 max | 1·0 max | 2·0 max | 0·045 max | 0·03 max | — | — | — | — |
| Blackburn (1972) | 18·4 | 9·7 | 0·1 | 0·07 | 0·5 | 0·9 | 0·02 | 0·01 | 0·05 | 0·013 | — | 0·0014 |
| Beckitt and Gladman (1971), (304L) | 17·7 | 9·99 | — | 0·015 | 0·31 | 0·57 | — | — | — | — | — | — |

creep and for diffusion, and the indifferent fit of the data to a power-law, this creep field must be regarded as a first approximation only.

The creep of 304 stainless is rather better described by a power-law (Fig. 8.13). The observed activation energies for creep and for diffusion are very similar; a single power-law with $n = 7·5$ describes most of Blackburn's data. The alloy studied by Beckitt and Gladman (1971) was 304L—the low-carbon form of 304. It creeps about an order of magnitude faster than 304, and has a slightly lower yield strength.

The reproducibility of creep data between different experimenters, for both 316 and 304 steels, using slightly different compositions, is reasonably good—better than for creep of pure metals. This evidently results from the solid-solution alloying which swamps out the effects of trace impurities.

For both 316 and 304 there is experimental evidence for diffusional flow; both Figs. 8.10 and 8.13 show a decrease in the exponent $n$ towards the value unity. This has importance in confirming experimentally the position of the field boundary between the diffusional flow and the power-law creep fields. The available evidence suggests that the position of the boundary is about right.

For both 316 and 304 the dislocation glide field is based on eqn. (2.9) with $\Delta F = 0·5 \, \mu b^3$ and $\hat{\tau} = 6·45 \times 10^{-3} \, \mu$. The value of $\hat{\tau}$ is chosen so that the yield stress at room temperature is equal to that quoted by Parr and Hanson (1965). A higher value of $\hat{\tau}$ should be used if the material is heavily worked.

Both steels show dynamic recrystallization above about 1000°C, at which temperature most carbides have redissolved. The shaded field is based on the observations of Nadai and Manjoine (1941) and of Wray, P. J. (1969) J. Appl. Phys. **40**, 4018.

### References for 316 and 304 stainless steels

Beckitt, F. R. and Gladman, T. (1971) British Steel, Special Steels Department Report No. PROD/PM/ 6041/1/71A.

Blackburn, L. D. (1972) "The Generation of Isochronous Stress-Strain Curves". ASME Winter Annual Meeting, New York.

Challenger, K. D. and Moteff, J. (1973) *Met. Trans.* **4A**, 749.

Garofalo, F., Richmond, O., Domis, W. F. and von Germminger, F. (1963) in *Joint International Conf. on Creep*. Int. Mech. Eng., London, p. 1.

Nadai, A. and Manjoine, M. J. (1941) *J. Appl. Mech.* **63**, A-77.

Parr, J. G. and Hanson, A. (1965) *An Introduction to Stainless Steel*. ASM, p. 54.

Perkins, R. A., Padgett, Jr., R. A. and Tunali, N. K. (1973) *Met. Trans.* **4A**, 2535.

Wray, P. J. (1969) *J. Appl. Phys.* **40**, 4018.

# CHAPTER 9

# THE COVALENT ELEMENTS, Si AND Ge

FOUR elements crystallize in the diamond cubic structures: silicon, germanium, α-tin and diamond itself. It can be thought of as an f.c.c. structure with two atoms associated with each f.c.c. atom position. Each atom has four nearest neighbours to which it is linked by four purely covalent bonds. The low co-ordination makes it a very open structure, with a low density. But despite this, the covalent elements are enormously strong, having a flow strength at which exceeds $\mu/20$ at 0 K and which remains high even at $0.5\ T_M$. Their mechanical properties are interesting because they typify the extreme behaviour associated with pure covalent bonding.

Maps and data for silicon and germanium are shown in Figs. 9.1 to 9.4. The parameters used to compute the maps are listed in Table 9.1.

## 9.1 GENERAL FEATURES OF THE MECHANICAL BEHAVIOUR OF THE DIAMOND CUBIC ELEMENTS

The mechanical behaviour of Si and Ge is dominated by their localized, strongly directional, covalent bonding. It creates an exceptionally large lattice resistance for slip on all slip systems, including the primary system $\{111\}\langle1\bar{1}0\rangle$ (Alexander and Haasen, 1968). The result is that the normalized flow strength at below half the melting point is larger than that for any other class of solid (except perhaps ice). It is so large that, in most modes of loading, covalent elements and compounds fracture before they flow. The hardness test allows the plastic behaviour to be studied in a simple way down to $0.2\ T_M$, when fracture becomes a problem here also.

The localized nature of the bonding has other consequences: the energy of formation and motion of point defects is large, so that diffusion (at a given homologous temperature) is slower than in other classes of solid. At high temperature ($>0.6\ T_M$) Si and Ge show power-law creep, often complicated by an inverse transient (in creep tests) or a yield drop (in constant strain-rate tests) during which the dislocation density increases towards a steady state. This steady state is rarely reached in the small

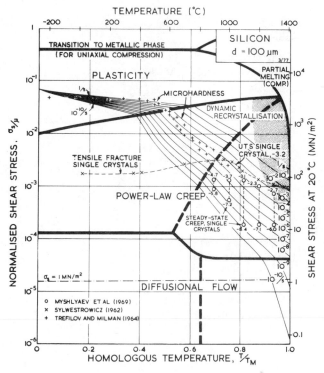

Fig. 9.1. Stress/temperature map for silicon of grain size 100 μm. Data are labelled with $\log_{10}(\dot{\gamma})$.

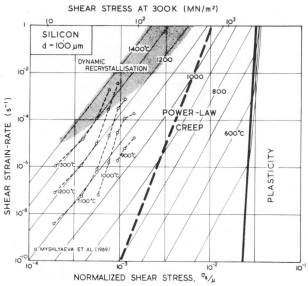

Fig. 9.2. Strain-rate/stress map for silicon of grain size 100 μm. Data are labelled with temperature (°C).

**TABLE 9.1   The covalent elements**

| Material | Silicon | | Germanium | |
|---|---|---|---|---|
| *Crystallographic and thermal data* | | | | |
| Atomic volume, $\Omega$ (m$^3$) | $2.00 \times 10^{-29}$ | | $2.26 \times 10^{-29}$ | |
| Burger's vector, $b$ (m) | $3.83 \times 10^{-10}$ | | $3.99 \times 10^{-10}$ | |
| Melting temperature, $T_M$ (K) | 1687 | | 1211 | |
| *Modulus* | | | | |
| Shear modulus at 300 K, $\mu_0$ (MN/m$^2$) | $6.37 \times 10^4$ | (a) | $5.20 \times 10^4$ | (g) |
| Temperature dependence of modulus, $\dfrac{T_M}{\mu_0}\dfrac{d\mu}{dT}$ | $-0.078$ | (b) | $-0.146$ | (b) |
| *Lattice diffusion (normal)* | | | | |
| Pre-exponential, $D_{0v}$ (m$^2$/s) | 0.9 | (c) | $7.8 \times 10^{-4}$ | (h) |
| Activation energy, $Q_v$ (kJ/mole) | 496 | | 287 | |
| *Boundary diffusion* | | | | |
| Pre-exponential, $\delta D_{0b}$ (m$^3$/s) | $10^{-15}$ | (d) | $10^{-17}$ | (d) |
| Activation energy, $Q_b$ (kJ/mole) | 300 | | 172 | |
| *Core diffusion* | | | | |
| Pre-exponential, $a_c D_{0c}$ (m$^4$/s) | $10^{-25}$ | (d) | $10^{-27}$ | (d) |
| Activation energy, $Q_c$ (kJ/mole) | 300 | | 172 | |
| *Power-law creep* | | | | |
| Exponent, $n$ | 5.0 | (e) | 5.0 | (i) |
| Dorn constant, $A$ | $2.5 \times 10^6$ | | $1.0 \times 10^8$ | |
| *Obstacle-controlled glide* | | | | |
| 0 K flow stress, $\hat{\tau}/\mu_0$ | — | | — | |
| Pre-exponential, $\dot{\gamma}_0$(s$^{-1}$) | — | | — | |
| Activation energy, $\Delta F/\mu_0 b^3$ | — | | — | |
| *Lattice-resistance-controlled glide* | | | | |
| 0 K flow stress, $\hat{\tau}_p/\mu_0$ | 0.07 | (f) | 0.06 | (f) |
| Pre-exponential, $\dot{\gamma}_p$(s$^{-1}$) | $10^{11}$ | | $10^{11}$ | |
| Activation energy, $\Delta F_p/\mu_0 b^3$ | 0.2 | | 0.2 | |

(a)  McSkimin (1953); Prasad and Wooster (1955);

$$\mu = [\tfrac{1}{2}c_{44}(c_{11} - c_{12})]^{\frac{1}{2}}$$

(b)  McSkimin (1959); McSkimin and Andreatch (1964);

$$\mu = \mu_0\left(1 + \left(\frac{T - 300}{T_M}\right)\left(\frac{T_M}{\mu_0}\frac{d\mu}{dT}\right)\right)$$

(c)  Masters and Fairfield (1966); $D_v = D_{0v} \exp - (Q_v/RT)$.
(d)  Estimated using $Q_b = 0.6 \ Q_v$ and $\delta D_{0b} = 2bD_{0v}$; $\delta D_b = \delta D_{0b} \exp - (Q_b/RT)$; $a_c D_c = a_c D_{0c} \exp - (Q_c/RT)$

(e)  Adjusted to fit Myschlyaev et al. (1969). The constant $A$ refers to tensile creep; in computing the maps $A_s = (\sqrt{3})^{n+1} A$ is used.
(f)  Adjusted to fit the hardness data of Trefilov and Mil'man (1964).
(g)  Fine (1953); McSkimin and Andreatch (1963).
(h)  Letaw et al. (1956).
(i)  Estimated by using the same creep exponent as silicon and fitting the constant $A$ to the data of Gerk (1975); see note (e).

strains of most published experiments, although Myshlyaev et al. (1969) report steady behaviour in silicon. We have assumed that steady-state creep can be described by the rate-equation used in earlier sections for metals (eqn. (2.21)) and that, at sufficiently low stresses, diffusional flow (as described by eqn. (2.29)) will appear. Both creep mechanisms are diffusion-controlled, and are slow because of the slow rates of diffusion. Thus, the covalent elements and the covalently bonded 3–5 and 2–6 compounds, as a class, are stronger (in terms of the normalized variable $\sigma_s/\mu$) at all homologous temperatures, than any other class of solid; the ultimate example, of course, is diamond. The maps are a tolerably good fit to the hardness data except below $0.2 \ T_M$ (when

fracture often occurs round the indenter) and to the high-temperature ($>0.8 \ T_M$) creep data. At intermediate temperatures, the maps match the creep data less well. Lower yield point and transient creep data, some of which are shown on Fig. 9.3, deviate widely from the predicted steady-state behaviour of the maps (as expected) and were ignored in deriving the parameters used in their construction.

The stress axes of the maps shown in this chapter have been extended upwards by a factor of 10 compared with the others in this book. This is because, for the reasons already given, the data lie at higher normalized stresses than those for most other materials; and because it allows us to show where certain pressure-induced phase trans-

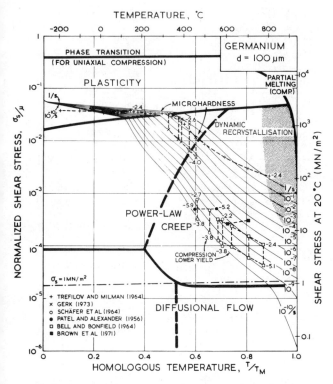

Fig. 9.3. Stress/temperature map for germanium of grain size 100 μm. Data are labelled with $\log_{10} \dot{\gamma}$.

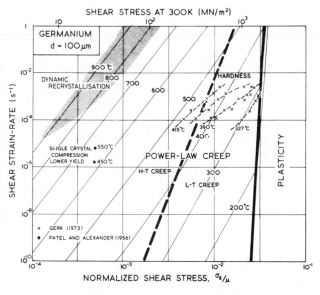

Fig. 9.4. Strain-rate/stress map for germanium of grain size 100 μm. Data are labelled with temperature (°C).

formations would occur in simple compression (using the pressure $-\frac{1}{3}\sigma_1$ that would appear in a uniaxial compression test). These phase transformations could become important in hardness tests, when the hydrostatic pressure at yield is roughly three times larger for the same equivalent shear stress, $\sigma_s$.

There is extensive literature on the mechanical properties of silicon and germanium. The interested reader should refer to Alexander and Haasen (1968) for a comprehensive review.

### References for Section 9.1

Alexander, H. and Haasen, P. (1968) *Solid State Physics* **22**, 27.
Myshlyaev, M. M., Nikitenko, V. I. and Nesterenko, V. I. (1969) *Phys. Stat. Sol.* **36**, 89.

## 9.2  DATA FOR SILICON AND GERMANIUM

The modulus and its temperature dependence were calculated from single-crystal data using:

$$\mu = \left[\tfrac{1}{2}c_{44}(c_{11} - c_{12})\right]^{\frac{1}{2}}$$

and the data of McSkimin (1953, 1959) and McSkimin and Andreatch (1963, 1964).

Dislocations and slip systems in diamond cubic crystals have been reviewed by Alexander and Haasen (1968). Under normal circumstances, slip occurs on $\{111\} \langle 1\bar{1}0 \rangle$, giving the five independent systems necessary for polycrystal plasticity.

The low-temperature yield strength ($T < 0.5\, T_M$) is based on the micro-hardness measurements of Trefilov and Mil'man (1964) corrected according to Marsh (1964). In both cases the temperature dependence of the hardness is less than can be explained using realistic values of $\Delta F_p$, and the rate-equation, using the constants of Table 9.1, does not match the measured hardness below $0.2\, T_M$ very well (this may be because fracture occurred at the indenter).

The power-law creep constants for silicon were fitted to the steady-state single-crystal creep data of Myshlyaev *et al* (1969). In the range of their measurements they observed an activation energy of between 400 and 520 kJ/mole; that for self-diffusion is 496 kJ/mole (Masters and Fairfield, 1966) which therefore provides an approximate description of their results. The use of a power-law, too, is an approximation (Fig. 9.2); our values of $n = 5$ and $A = 2.5 \times 10^6$ provide the best fit.

We have found no steady-state creep data for germanium. Brown *et al.* (1971), studying transient creep, found an activation energy of around 340 kJ/mole, which is rather larger than that for lattice diffusion of 287 kJ/mole (Letaw *et al.*, 1956). Useful information is contained in the high-temperature hardness measurements of Gerk (1975) and Trefilov and Mil'man (1964) which are plotted on Figs. 9.3 and 9.4. Gerk's data suggest a power-law with $n = 3$;

but because the hot-hardness test does not approach steady-state conditions, we have fitted the data to a power-law with the same exponent, $n$, as that for silicon (namely 5), adjusting the constant $A$ to fit Gerk's data. This choice leads to better agreement with the hardness data of Trefilov and Mil'man (1964) than the alternative choice of $n = 3$ suggested by Gerk (1975).

### References for silicon and germanium

Alexander, H. and Haasen, P. (1968) *Solid State Physics* **22**, 27.
Bell, R. L. and Bonfield, W. (1964) *Phil. Mag.* **9**, 9.
Brown, D., Chaudhari, G. and Feltham, P. (1971) *Phil. Mag.* **24**, 213.
Fine, M. E. (1953) *J.A.P.* **24**, 338.
Gerk, A. P. (1975) *Phil. Mag.* **32**, 355.

Letaw, H., Portnoy, W. M. and Slifkin, L. (1956) *Phys. Rev.* **102**, 636.
Marsh, D. M. (1964) *Proc. R. Soc.* **A279**, 420.
Masters, B. J. and Fairfield, J. M. (1966) *Appl. Phys. Letters* **8**, 280.
McSkimin, H. J. (1953) *J.A.P.* **24**, 988.
McSkimin, H. J. (1959) *J. Acoust. Soc. Am.* **31**, 287.
McSkimin, H. J. and Andreatch, P. Jr. (1963) *J.A.P.* **34**, 651.
McSkimin, H. J. and Andreatch, P. Jr. (1964) *J.A.P.* **35**, 2161.
Myshlyaev, M. M., Nikitenko, V. I. and Nesterenko, V. I. (1969) *Phys. Stat. Sol.* **36**, 89.
Patel, J. R. and Alexander, B. H. (1956) *Acta Met.* **4**, 385.
Prasad, S. C. and Wooster, W. A. (1955) *Acta Cryst.* **8**, 361.
Schäfer, S., Alexander, H. and Haasen, P. (1964) *Phys. Stat. Sol.* **5**, 247.
Sylwestrowicz (1962) *Phil. Mag.* **7**, 1825.
Trefilov, V. I. and Mil'man, Yu V. (1964) *Soviet Physics Doklady* **8**, 1240.

# CHAPTER 10

# THE ALKALI HALIDES: NaCl AND LiF

SODIUM CHLORIDE and lithium fluoride typify a large group of alkali halides with the rock-salt structure. The unit cell can be thought of as two inter-penetrating face-centred cubic sublattices, each ion surrounded octahedrally by six ions of the other charge. The unit cell contains four ions of each type.

More than 200 compounds with this structure are known. It occurs most commonly among alkali metal halides, the oxides of divalent transition metals, the alkaline earth oxides and calco-genides, and the transition metal carbides, nitrides and hydrides. But at least four mechanically-distinguishable subgroups with this structure exist (see Chapter 17). Three are examined in this book: the alkali halides (this chapter), the transition metal carbides (Chapter 11) and the simple oxides (Chapter 12). The bonding of each subgroup is different, so that despite the similarity in structure, the mechanical strengths differ. The alkali halides are purely ionic in their bonding. Their maps are shown in Figs. 10.1 to 10.4. The parameters used to construct them are listed in Table 10.1.

## 10.1 GENERAL FEATURES OF THE DEFORMATION OF ALKALI HALIDES

Like other polycrystalline materials, the plastic deformation of ionic solids involves a number of distinct mechanisms: low-temperature plasticity, power-law creep and diffusional flow. All three have been observed in alkali halides, among which LiF and NaCl are the most extensively studied.

At low temperatures, NaCl and LiF slip most easily on $\{110\}$ planes in the $\langle 1\bar{1}0 \rangle$ directions—but these provide only two independent systems. For polycrystal plasticity, slip on the harder $\{001\}$ and $\{111\}$ planes in the $\langle 110 \rangle$ directions is required, and it is this which determines the flow strength (Gutmanas and Nadgornyi, 1970; see also Chapter 2, Section 2.2). Even at 250°C, flow is governed by the hard slip systems (Carter and Heard, 1970).

Impurities influence the defect structure of ionic solids (and thus the rates of the mechanisms) in ways which are more complicated than in metals (see, for example, Greenwood, 1970). First, divalent impurities alter the concentration of vacancies (and other point defects) and thus affect the diffusion coefficients. A divalent impurity at concentration $C_0$, for instance, stabilizes a cation vacancy concentration $C_0$. At temperatures above that at which the thermal concentration of vacancies is $C_0$, the impurities have little effect and cation diffusion is said to be intrinsic. At lower temperatures the impurity maintains the cation vacancy concentration at $C_0$, and (by the law of mass-action) depresses the anion vacancy concentration to less than its thermal-equilibrium value in the pure solid. Diffusion is then influenced by the impurities, and is said to be extrinsic. Since anion diffusion usually controls the rate of mass-transport in the alkali halides, impurities of this sort can slow the rate of power-law creep and diffusional flow greatly.

A second effect of impurities is to alter the electrical charge carried by dislocations and grain boundaries (see, for example, Kliewer and Koehler, 1965). In sodium chloride at high temperatures, for example, dislocations carry a positive charge. Divalent cation impurities cause the ions to distribute themselves such that, at the isoelectric temperature $T_I$, the charge changes sign. Above $T_I$, each dislocation core carries an excess of $Na^+$ ions, and a negative space charge caused by an excess of $Na^+$ vacancies surrounds it. But below $T_I$, the core carries an excess of $Cl^-$ ions surrounded by a positive space charge. For reasons which are not completely understood, this seems to enhance diffusion at or near the core, and may explain certain aspects of creep in NaCl discussed in Section 10.2 (below).

There is a third effect of impurities: divalent solutes cause rapid solution-hardening, raising the yield stress in the dislocation glide regime. Johnston (1962), and Guiu and Langdon (1974), found that the critical resolved shear stress of $Mg^{2+}$–doped LiF single crystals was higher than that of purer ($<10$ ppm Mg) ones, probably because of the inter-actions of edge dislocations with tetragonal lattice

**TABLE 10.1  The alkali halides**

| Material | Sodium chloride | | Lithium fluoride | |
|---|---|---|---|---|
| *Crystallographic and thermal data* | | | | |
| Atomic volume, $\Omega$ (m$^3$) | $4\cdot49 \times 10^{-29}$ | (a) | $1\cdot63 \times 10^{-29}$ | (a) |
| Burger's vector, $b$ (m) | $3\cdot99 \times 10^{-10}$ | | $2\cdot84 \times 10^{-10}$ | |
| Melting temperature, $T_m$ (K) | 1070 | | 1140 | |
| *Modulus* | | | | |
| Shear modulus at 300 K, $\mu_0$ (MN/m$^2$) | $1\cdot5 \times 10^4$ | (b) | $4\cdot58 \times 10^4$ | (b) |
| Temperature dependence of modulus, $\dfrac{T_m}{\mu_0}\dfrac{d\mu}{dT}$ | $-0\cdot73$ | | $-0\cdot8$ | |
| *Lattice diffusion (anion)* | | | | |
| Pre-exponential, $D_{0v}$ (m$^2$/s) | $2\cdot5 \times 10^{-2}$ | (c) | $7\cdot4 \times 10^{-3}$ | (g) |
| Activation energy, $Q_v$ (kJ/mole) | 217 | | 214 | |
| *Boundary diffusion (anion)* | | | | |
| Pre-exponential, $\delta D_{0b}$ (m$^3$/s) | $6\cdot2 \times 10^{-10}$ | (d) | — | |
| Activation energy, $Q_b$ (kJ/mole) | 155 | | — | |
| *Core diffusion (anion)* | | | | |
| Pre-exponential, $a_c D_{0c}$ (m$^4$/s) | $4\cdot9 \times 10^{-19}$ | (d) | — | |
| Activation energy, $Q_c$ (kJ/mole) | 155 | | — | |
| *Power-law creep* | | | | |
| Exponent, $n$ | $3\cdot6$ | (e) | $6\cdot6$ | (h) |
| Dorn constant, $A$ | $6\cdot6 \times 10^2$ | | $2\cdot6 \times 10^{13}$ | |
| *Obstacle-controlled glide* | | | | |
| 0 K flow stress, $\hat{\tau}/\mu_0$ | $2\cdot5 \times 10^{-3}$ | | $1\cdot87 \times 10^{-2}$ | |
| Pre-exponential, $\dot{\gamma}_0$ (s$^{-1}$) | $10^6$ | (f) | $10^6$ | (i) |
| Activation energy, $\Delta F/\mu_0 b^3$ | $0\cdot5$ | | $0\cdot5$ | |
| *Lattice-resistance-controlled glide* | | | | |
| 0 K flow stress, $\hat{\tau}_p/\mu_0$ | $1\cdot8 \times 10^{-2}$ | | $3\cdot7 \times 10^{-2}$ | |
| Pre-exponential, $\dot{\gamma}_p$ (s$^{-1}$) | $2\cdot5 \times 10^{11}$ | (f) | $2\cdot5 \times 10^{11}$ | (i) |
| Activation energy, $\Delta F_p/\mu_0 b^3$ | $0\cdot12$ | | $0\cdot11$ | |

(a) *Handbook of Chemistry and Physics* (1973), with $b = a/\sqrt{2}$.

(b) Calculated from single-crystal constants from the *Landolt–Bornstein Tables* 3–1 (1966) and 3–2 (1969); $\mu_0 = (\tfrac{1}{2} c_{44} (c_{11} - c_{12}))^{\frac{1}{2}}$; $\mu = \mu_0 \left(1 + \dfrac{T - 300}{T_M}\left(\dfrac{T_M}{\mu_0} . \dfrac{d\mu}{dT}\right)\right)$.

(c) Best fit to diffusion plot for Cl$^-$ shown in text; $D_v = D_{0v} \exp - (Q_v/RT)$.

(d) Burke (1968), Cl$^-$ ion diffusion; $\delta D_b = \delta D_{0b} \exp - (Q_b/RT)$, $a_c D_c = a_c D_{0c} \exp -\left(\dfrac{Q_c}{RT}\right)$.

(e) Burke (1968). His specimens contained the following impurities in less than 10 ppm: Ag, Al, Ba, Ca, Co, Cr, Cu, Fe, Mg, Mn, Mo, Ni, Pb, Si, Sn, Ti, V, Zn. Also Cd was present in less than 25 ppm. The value of $A$ is for tensile creep; the maps are constructed using $A_s = (\sqrt{3})^{n+1} A$.

(f) Parameters obtained by fitting eqns. (2.9) and (2.12) to data from Aldag *et al.* (1970) (100 $\mu$m grain size polycrystals of NaCl in compression at room temperature and pressures from atmospheric to 10 kbar at a strain-rate of $10^{-4}$/s) Stokes (1966) (100 $\mu$m grain size polycrystals at $5 \times 10^{-10}$/s; and temperatures of 150°C, 250°C and 350°C); Gilman (1961); and Verrall *et al.* (1977).

(g) Matzke, Hj. (1971), and Naumov and Ptashnik (1968); see also note (c).

(h) Cropper and Langdon (1968). The impurities in ppm, reported as oxides of the elements indicated, were: Mg: 10, Ag: 5, Cu: 5, Al: 10, Cu: 4, Si: 150. The value of $A$ refers to tensile stress and strain-rate (see Section 2.4); see also note (e).

(i) Parameters obtained by fitting eqns. (2.9) and (2.12) to data from Verrall *et al.* (1977).

distortions produced by Mg$^{2+}$–vacancy pairs. The hardness also varies with impurity concentration (Chin *et al.*, 1972 and Suszynska, 1971), but slowly; so at low impurity levels the yield stress is almost independent of the impurity concentration.

For all these reasons, there is a great deal of scatter in published data for alkali halides. We have fitted the equations to the available data, giving most weight to data from pure polycrystalline specimens.

We have calculated the shear modulus and its temperature dependence from:

$$\mu = (\tfrac{1}{2} c_{44} (c_{11} - c_{12}))^{\frac{1}{2}}$$

using the single-crystal constants of the Landolt–Bornstein Tables (1966 and 1969).

A recent study of the microhardness of nine alkali halides (Brown and Ashby, 1981) has shown that, when plotted on the normalized axes $\sigma_s/\mu$ and $T/T_M$, the mechanical strengths of all nine are brought into coincidence, so that the maps shown here can be treated as broadly typical of all alkali halides. This scaling of mechanical properties is discussed further in Chapter 18.

### References for Section 10.1

Brown, A. M. and Ashby, M. F. (1981) in: *Deformation of Polycrystals: Mechanisms and Microstructures* (eds. Leffers, T. and Hanson, N.). Second Risø Int. Symp., Risø Nat. Lab., Denmark. p. 1.

Carter, N. L. and Heard, H. C. (1970) *Am. J. Sci.* **269**, 193.

Chin, G. Y., Van Uitert, L. G., Green, M. L. and Zydzik, C. (1972) *Scripta Met.* **6**, 475.

Greenwood, N. N. (1970) *Ionic Crystals, Lattice Defects and Non-Stoichiometry.* Butterworths, London.

Guiu, F. and Langdon, T. G. (1974) *Phil. Mag.* **24**, 145.

Gutmanas, E. Yu. and Nadgornyi, E. M. (1970) *Phys. Stat. Sol.* **38**, 777.

Johnston, W. G. (1962) *J. Appl. Phys.* **33**, 2050.

Kliewer, K. and Koehler, J. (1965) *Phys. Rev.* **A140**, 1226.

*Landolt–Bornstein Tables*, 3–1 (1966) and 3–2 (1969). Springer-Verlag, Berlin.

Suszynska, M. (1971) *Phys. Stat. Sol.* (a) **6**, 79.

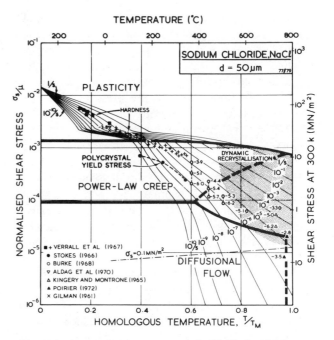

Fig. 10.1. A stress/temperature map for NaCl of grain size 50 μm. Data are labelled with $\log_{10} \dot{\gamma}$.

## 10.2  DATA FOR SODIUM CHLORIDE, NaCl

Figs. 10.1 and 10.2 show maps for NaCl with a grain size of 50 μm. They are based on data plotted on the figures, and summarized in Table 10.1.

The crystallographic, thermal and elastic data are from standard sources, listed on Table 10.1. Diffusion measurements show much scatter; in making the maps, it is best to replot the available data (rejecting any that is suspect because of dubious purity or poor technique), and then to choose diffusion constants that best fit all the data, as shown in Fig. 10.5. The anion diffuses more slowly than the cation at all temperatures. It is the anion diffusion data that appears in Table 10.1.

The scope of the data for glide and creep is shown in Fig. 10.1. The data are confused, and our choice of parameters requires explanation.

Burke (1968) compressed polycrystalline NaCl to steady state in the power-law creep regime, between 330°C and 740°C. Above 530°C, he found an activation energy off 205 ± 25 kJ/mole, close to that for lattice diffusion of chlorine, the slower diffusing species in NaCl. Below 530°C he observed a drop in activation energy to 155 ± 25 kJ/mole—which matches inferred data on core diffusion (Barr *et al.*, 1960)—and an increase in the stress exponent *n* by about 2. This, we think, reflects the influence of impurities on the isoelectric temperatures (and thus

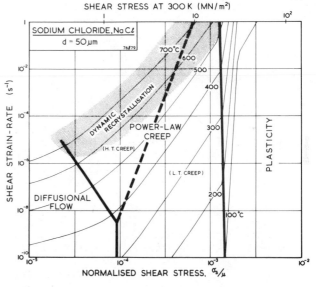

Fig. 10.2. A strain-rate/stress map for NaCl of grain size 50 μm.

on core diffusion) discussed in Section 10.1. On this assumption we have fitted Burke's data to eqn. (2.21), using an effective diffusion coefficient (eqn. (2.20)) which includes the core-diffusion controlled term: the resulting parameters are listed in Table 10.1.

There is some additional information for NaCl. Poirier (1972) compressed pure single crystals in the higher temperature range from 750 to 795°C. Though his measurements of the activation energy for creep disagree with those of Burke, a corrected analysis by Rothman *et al.* (1972) support Burke's results. Further data exist in the core-diffusion controlled creep regime. Heard (1972), using comparatively impure Baker reagent grade salt (including about 40 ppm $H_2O$), extended polycrystalline specimens under a confining pressure of 2 kbar. The observed shear strain rates ($3 \cdot 2 \times 10^{-1}$ to $2 \times 10^{-8}$/s between the temperatures 23°C and 400°C) were $10^2$ to $10^4$ times faster than indicated on the maps. The activation energy measured was only 99·5 kJ/mole (cf. 155 kJ/mole used in the low-temperature creep regime). The increased strain-rates and altered activation energy may be caused by the presence of water: Benard and Cabane, 1967, for example, report enhanced boundary diffusion in the presence of water. Some of this additional information is plotted on the figures, but was ignored in deriving the parameters of Table 10.1.

Kingery and Montrone (1965) measured diffusional flow in polycrystalline NaCl at 740°C. Their material was relatively impure analytical reagent-grade NaCl, also containing 1 $\mu$m particles of alumina and about 4% porosity. The data are included in Fig. 10.1.

The parameters describing glide (Table 10.1) are based on microhardness measurements (Verrall *et al.*, 1977, and Ashby and Brown, 1981).

### References for NaCl

Aldag, E., Davis, L. A. and Gordon, R. B. (1970) *Phil. Mag.* **21**, 469.

Ashby, M. F. and Brown, A. M. (1981) Proc. 2nd Risø Int. Symp. on *Deformation of Polycrystals* (eds. Hansen, N., Horsewell, A., Leffers, T. and Lillholt, H.) p. 1.

Barr, L. W., Hoodless, I. M., Morrison, J. A. and Rudham, R. (1960) *Trans. Faraday Soc.* **56**, 697.

Barr, L. W., Morrison, J. A. and Schroeder, P. A. (1965) *J. Appl. Phys.* **36**, 624.

Benard, J. and Cabane, J. (1967) *Sintering* (ed. Kuczynski, G. C.). Gordon & Breach.

Benière, F., Benière, M. and Chemla, M. (1970) *J. Phys. Chem. Solids* **31**, 1205.

Burke, P. M. (1968) Ph.D. thesis, Stanford University, Dept. of Engineering and Metallurgy.

Downing, H. L. and Friauf, R. J. (1970) *J. Phys. Chem. Solids* **31**, 845.

Gilman, J. J. (1961) *Prog. Ceram. Sci.* **1**, 146.

*Handbook of Chemistry and Physics*, 54th edn. (1973); Chemical Rubber Co., 18901 Cranwood Parkway, Cleveland, Ohio 44128.

Heard, H. C. (1972) The Griggs Volume, *American Geophysical Union Monograph* **16**.

Kingery, W. D. and Montrone, E. B. (1965) *J. Appl. Phys.* **36**, 2412.

*Landolt–Bornstein Tables*, 3–1 (1966) and 3–2 (1969). Springer-Verlag, Berlin.

Nelson, V. C. and Friauf, R. J. (1970) *J. Phys. Chem. Solids* **31**, 825.

Poirier, J. P. (1972) *Phil. Mag.* **26**, 701.

Rothman, S. J., Peterson, N. L., Laskar, A. L. and Robinson, L. C. (1972) *J. Phys. Chem. Solids* **33**, 1061.

Stokes, R. J. (1966) *Proc. Br. Ceram. Soc.* **6**, 189.

Verrall, R. A., Fields, R. J. and Ashby, M. F. (1977) *J. Am. Ceram. Soc.* **60**, 211.

### 10.3   DATA FOR LITHIUM FLUORIDE, LiF

Figs. 10.3 and 10.4 show maps for LiF with a grain size of 50 $\mu$m. They are based on data plotted

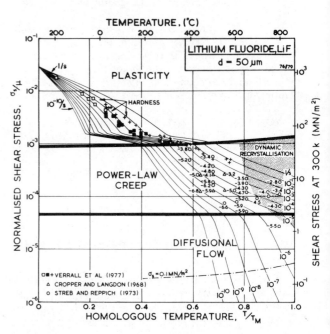

Fig. 10.3. A stress/temperature map for LiF of grain size 50 $\mu$m. Data are labelled with $\log_{10} \dot{\gamma}$.

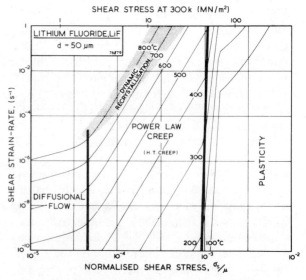

Fig. 10.4. A strain-rate/stress map for LiF of grain size 50 $\mu$m.

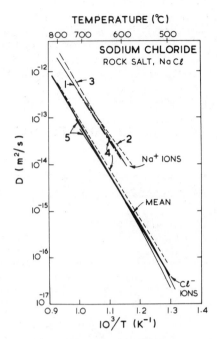

Fig. 10.5. Anion and cation diffusion in NaCl. The data are from: (1) Rothman *et al.* (1972); (2) Downing and Friauf (1970); (3) Nelson and Friauf (1970); (4) Benière *et al.* (1970); (5) Barr *et al.* (1965).

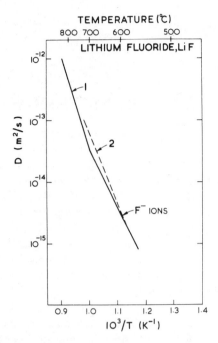

Fig. 10.6. Anion diffusion in LiF. The data are from: (1) Matzke (1971); (2) Eisenstadt (1963).

on the figures, and summarized in Table 10.1. The crystallographic, thermal and elastic data are from standard sources listed on Table 10.1.

Cropper and Langdon (1968) observed power-law creep in polycrystalline LiF between 450°C and 550°C, tested in compression. Their data appear to be the most reliable, and the parameters of Table 10.1, and the maps, are based on it. A substantial body of single-crystal data is available (Streb and Reppich, 1977). It is shown in the data plots of Fig. 10.3; it is not corrected in any way to make it comparable with the polycrystal data, nor was it used in the construction of the maps.

Data for lattice diffusion of the anion $F^-$ are shown in Fig. 10.6. Neither dislocation-enhanced nor grain boundary-enhanced diffusion have been observed in lithium fluoride. Impurity effects (discussed in Section 10.1) may provide an explanation for this difference with sodium chloride. As a result, however, no low temperature power-law creep regime or Coble-creep regime appear in the maps for lithium fluoride.

The glide parameters (Table 10.1) are based on microhardness measurements of Verrall *et al.* (1977) and Ashby and Brown (1981).

### References for LiF

Ashby, M. F. and Brown, A. M. (1981) and Risø Int. Symp. on the *Deformation of Polycrystals* (eds. Hansen, N., Horsewell, A., Leffers, T. and Lilholt, H.), p. 1.

Cropper, D. R. and Langdon, T. G. (1968) *Phil. Mag.* **18**, 1181.

*Handbook of Physics and Chemistry*, 54th edn. (1973). Chemical Rubber Co., 18901, Cranwood Parkway, Cleveland, Ohio, 44128.

*Landolt–Bornstein Tables*, 3–1 (1966) and 3–2 (1969). Springer-Verlag, Berlin.

Matzke, Hj. (1971) *J. Phys. Chem. Solids* **32**, 437.

Naumov, A. N. and Ptashnik, V. B. (1968) *Fiz. Tverd. Tela* **10**, 3710 (see *Diffusion Data* (1969) 3, 332).

Streb, G. and Reppich, B. (1972) *Phys. Stat. Sol. (a)* **16**, 493.

Verrall, R. A., Fields, R. J. and Ashby, M. F. (1977) *J. Am. Ceram. Soc.* **60**, 211.

# CHAPTER 11

# THE TRANSITION-METAL CARBIDES: ZrC AND TiC

ZIRCONIUM CARBIDE and titanium carbide typify a large group of refractory carbides with the rock-salt structure (examples are TiC, ZrC, HfC, VC, NbC, TaC). They form part of a larger group of over 80 hydrides, borides and carbides with this structure. They differ from alkali halides in that the bonding is less ionic in character: it is sometimes helpful to think of them as interstitial compounds based on metallic zirconium, titanium, etc. Almost all of them are very hard: they are used for cutting tools, abrasives and dies. Mechanical data are available for only a few; we consider TiC and ZrC as typical.

Maps and data for ZrC are shown in Figs. 11.1 and 11.2. Those for TiC are so nearly identical that, although the data are discussed below, no additional maps are given. The parameters used to compute the maps are listed in Table 11.1.

## 11.1 GENERAL FEATURES OF THE MECHANICAL BEHAVIOUR OF TRANSITION-METAL CARBIDES

These carbides are characterized by large elastic moduli ($\mu_0 \approx 200$ GN/m²) and high melting temperatures (*circa* 3500 K). At low temperatures they are extremely hard, among the hardest known materials, with yield strengths which (at 0 K) approach the ideal strength (Fig. 11.1). Yet at high homologous temperatures they are very soft; or, more accurately, they creep exceptionally rapidly. This behaviour is related to two important properties of these carbides: at low temperatures they have an exceptionally high lattice resistance to dislocation glide: but at high temperatures the diffusivity of the metal ions is exceptionally rapid.

The primary glide system in single crystals is $\{111\} \langle 1\bar{1}0 \rangle$, with a Burger's vector of $\frac{1}{2} a \langle 1\bar{1}0 \rangle$; slip on $\{100\}$ and $\{110\}$ is also observed (Williams and Schaal, 1962; Hollox and Smallman, 1966; Lee and Haggerty, 1969). This implies (Williams, 1964) that there is little ionic contribution to the bonding, since charged ions would enforce $\{110\} \langle 1\bar{1}0 \rangle$ slip. Part, at least, of the bonding is metallic. The hardness of these materials falls rapidly between 0·2 and

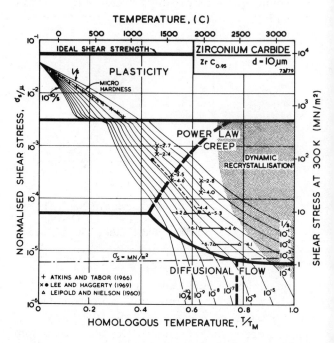

Fig. 11.1. A stress/temperature map for ZrC$_{0.95}$ of grain size 10 $\mu$m. Data are labelled with $\log_{10} (\dot{\gamma})$.

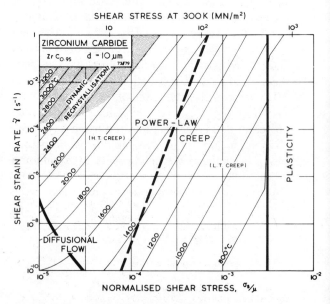

Fig. 11.2. A strain-rate/stress map for ZrC$_{0.95}$ of grain size 10 $\mu$m.

$0.4 T_M$, and at $0.5 T_M$ they are as soft, or softer, than pure copper at room temperature. (By contrast the high hardness of the diamond cubic elements Si and Ge, the covalent compounds SiC and $Si_3N_4$, and hydrogen-bonded ice, first starts to decrease at $0.5 T_M$.) For this reason, the transition-metal carbides are not attractive as high-temperature materials: they creep rapidly at comparatively low temperatures. Dispersion strengthening by doping with borides to increase their creep strength remains a possibility.

There is a considerable literature on the transition-metal carbides. The most important general references are: Atkins and Tabor (1966) (hardness data); Lee and Haggerty (1969) (slip and creep in ZrC); Hollox (1968) and Hollox and Smallman (1966) (slip and creep in TiC); Williams (1964) (slip in single crystals of TiC); and the review by Williams (1971). Data from all these sources are included in Figs. 11.1, 11.2 and 11.3.

**TABLE 11.1   The transition-metal carbides**

| Material | $ZrC_{0.95}$ | | $TiC_{0.97}$ | |
|---|---|---|---|---|
| *Crystallographic and thermal data* | | | | |
| Atomic volume, $\Omega$ (m$^3$) | $2.64 \times 10^{-29}$ | (a) | $2.0 \times 10^{-29}$ | (a) |
| Burger's vector, $b$ (m) | $3.34 \times 10^{-10}$ | (b) | $3.05 \times 10^{-10}$ | (i) |
| Melting temperature, $T_m$ (K) | 3530 | (c) | 3523 | (c) |
| *Modulus* | | | | |
| Shear modulus at 300 K, $\mu_0$ (MN/m$^2$) | $1.73 \times 10^5$ | (d) | $1.93 \times 10^5$ | (d,f) |
| Temperature dependence of modulus, $\dfrac{T_m}{\mu_0}\dfrac{d\mu}{dT}$ | $-0.28$ | (d) | $-0.18$ | (d,j) |
| *Lattice diffusion* | | | | |
| Pre-exponential, $D_{0v}$ (m$^2$/s) | $1.03 \times 10^{-1}$ | (e) | 4.4 | (e) |
| Activation energy, $Q_v$ (kJ/mole) | 720 | (e) | 740 | (e) |
| *Boundary diffusion* | | | | |
| Pre-exponential, $\delta D_{0b}$ (m$^3$/s) | $5 \times 10^{-12}$ | (e) | $2 \times 10^{-9}$ | (e) |
| Activation energy, $Q_b$ (kJ/mole) | 468 | (e) | 543 | (e) |
| *Core diffusion* | | | | |
| Pre-exponential, $a_c D_{0c}$ (m$^4$/s) | $1 \times 10^{-21}$ | (e) | $1 \times 10^{-21}$ | (e) |
| Activation energy, $Q_c$ (kJ/mole) | 468 | (e) | 543 | (e) |
| *Power-law creep* | | | | |
| Exponent, $n$ | 5.0 | (f) | 5.0 | (k) |
| Dorn constant, $A$ | $1 \times 10^{12}$ | | $2 \times 10^{13}$ | |
| *Obstacle-controlled glide* | | | | |
| 0 K flow stress, $\hat{\tau}/\mu_0$ | $3.34 \times 10^{-3}$ | (g) | $3.3 \times 10^{-3}$ | |
| Pre-exponential, $\dot{\gamma}_0$ (s$^{-1}$) | $10^8$ | | $10^8$ | (e) |
| Activation energy, $\Delta F/\mu_0 b^3$ | 1.0 | | 1.0 | |
| *Lattice-resistance-controlled-glide* | | | | |
| 0 K flow stress, $\hat{\tau}_p/\mu$ | $4.0 \times 10^{-2}$ | | $5.2 \times 10^{-2}$ | (h) |
| Pre-exponential, $\dot{\gamma}_p$ (s$^{-1}$) | $10^{11}$ | (h) | $10^{11}$ | |
| Activation energy, $\Delta F_p/\mu_0 b^3$ | 0.05 | | 0.05 | |

(a) Pearson (1967) and Gilman (1970).

(b) Lee and Haggerty (1969) and Williams and Schall (1962).

(c) Storms (1967); Williams (1971); Routbourt (1971). (There is some question about the values of $T_M$. Atkins and Tabor, 1966, give 3803 K for TiC, other sources give 3693 K for ZrC.)

(d) From the single-crystal constants of Chang and Graham (1966) using $\mu = (\frac{1}{2} c_{44} (c_{11} - c_{12}))^{\frac{1}{3}}$, with data for $ZrC_{0.94}$; $\mu = \mu_0 (1 + (T - 300)/T_M (T_M/\mu_0 \cdot d\mu/dT)$.

(e) See diffusion plot and discussion in text; $D_v = D_{0v} \exp - (Q_v/RT)$, $\delta D_b = \delta D_{0b} \exp - (Q_b/RT)$ and $a_c D_c = a_c D_{0c} \exp - (Q_c/RT)$.

(f) Based on data of Lee and Haggerty (1969) and Liepold and Nielsen (1964). The value of $A$ refers to tensile creep. The maps are computed using $A_s = (\sqrt{3})^{n+1} A$.

(g) Based on the plateau in the critical shear stress data of Lee and Haggerty (1969) and of Williams (1964).

(h) See text.

(i) Hollox and Smallman (1966).

(j) Gilman and Roberts (1961).

(k) Keihn and Kebler (1964); see also Hollox (1968), and note (f).

(l) Based on the plateau in the critical resolved shear stress data of Williams (1964).

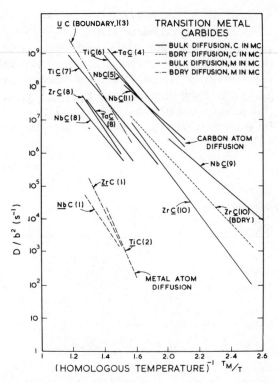

Fig. 11.3. Metal-ion and carbide diffusion in transition metal carbides. The diffusion coefficient has been normalized by the square of the Burger's vector. (1) Andrievskii *et al.* (1971); (2) Sarian (1968); (3) Villane (1968); (4) Brizes (1968); (5) Brizes *et al.* (1966); (6) Kohlstedt *et al.* (1970); (7) Sarian (1969); (8) Andrievskii *et al.* (1969); (9) Bornstein *et al.* (1965); (10) Sarian and Criscione (1967); (11) Lidner *et al.* (1967).

## 11.2   ORIGINS OF DATA FOR TRANSITION-METAL CARBIDES

Much of the data used here for ZrC is from Lee and Haggerty (1969) and refers to $ZrC_{0.945}$. Where possible, the TiC data refer to $TiC_{0.97}$. Moduli were calculated from single-crystal data using $\mu = (\frac{1}{2}c_{44}\,(c_{11} - c_{12}))^{\frac{1}{2}}$.

Figs. 11.1 and 11.2 show maps for $ZrC_{0.95}$ with a grain size of 10 $\mu$m. They are based on data shown on Fig. 11.1, and on data inferred from studies of $TiC_{0.97}$. The parameters used to compute the maps are listed in Table 11.1, together with relevant information for $TiC_{0.97}$.

Diffusion data for carbides are plotted (on normalized axes) in Fig. 11.3. The metal ion diffuses much more slowly than does the carbon (typically, by a factor of $10^{-3}$); it is metal-ion diffusion which will determine the rate of dislocation climb or of diffusional flow. Most of the transition metal carbides exist in the hypostoichiometric (C-deficient) state. The carbon deficiency is taken up by vacancies on the carbon sub-lattice (which, under certain circumstances, may order). The transport of carbon then depends on composition; but as far as is known, diffusion of the metal ion is by a thermal-vacancy mechanism, and is independent of stoichiometry (Sarian, 1968, 1969). We have selected values for $D_{0v}$ and $Q_v$ for the metal ion which are consistent with the data plotted in Fig. 11.3, using data from Andrievskii *et al.* (1971) for Zr and from Sarian (1968) for Ti.

There are very little data for boundary and core diffusion. What little there is appears on Fig. 11.3. Data for UC (Villane, 1968) and TiC (Hollox and Smallman, 1966) suggest, respectively, that enhanced boundary and core diffusion occur in the carbides. Our values (Table 11.1) are chosen to be consistent with these data and with observations of creep rates (see below).

Creep data for the transition metal carbides are limited to high temperatures (Keihn and Kebler, 1964; Hollox, 1968; Lee and Haggerty, 1969; and Liepold and Nielson, 1964). Microstructural studies suggest a creep process very like that in pure metals: cells form, and (as far as one can judge from the meagre data) the activation energy is close to that of self-diffusion. Single crystals of ZrC show a well-developed power-law regime, with an exponent $n$ of 5 (Lee and Haggerty, 1969). Polycrystals, too, show power-law creep and there is evidence of diffusional flow at low stresses (Liepold and Nielson, 1964).

For TiC the creep data are more limited. Keihn and Kebler (1964) give an activation energy of 732 kJ/mole for creep above 2000°C, and 544 kJ/mole below. This seems self-consistent, since 732 kJ/mole is almost exactly the activation energy for bulk diffusion of Ti. We have assumed $n = 5$ (the value observed for ZrC) and calculated the constant $A$ from the best fit to their data. The detailed fitting process by which we make final adjustments to the parameters $A$ and $n$ (Chapter 3) has not been performed for TiC; the raw data are presented here for comparison with those for ZrC.

The lattice-resistance controlled glide parameters are based on microhardness measurements, extrapolated to 0 K, and on measurements of critical resolved shear strength. For both ZrC and TiC we used the microhardness data of Westbrook and Stover (1967), Atkins and Tabor (1966) and Kornilov (1966). These data, treated as described in Chapter 3, using Marsh's (1964) correction, combined with critical resolved shear stress measurements (Williams, 1964) lead to values of $\hat{\tau}_p$ and $\Delta F_p$ given in Table 11.1.

## References for Chapter 11

Andrievskii, R. A., Khormov, Yu. F. and Alekseeva, I. S. (1971) *Fiz. Metal Metalloved* **32**, 664 (*Diffusion Data* (1972) **6**, 147).

Andrievskii, R. A., Klimenko, V. V. and Khormov, Yu. F. (1969) *Fiz. Metal Metalloved* **28**, 298.

Atkins, A. G. and Tabor, D. (1966) *Proc. R. Soc.* **A292**, 441.

Bornstein, N. S., Hirakis, E. C. and Friedrich, L. A. (1965) *Pratt and Whitney Report* No. TIM-927.

Brizes, W. F. (1968) *J. Nucl. Mat.* **26**, 227.

Brizes, W. F., Cadoff, L. H. and Tobin, L. M. (1966) *J. Nucl. Mat.* **20**, 57.

Chang, R. and Graham, L. J. (1966) *J.A.P.* **37**, 2778.

Gilman, J. J. (1970) *J.A.P.* **41**, 1664.

Gilman, J. J. and Roberts, B. W. (1961) *J.A.P.* **32**, 1405.

Hollox, G. E. (1968) *Mat. Sci. and Eng.* **3**, 121.

Hollox, G. E. and Smallman, R. E. (1966) *J.A.P.* **37**, 818.

Hollox, G. E. and Venables, J. D. (1967) Int. Cong. on Strength of Metals and Alloys, Tokyo, Japan Inst. Met.

Keihn, F. and Kebler, R. (1964) *J. Less Common Metals* **6**, 484.

Kohlstedt, D. L., Williams, W. S. and Woodhouse, J. B. (1970) *J.A.P.* **41**, 4476.

Kornilov, I. I. (1966) *The Chemistry of Metallides*, translated by Loweberg, J. W., Consultants Bureau, NY.

Lee, D. W. and Haggerty, J. S. (1969) *J. Am. Ceram. Soc.* **52**, 641.

Liepold, M. H. and Nielson, T. H. (1964) *J. Am. Ceram. Soc.* **47**, 419.

Lidner, R., Riemer, G. and Scherff, H. L. (1967) *J. Nucl. Mat.* **23**, 222.

Marsh, D. M. (1964) *Proc. R. Soc.* **A279**, 420.

Pearson, W. B. (1967) *Handbook of Lattice Spacings and Structures of Metals*. Pergamon, p. 422 *et seq.*

Routbourt, J. L. (1971) *J. Nucl. Mat.* **40**, 17.

Sarian, S. (1968) *J.A.P.* **39**, 5036.

Sarian, S. (1969) *J.A.P.* **40**, 3515.

Sarian, S. and Criscione, J. M. (1967) *J.A.P.* **38**, 1794.

Storms, E. J. (1967) *The Refractory Oxides*. Academic Press.

Villane, P. (1968) *C.E.A. Report* No. 3436.

Westbrook, J. H. and Stover, E. R. (1967) *High Temperature Materials and Technology* (eds. Campbell, I. E. and Sherwood, E. M.). Wiley, NY.

Williams, W. S. (1964) *J.A.P.* **35**, 1329.

Williams, W. S. (1971) *Prog. Solid State Chem.* **6**, 57.

Williams, W. S. and Schall, R. D. (1962) *J.A.P.* **37**, 818.

# CHAPTER 12

## OXIDES WITH THE ROCK-SALT STRUCTURE: MgO, CoO AND FeO

A NUMBER of simple oxides crystallize with the rock-salt structure; they include MgO, CaO, CdO, MnO, FeO, NiO, CoO, SrO and BaO. Some, like magnesia, are stable as the stoichiometric oxide MgO. Others, like cobalt monoxide, exist only as the hyper-stoichiometric oxide $Co_{1-x}O$, always oxygen-rich because of the presence of $Co^{3+}$ ions. Wüstite, too, is always hyper-stoichiometric at atmospheric pressure, having the composition $Fe_{1-x}O$ with $0.05 < x < 0.15$. In all these oxides the oxygens are the larger ions. They are packed in an f.c.c. array, with the metal ions occupying the octahedral interstices.

Maps for MgO, CoO and FeO are shown in Figs. 12.1 to 12.10. The parameters used to construct them are listed in Table 12.1.

### 12.1 GENERAL FEATURES OF THE DEFORMATION OF THE ROCK-SALT STRUCTURED OXIDES

The bonding in rock-salt structured oxides is largely ionic. Like the alkali halides of Chapter 10, but unlike the transitional-metal carbides (Chapter 11), they slip easily on $\langle 1\bar{1}0 \rangle \{110\}$, though this provides only two independent slip systems. Poly-crystal plasticity is made possible by slip on $\langle 0\bar{1}1 \rangle \{100\}$. But despite this similarity with the alkali halides, these oxides form a distinct isomechanical group: their melting points and moduli are much higher; their strength at 0 K is a little larger (about $\mu/30$ compared with $\mu/50$); and this low-temperature strength is retained to slightly higher homologous temperatures.

Above about $0.4 \, T_M$ the oxides start to creep. Once they do, their strength falls as fast or faster than that of the alkali halides. When creep of a compound is diffusion-controlled, both components—oxygen and metal in these oxides—must move. The creep rate is then controlled by a weighted mean of the diffusivities of the components. But diffusion may be intrinsic or extrinsic, and alternative diffusion paths (lattice and grain-boundary paths, for instance) may be available. As a general rule, the creep rate is determined by the fastest path of the slowest species—and this may change with temperature and grain size. The result is a proliferation of different diffusional flow fields—six or more are possible (Stocker and Ashby, 1973).

This proliferation is illustrated below for magnesia. As the grain size decreases from 100 μm to 1 μm, new diffusional flow fields appear as new diffusion paths become rate-controlling. Even for this well-studied oxide the diffusion data are ambiguous and incomplete; we have had to guess values for some of the coefficients. But despite this, the picture is full enough to give a fair idea of how this, and the other oxides, should behave. All compounds, potentially, can show such complications; the figures for α-alumina (Chapter 14 and Section 14.1) give a further example. They are not shown on the other figures in this part of the book only because diffusion data are too meagre to allow them to be predicted with any precision.

A general review of the mechanical behaviour of ceramics, including MgO, is given by Evans and Langdon (1976). It is recommended for further background.

### References for Section 12.1

Evans, A. G. and Langdon, T. G. (1976) Structural Ceramics, *Prog. Mat. Sci.* **21**.
Stocker, R. L. and Ashby, M. F. (1973) *Rev. Geophys. Space Phys.* **11**, 391.

### 12.2 DATA FOR MgO

The shear modulus and its temperature dependence were calculated from single-crystal data, using $\mu_0 = [\frac{1}{2}c_{44}(c_{11} - c_{12})]^{\frac{1}{2}}$ together with the measurements of Speltzler and Anderson (1971) between 500 and 1000 K. The normalized temperature dependence $(T_M/\mu_0 . d\mu/dT)$ is higher than that for other oxides; but it was calculated from the low-temperature data, where relaxation should be minimal.

The easy slip systems in MgO are those of the

**TABLE 12.1  Rock-salt structured oxides**

| Material | MgO | | CoO | | FeO | |
|---|---|---|---|---|---|---|
| *Crystallographic and thermal data* | | | | | | |
| Atomic volume, $\Omega$ (m³) | $1 \cdot 87 \times 10^{-29}$ | | $1 \cdot 93 \times 10^{-29}$ | | $2 \cdot 00 \times 10^{-29}$ | |
| Burger's vector, $b$ (m) | $2 \cdot 98 \times 10^{-10}$ | | $3 \cdot 01 \times 10^{-10}$ | | $3 \cdot 05 \times 10^{-10}$ | |
| Melting temperature, $T_m$ (K) | 3125 | (a) | 2083 | (a) | 1643 | (o) |
| *Modulus* | | | | | | |
| Shear modulus at 300 K, $\mu_0$ (MN/m²) | $1 \cdot 26 \times 10^5$ | (b) | $6 \cdot 97 \times 10^4$ | (k) | $5 \cdot 57 \times 10^4$ | (p) |
| Temperature dependence of modulus, $\dfrac{T_m}{\mu_0}\dfrac{d\mu}{dT}$ | $-0 \cdot 68$ | (b) | $-0 \cdot 3$ | (g) | $-0 \cdot 3$ | (g) |
| *Lattice diffusion: oxygen, intrinsic* | | | | | | |
| Pre-exponential, $D_{0v}$ (m²/s) | $1 \cdot 37 \times 10^{-6}$ | (c) | $5 \times 10^{-3}$ | (l) | $1 \cdot 0 \times 10^{-2}$ | (q) |
| Activation energy, $Q_v$ (kJ/mole) | 460 | (c) | 398 | (l) | 326 | (q) |
| *Lattice diffusion: oxygen, extrinsic* | | | | | | |
| Pre-exponential, $D_{0v}$ (m²/s) | $2 \cdot 5 \times 10^{-10}$ | (d) | — | | — | |
| Activation energy, $Q_v$ (kJ/mole) | 261 | (d) | — | | — | |
| *Lattice diffusion: metal, extrinsic* | | | | | | |
| Pre-exponential, $D_{0v}$ (m²/s) | $4 \cdot 2 \times 10^{-8}$ | (e) | — | | — | |
| Activation energy, $Q_v$ (kJ/mole) | 266 | (e) | — | | — | |
| *Boundary diffusion: oxygen* | | | | | | |
| Pre-exponential, $\delta D_{0b}$ (m³/s) | $1 \cdot 36 \times 10^{-15}$ | (f) | $1 \times 10^{-13}$ | (m) | $1 \cdot 0 \times 10^{-13}$ | (g) |
| Activation energy, $Q_b$ (kJ/mole) | 230 | (f) | 200 | (m) | 195 | (g) |
| *Boundary diffusion: metal* | | | | | | |
| Pre-exponential, $\delta D_{0b}$ (m³/s) | $3 \times 10^{-17}$ | (g) | — | | — | |
| Activation energy, $Q_b$ (kJ/mole) | 200 | (g) | | | | |
| *Core diffusion: oxygen* | | | | | | |
| Pre-exponential, $a_c D_{0c}$ (m⁴/s) | $7 \cdot 5 \times 10^{-26}$ | (h) | $1 \times 10^{-23}$ | (g) | $1 \cdot 0 \times 10^{-24}$ | (g) |
| Activation energy, $Q_c$ (kJ/mole) | 252 | (h) | 238 | (g) | 195 | (g) |
| *Power-law creep* | | | | | | |
| Exponent, $n$ | $3 \cdot 3$ | (i) | $3 \cdot 5$ | (n) | $4 \cdot 2$ | (r) |
| Dorn constant, $A$ | 20 | (i) | $1 \cdot 07 \times 10^4$ | (n) | $2 \cdot 4 \times 10^5$ | (r) |
| Activation energy, $Q_{cr}$ (kJ/mole) | — | | — | | — | |
| *Obstacle-controlled glide* | | | | | | |
| 0 K flow stress, $\hat{\tau}/\mu_0$ | $5 \times 10^{-3}$ | | $3 \times 10^{-3}$ | | $1 \times 10^{-2}$ | |
| Pre-exponential, $\dot{\gamma}_0$ (s⁻¹) | $10^6$ | | $10^6$ | | $10^6$ | |
| Activation energy, $\Delta F/\mu_0 b^3$ | $0 \cdot 5$ | | $0 \cdot 5$ | | $0 \cdot 5$ | |
| *Lattice-resistance-controlled-glide* | | | | | | |
| 0 K flow stress, $\hat{\tau}_p/\mu_0$ | $3 \cdot 5 \times 10^{-2}$ | (j) | $1 \cdot 6 \times 10^{-2}$ | (j) | $6 \times 10^{-2}$ | (j) |
| Pre-exponential, $\dot{\gamma}_p$ (s⁻¹) | $10^{11}$ | | $10^{11}$ | | $10^{11}$ | |
| Activation energy, $\Delta F_p/\mu_0 b^3$ | $0 \cdot 08$ | | $0 \cdot 18$ | | $0 \cdot 18$ | |

(a) Shunk (1969).
(b) Speltzler and Anderson (1971); $\mu_0 = (\frac{1}{2}c_{44}(c_{11} - c_{12}))^{\frac{1}{2}}$;

$$\mu = \mu_0\left(1 + \frac{T - 300}{T_M}\left(\frac{T_M}{\mu_0}\frac{d\mu}{dT}\right)\right).$$

(c) Narayan and Washburn (1973); $D_v = D_{0v} \exp - (Q_v/RT)$.
(d) Oishi and Kingery (1960). See diffusion plot, text and note (c).
(e) Wuensch et al. (1973). See diffusion plot, text and note (c).
(f) A fit to McKenzie et al. (1971) and Passmore et al. (1966); $\delta D_b = \delta D_{0b} \exp - (Q_b/RT)$.
(g) An estimate.
(h) Narayan and Washburn (1972); $a_c D_c = a_c D_{0c} \exp - (Q_c/RT)$.
(i) Fitted to Langdon and Pask (1970), assuming control by

extrinsic oxygen diffusion. The value of $A$ is for tensile creep. The maps were computed using $A_s = (\sqrt{3})^{n+1} A$.

(j) See text and data plot.
(k) Aleksandrov et al. (1968); see also note (b).
(l) Chen and Jackson (1969). See diffusion plot. text and note (c).
(m) An estimate, consistent with Clauer et al. (1971b).
(n) Chosen to match the data of Krishnamachari and Notis (1977) at 1100°C; see also note (i).
(o) The value for $Co_{0.95}O$.
(p) Polycrystal shear modulus of Akimoto (1972).
(q) See text and note (c).
(r) Ilschner et al. (1964) for $x = 0 \cdot 054$; see also note (i).

family $\langle 1\bar{1}0 \rangle$ $\{110\}$ (Parker, 1961). Slip on the harder family $\langle 0\bar{1}1 \rangle$ $\{100\}$ is observed in single crystals loaded along a $\langle 111 \rangle$ axis, and in polycrystals at high temperatures, and it is presumed to occur in test on polycrystals at low temperatures (Hulse and Pask, 1960; Hulse *et al.*, 1963; Day and Stokes, 1966). It is this system which determines the low-temperature strength.

Table 12.1 includes parameters for extrinsic and intrinsic oxygen diffusion, and for the diffusion of magnesium ions. The rate of diffusional flow and

creep can be controlled by any one of these, as well as by boundary diffusion, as illustrated by Figs. 12.1 to 12.6. Diffusion data are assembled in Fig. 12.11. The parameters were arrived at in the following way.

According to Wuensch *et al.* (1973) lattice diffusion of $Mg^{2+}$ in MgO is almost always extrinsic; accordingly, the extrinsic $Mg^{2+}$ diffusion constants are based on their data. The rate of $O^{2-}$ diffusion is much lower than that of $Mg^{2+}$. We have attributed the $O^{2-}$ diffusion rates measured by Oishi and

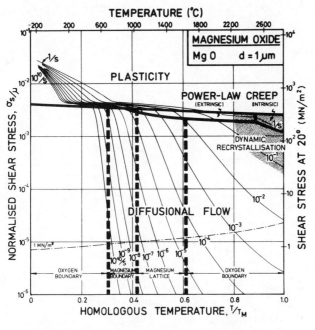

Fig. 12.1.  A stress/temperature map for MgO with a grain size of 1 $\mu$m. Data are shown on Fig. 12.3.

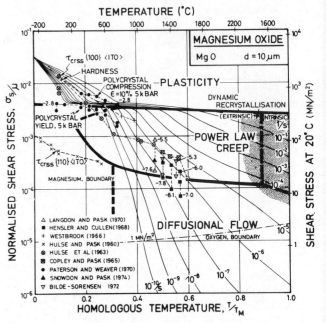

Fig. 12.3.  A stress/temperature map for MgO with a grain size of 10 $\mu$m. Data are labelled with $\log_{10} \dot{\gamma}$.

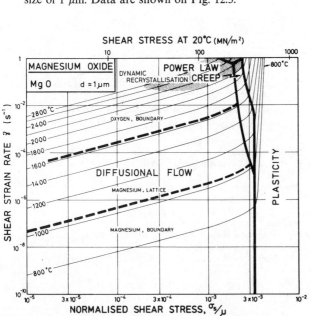

Fig. 12.2.  A strain-rate/stress map for MgO with a grain size of 1 $\mu$m. Data are shown on Fig. 12.4.

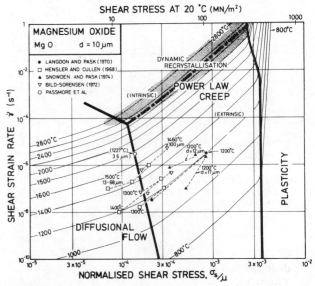

Fig. 12.4.  A strain-rate/stress map for MgO with a grain size of 10 $\mu$m. Data are labelled with the temperature °C.

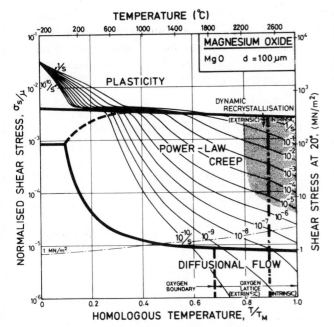

Fig. 12.5. A stress/temperature map for MgO with a grain size of 100 μm. Data are shown on Fig. 12.3.

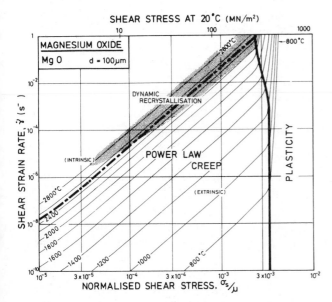

Fig. 12.6. A strain-rate/stress map for MgO with a grain size of 100 μm. Data are shown on Fig. 12.4.

Kingery (1960) (who found $Q_v = 261$ kJ/mole) to extrinsic diffusion, and those of Narayan and Washburn (1973) (who found $Q_v = 460$ kJ/mole) to intrinsic diffusion. Using the parameters listed in Table 12.1, $O^{2-}$ diffusion becomes intrinsic above $0.9\,T_M$. The maps show a dash–dot line at $0.9\,T_M$, separating the regime of extrinsic from that of intrinsic lattice diffusion of oxygen.

The activation energy for boundary diffusion of $O^{2-}$ is taken as 230 kJ/mole, which matches the Coble-creep data of Passmore et al. (1966). The pre-

exponential is derived from data given by McKenzie et al. (1971). The boundary diffusion parameters for $Mg^{2+}$ are merely an educated guess. Core diffusion data for $O^{2-}$ are from Narayan and Washburn's (1973) observations of shrinking loops.

This set of diffusion parameters for MgO leads to the subdivision of the diffusional flow field into a number of subfields, the size of which varies with grain size. In general, the creep rate is controlled by the slower species moving on the fastest path available to it. When both $Mg^{2+}$ and $O^{2-}$ move by lattice diffusion, it is the oxygen which is slower, though even then the activation energy will change at the intrinsic to extrinsic transition. When boundaries offer a faster path for oxygen, the lattice diffusion of $Mg^{2+}$ may become rate-controlling. Finally, if $Mg^{2+}$ moves predominantly on boundaries, the boundary diffusion of oxygen becomes controlling.

This sequence, and the way it changes with grain size, are illustrated by Figs. 12.1 to 12.6. They were computed by using a new effective diffusion coefficient $D_{eff}$ in place of eqn. (2.30), defined by:

$$D_{Oxy} = (D_v)_{Oxy} + \frac{\pi\delta(D_b)_{Oxy}}{d}$$

$$D_{Mg} = (D_v)_{Mg} + \frac{\pi\delta(D_b)_{Oxy}}{d}$$

$$D_{eff} = \frac{D_{Mg}D_{Oxy}}{D_{Mg} + D_{Oxy}}$$

Throughout, the faster of extrinsic diffusion and intrinsic was used. The rate-controlling species is best illustrated by a plot of $D_v$ and $\pi\delta D_b/d$ for each species, against $1/T$ (Stocker and Ashby, 1973). Fig. 12.12 shows such a plot for two grain sizes: 1 μm and 100 μm. The size of the fields on the maps for these two grain sizes (Figs. 12.1 and 12.5) are determined by the intersection of the diffusion rates on Fig. 12.12, remembering that it is always the faster path of the slower species which is important.

The power-law creep parameters are based on the data of Langdon and Pask (1970) at 1200°C, calculating $A$ on the assumption that the creep rate at this temperature is limited by the extrinsic diffusion of oxygen. These parameters give a reasonable fit with other published data (see Figs. 12.3 and 12.4).

The low-temperature plasticity is based on the polycrystal yield data of Paterson and Weaver (1970), who prevented fracture by applying a confining pressure of 500 MN/m². The results are consistent with the hardness data of Westbrook (1966), and, as expected, are close to the shear stress required to operate the "hard" slip system {100} ⟨110⟩, data for which are shown on Fig. 12.3.

**References for MgO**

Bilde-Sörensen, J. B. (1972) *J. Am. Ceram. Soc.* **55**, 606.

Copley, S. M. and Pask, J. A. (1965) *J. Am. Ceram. Soc.* **48**, 139.

Day, R. B. and Stokes, R. J. (1966) *J. Am. Ceram. Soc.* **49**, 345.

Harding, B. C. and Price, D. M. (1972) *Phil. Mag.* **26**, 253.

Harding, B. C., Price, D. M. and Mortlock, A. J. (1971) *Phil. Mag.* **23**, 399.

Hensler, J. H. and Cullen, G. V. (1968) *J. Am. Ceram. Soc.* **51**, 557.

Hulse, C. O., Copley, S. M. and Pask, J. A. (1963) *J. Am. Ceram. Soc.* **46**, 317.

Hulse, C. O. and Pask, J. A. (1960) *J. Am. Ceram. Soc.* **43**, 373.

Langdon, T. G. and Pask, J. A. (1970) *Acta Met.* **18**, 505.

Lindler, R. and Parfitt, G. D. (1957) *J. Chem. Phys.* **26**, 182.

Narayan, J. and Washburn, J. (1972) *Phil. Mag.* **26**, 1179.

Narayan, J. and Washburn, J. (1973) *Acta Met.* **21**, 533.

Oishi, Y. and Kingery, W. D. (1960) *J. Chem. Phys.* **33**, 905.

Parker, E. R. (1961). *Mechanical Properties of Engineering Ceramics* (eds. Kriegel, W. W. and Palmour, H.), Interscience, NY, p. 61.

Passmore, E. M., Duff, R. H. and Vasilov, T. (1966) *J. Am. Ceram. Soc.* **49**, 594.

Paterson, M. S. and Weaver, C. W. (1970) *J. Am. Ceram. Soc.* **53**, 463.

Rovner, L. H. (1966) Ph.D. thesis, Cornell University.

Shirasaki, S. and Hama, M. (1973) *Chem. Phys. Lett.* **20**, 361.

Shirasaki, S. and Oishi, Y. (1971) *Jap. J. Appl. Phys.* **10**, 1109.

Shirasaki, S. and Yamamura, H. (1973) *Jap. J. Appl. Phys.* **12**, 1654.

Shunk, F. A. (1969) *Constitution of Binary Alloys*, 2nd Supplement. McGraw-Hill.

Snowden, W. E. and Pask, J. A. (1974) *Phil. Mag.* **29**, 441.

Speltzler, H. A. and Anderson, D. L. (1971) *J. Am. Ceram. Soc.* **54**, 520.

Stocker, R. L. and Ashby, M. F. (1973) *Rev. Geophys. Space Phys.* **11**, 391.

Vasilos, T., Wuensch, B. J., Gruber, P. E. and Rhodes, W. H. (1969) Avco. Corp. Rpt. AVSO-0067-70-CR 15/12/69.

Westbrook, J. H. (1966) *Rev. Hautes Temper. et Refract.* **3**, 47.

Wuensch, B. J., Steele, W. C. and Vasilos, T. (1973) *J. Chem. Phys.* **58**, 5258.

## 12.3  DATA FOR $Co_{1-x}O$

Data for the moduli and other properties of CoO are complicated by its deviations from stoichiometry, and by the transition from anti-ferromagnetic to paramagnetic behaviour at 289 K (Aleksandrov *et al.*, 1968). Below this temperature, there is a slight tetragonal distortion of the structure, and the shear modulus decreases substantially.

We have neglected both, and calculated $\mu_0$ (using $\mu_0 = [\frac{1}{2}c_{44}(c_{11} - c_{12})]^{\frac{1}{2}}$) from single-crystal constants at 300 K of Aleksandrov *et al.* (1968). Since its temperature dependence has not been measured with useful precision, we have adopted the "typical" value

$$\frac{T_M}{\mu_0}\frac{d\mu}{dT} = -0.3.$$

Like magnesia, easy slip in CoO is found on the soft systems $\langle1\bar{1}0\rangle\{110\}$, with additional slip on the hard systems $\langle0\bar{1}1\rangle\{100\}$ when compatibility requires it (Clauer *et al.*, 1971a; Krishnamachari and Jones, 1974).

As in MgO, oxygen is the slower diffusing species. Lattice diffusion of $Co^{2+}$ in $Co_{1-x}O$ depends on stoichiometry. The dominant defects are cobalt-ion vacancies, with a concentration $x$ which decreases with decreasing oxygen partial pressure. Partly because of this, $Co^{2+}$ diffuses at least 100 times faster than does $O^{2-}$, and with a much lower activation energy. From the data plotted in Fig. 12.13 we selected the oxygen diffusion data of Chen and Jackson (1969) obtained under an oxygen partial pressure of 0.2 bars (one atmosphere of air); these conditions do not lead to stoichiometric CoO, but they are typical of much of the creep data. Boundary diffusion in CoO has not been measured directly; but creep data reviewed below suggest the values listed in the table. We have assumed an activation energy for core diffusion of $Q_c = 0.6\ Q_v$, giving a value of $Q_c/RT_M$ very close to that measured for MgO.

The most extensive body of creep data for polycrystalline CoO is that of Krishnamachari and Notis (1977). They measured an activation energy of 301 kJ/mole, substantially lower than the preferred activation energy for oxygen diffusion (398 kJ/mole), and lower also than that measured in creep by Clauer *et al.* (1971a) whose data, however, are for primary creep in single crystals. Using this higher activation energy, it is not possible to match Krishnamachari and Notis' data over a wide range of temperature. We have chosen creep parameters to give an exact match at 1100°C.

At low stress and high temperature, Krishnamachari and Notis (1977) observed more or less Newtonian creep with an activation energy of 180 kJ/mole, suggesting Coble-creep. This is further supported by the observations of Clauer *et al.* (1971b) and by Strafford and Gartside (1969), of an activation energy of 213 kJ/mole during the low-stress creep of polycrystalline CoO. Considering these observations, we have chosen a value of $Q_b$ of 200 kJ/mole (0.5 $Q_V$) and a value of $\delta D_{0b}$ which matches the Coble-creep equation (eqn. (2.29)) to the low-stress data of Krishnamachari and Notis (1977). Their data suggested a grain size dependence of $1/d^2$ which they interpret as Nabarro–Herring, rather than Coble-creep. But lattice diffusion of oxygen is much too slow to account for the rates they observed, and if lattice diffusion of cobalt is rate-controlling, then boundary diffusion of oxygen must be even faster than we have assumed it to be (see the discussion for MgO). The unexpected grain size dependence may be caused by the proximity

of the field boundary with dislocation creep.

Dislocation glide parameters are deduced from the hardness data of Westbrook (1966), with the Marsh (1964) correction.

The two maps (Figs. 12.7 and 12.8) are broadly consistent with the limited data. The real behaviour may be complicated by transitions from intrinsic to extrinsic diffusion, and by oxygen transport by boundary diffusion while cobalt still moves by lattice diffusion. The effect of such changes is illustrated in the section on MgO.

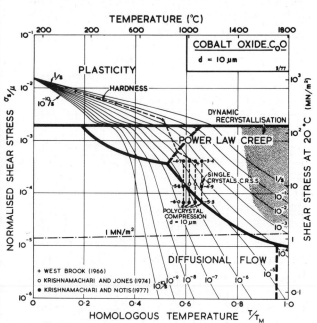

Fig. 12.7. A stress/temperature map for CoO with a grain size of 10 $\mu$m. Data are labelled with $\log_{10} \dot{\gamma}$.

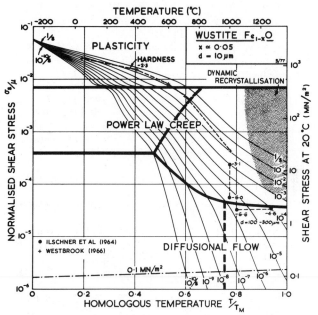

Fig. 12.9. A stress/temperature map for FeO with a grain size of 10 $\mu$m. Data are labelled with $\log_{10} \dot{\gamma}$.

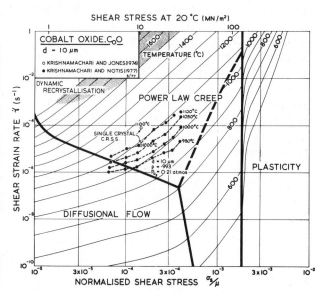

Fig. 12.8. A strain-rate/stress map for CoO with a grain size of 10 $\mu$m. Data are labelled with the temperature in °C.

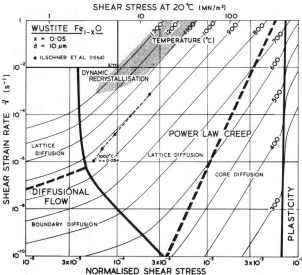

Fig. 12.10. A strain-rate/stress map for FeO with a grain size of 10 $\mu$m. Data are labelled with the temperature in °C.

## References for CoO

Aleksandrov, K. S., Shabanova, L. A. and Reshchikova, L. M. (1968) *Sov. Phys. Solid State* **10**, 1316.

Asanti, P. and Kohlmeyer, E. J. (1951) *Z. Anorg. Chemie* **265**, 90.

Carter, R. E. and Richardson, F. D. (1954) *J. Metals* **6**, 1244.

Chen, W. K. and Jackson, R. A. (1969) *J. Phys. Chem. Solids* **30**, 1309.

Chen, W. K., Peterson, N. L. and Reeves, W. T. (1969) *Phys. Rev.* **186**, 887.

Clauer, A. H., Seltzer, M. S. and Wilcox, B. A. (1971a) *J. Mat. Sci.* **4**, 1379.

Clauer, A. H., Seltzer, M. S. and Wilcox, B. A. (1971b) in *Ceramics in Severe Environments* (eds. Kriegel, W. W. and Palmour, H.). Plenum, p. 361.

Crow, W. B. (1969) Ph.D. thesis, Ohio State University.

Holt, J. B. (1967) *Proc. Br. Ceram. Soc.* **9**, 157.

Krishnamachari, V. and Jones, J. T. (1974). *J. Am. Ceram. Soc.* **57**, 506.

Krishnamachari, V. and Notis, M. R. (1977) *Acta. Met.* **25**, 1025.

Maiya, P. S., Chen, W. K. and Peterson, N. L. (1970) *Met. Trans.* **1**, 801.

Marsh, D. M. (1964) *Proc. R. Soc.* **A279**, 420.

Mrowec, S. (1967) *Bull. Acad. Pol. Sci. Ser. Sci. Chim.* **15**, 373.

Shunk, F. A. (1969) *Constitution of Binary Alloys*, 2nd Supplement. McGraw-Hill.

Strafford, K. N. and Gartside, H. (1969) *J. Mat. Sci.* **4**, 760.

Thompson, B. A. (1962) Ph.D. thesis, Rensselaer Polytech. Instit.

Wartenberg, H. V. and Gurr, W. (1931) *Z. Anorg. Chemie* **196**, 377.

## 12.4    DATA FOR $Fe_{1-x}O$

No single-crystal elastic constants are available for FeO. Table 12.1 lists the polycrystal shear modulus of Akimoto (1972), and the typical temperature dependence of $-0.3$. By analogy with the other oxides of this chapter, slip in FeO is expected to occur on the easy system $\langle 1\bar{1}0 \rangle \{110\}$ with slip on the hard system $\langle 011 \rangle \{100\}$ appearing in polycrystals.

As in $Co_{1-x}O$, deviations from stoichiometry influence the properties of $Fe_{1-x}O$. Its melting point, for example, depends on composition; we have used the value for $x = 0.05$. More important, diffusion rates depend on composition. As far as we know, lattice diffusion of $O^{2-}$ in $Fe_{1-x}O$ has not been measured. The oxide is always hyperstoichio-

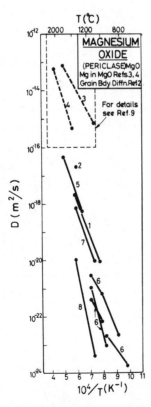

Fig. 12.11. Diffusion data plot for MgO. Data from (1) Shirasaki and Hama (1973); (2) Shirasaki and Oishi (1971); (3) Harding *et al.* (1971); (4) Vasilov *et al.* (1969); (5) Oishi and Kingery (1960); (6) Rovner (1966); (7) Shirasaki and Yamamura (1973); (8) Narayan and Washburn (1973); (9) Harding and Price (1972).

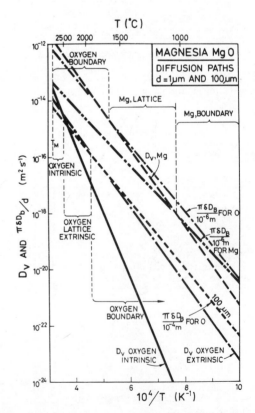

Fig. 12.12. Selected diffusion coefficients for MgO illustrating the selection of the slowest species diffusing by the fastest path.

metric, and the metal ion is smaller than the oxygen ion. For both reasons we expect that the diffusion of $O^{2-}$ should be much slower than that of $Fe^{2+}$, and should control creep rates. The creep data of Ilschner et al. (1964) support this view: they observed an activation energy for creep of 326 kJ/mole, compared with about 125 kJ/mole for the lattice diffusion of $Fe^{2+}$. We have therefore assumed that $Q_v (O^{2-})$ is 326 kJ/mole even though it leads to a normalized value, $Q_v/RT_M$, of 24 which is higher than that for most other oxides. We have combined this with a pre-exponential $D_{0v} (O^{2-})$ of $10^{-2}$ m²/s to give a melting point diffusivity in general agreement with that for other oxides. There are no measurements of grain-boundary or core diffusion for FeO. The data given in the table are estimates.

The parameters describing power-law creep are chosen to fit the data of Ilschner et al. (1964) for $x = 0.054$. The creep rate of both $Co_{1-x}O$ and $Fe_{1-x}O$ increases with $x$, an observation not easy to explain if the creep is controlled by oxygen diffusion on its own sublattice. It is possible that complex oxygen defects are involved.

The parameters for lattice-resistance controlled glide are chosen to fit the hardness data of West-brook (1966), adjusted according to Marsh (1964).

The simple maps of Figs. 12.9 and 12.10 are consistent with this very limited data. One must anticipate that the real behaviour may be complicated in the ways described for MgO.

### References for FeO

Akimoto, S. (1972) *Tectonophysics* **13**, 161.

Campbell, R. H. (1969) Ph.D. thesis, Arizona State University, p. 95. Univ. Microf. No. 69–5710.

Carter, R. E. and Richardson, F. D. (1954) *J. Metals* **6**, 1244.

Chen, W. K. and Peterson, N. L. (1973) *J. Phys. (Paris), Colloq.* **9**, 303.

Chu, W. F. (1973) *Proc. Nato Conf. Sept., Belgirate, Italy.* North Holland Publ. Co., p. 181.

Fujii, C. T. and Meussner, R. A. (1967) *Rept. Nav. Res. Lab. Progr.*, p. 27.

Greenwood, N. N. and Howe, A. T. (1972) *J. Chem. Soc. Dalton Trans.* **1**, 122.

Hembree, P. and Wagner, J. B. Jr. (1969) *Trans. AIME* **245**, 1547.

Himmel, L., Mehl, R. F. and Birchenall, C. E. (1953) *J. Metals* **5**, 827.

Ilschner, B., Reppich, B. and Riecke, E. (1964) *Faraday Soc. Disc.* **38**, 243.

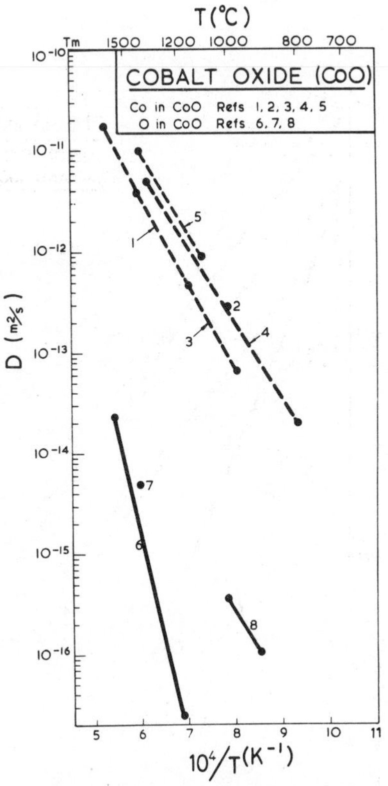

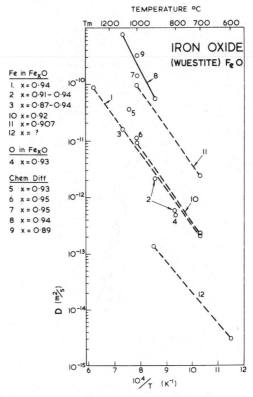

Fig. 12.13. Diffusion data plot for CoO. Data from (1) Maiya et al. (1970); (2) Mrowec (1967); (3) Chen et al. (1969); (4) Carter and Richardson (1954); (5) Crow (1969); (6) Chen and Jackson (1969); (7) Holt (1967); (8) Thompson (1962).

Fig. 12.14. Diffusion data plot for FeO. Data from (1) Chen and Peterson (1973); (2) Greenwood and Howe (1972); (3) and (4) Hembree and Wagner (1969); (5) Campbell (1969); (6) Fujii and Meussner (1967); (7) Levin and Wagner (1965); (8) Rickert and Weppner (1974); (9) Chu (1973); (10) Carter and Richardson (1954); (11) Himmel et al. (1953); (12) Valov (1970).

Levin, R. L. and Wagner, J. B. Jr. (1965) *Trans. AIME* **233**, 159.

Marsh, D. M. (1964) *Proc. R. Soc.* **A279**, 420.

Rickert, H. and Weppner, W. (1974) *Z. Naturforsch.* **A29**, 1849.

Valov, P. M. (1970) *J. Solid State Chem.* **1**, 215.

Westbrook, J. H. (1966) *Rev. Hautes Temper. et Refract.* **3**, 47.

# CHAPTER 13

# OXIDES WITH THE FLUORITE STRUCTURE: $UO_2$ AND $ThO_2$

BOTH uranium dioxide and thorium dioxide crystallize with the fluorite ($CaF_2$) structure. They differ from most other oxides in that the metal ion is larger than the oxygen ion and, partly for this reason, diffuses more slowly. Both oxides can be thought of as an f.c.c. stacking of metal ions, with oxygen ions contained in the tetrahedral holes. The oxides $PuO_2$ and $CeO_2$ are like $UO_2$ and $ThO_2$ in the size and packing of their ions. Their mechanical behaviour should closely resemble that described below.

All these oxides exist over a range of composition. They are usually hyperstoichiometric with the composition $UO_{2+x}$ and $ThO_{2+x}$. Throughout we have tried to select data for oxides close to stoichiometry.

Data and maps for the two oxides are shown in Figs. 13.1 to 13.4. The parameters used to construct them are listed in Table 13.1.

## 13.1 GENERAL FEATURES OF THE DEFORMATION OF FLUORITE-STRUCTURED OXIDES

At low temperatures the fluorite-structured oxides show a substantial lattice resistance, which dominates the mechanical strength up to about 0·4 $T_M$. Hardness studies and compression tests both indicate a flow strength at 0 K of about $\mu/50$.

Above 0·4 $T_M$, the oxides creep. Data for $UO_2$ and $ThO_2$ are detailed on the maps shown as Figs. 13.1 to 13.4. At the higher stresses and larger grain sizes, both oxides exhibit power-law creep, with a creep exponent of about 4. Even though the metal ion is the slower diffusing species, the creep rate depends on the oxygen partial-pressure in the surrounding atmosphere. Seltzer et al. (1971), who have reviewed the creep data for this class of oxides, report that the creep rate of $UO_{2+x}$ increases, and its apparent activation energy decreases, with increasing amounts of excess oxygen ($x$)—probably because it

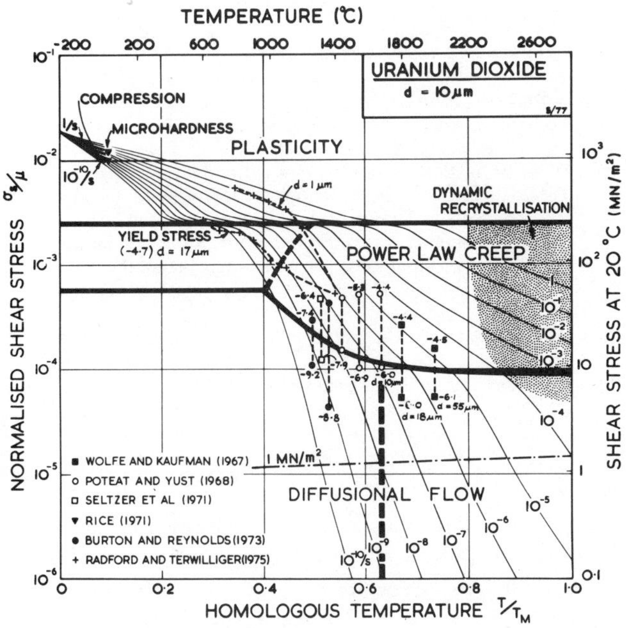

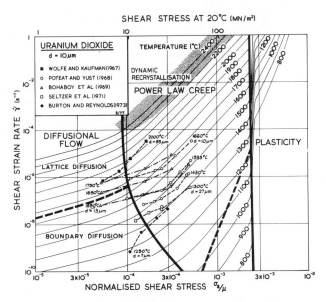

Fig. 13.1. A stress/temperature map for $UO_{2·00}$ with a grain size of 10 $\mu$m. Data are labelled with $\log_{10} \dot{\gamma}$.

Fig. 13.2. A strain-rate/stress map for $UO_{2·00}$ with a grain size of 10 $\mu$m. Data are labelled with the temperature in °C.

93

**TABLE 13.1    Oxides with the fluorite structure**

| Material | $UO_{2\cdot00}$ | | $ThO_{2\cdot00}$ | |
|---|---|---|---|---|
| *Crystallographic and thermal data* | | | | |
| Atomic volume, (m³) | $4\cdot09 \times 10^{-29}$ | (a) | $4\cdot39 \times 10^{-29}$ | (a) |
| Burger's vector, $b$ (m) | $3\cdot87 \times 10^{-10}$ | (a) | $3\cdot96 \times 10^{-10}$ | (a) |
| Melting temperature, $T_m$ (K) | 3080 | (b) | 3520 | (b) |
| *Modulus* | | | | |
| Shear modulus at 300 K, $\mu_0$ (MN/m²) | $9\cdot39 \times 10^4$ | (c) | $1\cdot02 \times 10^5$ | (k) |
| Temperature dependence of modulus, $\dfrac{T_m}{\mu_0}\dfrac{d\mu}{dT}$ | $-0\cdot35$ | (d) | $-0\cdot48$ | (l) |
| *Lattice diffusion* | | | | |
| Pre-exponential, $D_{0v}$ (m²/s) | $1\cdot2 \times 10^{-5}$ | (e) | $3\cdot5 \times 10^{-5}$ | |
| Activation energy, $Q_v$ (kJ/mole) | 452 | | 625 | (m) |
| *Boundary diffusion* | | | | |
| Pre-exponential, $\delta D_{0b}$ (m³/s) | $2 \times 10^{-15}$ | (f) | $10^{-14}$ | (n) |
| Activation energy, $Q_b$ (kJ/mole) | 293 | | 375 | (n) |
| *Core diffusion* | | | | |
| Pre-exponential, $a_c D_{0c}$ (m⁴/s) | $10^{-25}$ | (g) | $10^{-25}$ | (g) |
| Activation energy, $Q_c$ (kJ/mole) | 293 | | 375 | |
| *Power-law creep* | | | | |
| Exponent, $n$ | $4\cdot0$ | | $4\cdot0$ | |
| Dorn constant, $A$ | $3\cdot88 \times 10^3$ | (h) | $5\cdot2 \times 10^6$ | (o) |
| *Obstacle-controlled glide* | | | | |
| 0 K flow stress, $\hat{\tau}/\mu_0$ | $3 \times 10^{-3}$ | | $3 \times 10^{-3}$ | |
| Pre-exponential, $\dot{\gamma}_0$ (s⁻¹) | $10^6$ | (i) | $10^6$ | (p) |
| Activation energy, $\Delta F/\mu_0 b^3$ | $0\cdot5$ | | $0\cdot5$ | |
| *Lattice-resistance-controlled-glide* | | | | |
| 0 K flow stress, $\hat{\tau}_p/\mu_0$ | $2 \times 10^{-2}$ | | $2 \times 10^{-2}$ | |
| Pre-exponential, $\dot{\gamma}_p$ (s⁻¹) | $10^{11}$ | (j) | $10^{11}$ | (q) |
| Activation energy, $\Delta F_p/\mu_0 b^3$ | $0\cdot08$ | | $0\cdot08$ | |

(a)    $\Omega = a^3/4$ is the volume of one $UO_2$ or $ThO_2$ molecule, and $b = \frac{1}{2}a\langle 110 \rangle$ where $a$ is the lattice parameter.

(b)    Shunk (1969).

(c)    Wachtman *et al.* (1965); $\mu_0 = (\frac{1}{2}c_{44}(c_{11} - c_{12}))^{\frac{1}{3}}$.

(d)    Marlowe and Kaznoff (1976);

$$\mu = \mu_0\left(1 + \frac{T-300}{T_M}\left(\frac{T_M}{\mu_0}\frac{d\mu}{dT}\right)\right).$$

(e)    Diffusion of uranium ions; see also the diffusion plot for $UO_2$, Fig. 13.5. These parameters correspond to the measurements of McNamara (1963). They were chosen because they lie centrally through the data shown in the diffusion plot; $D_v = D_{0v} \exp -(Q_v/RT)$.

(f)    Diffusion of uranium ions; Alcock *et al.* (1966); $\delta D_b = \delta D_{0b} \exp -(Q_b/RT)$.

(g)    A dubious estimate, based on the data for $D_b$; $a_c D_c = a_c D_{0c} \exp -(Q_c/RT)$.

(h)    Based on data of Poteat and Yust (1968). These parameters

give an exact fit at 1535°C; *see* Fig. 13.2. The value of $A$ is for tensile creep; the maps were computed using $A_s = (\sqrt{3})^{n+1} A$.

(i,j)    Data are sparse. These parameters give a tolerable fit to the measurements of Radford and Terwilliger (1975) and of Rice (1971).

(k)    Bechmann and Hearmon (1969); see also note (c).

(l)    Spinner *et al.* (1963); see also note (d).

(m)    Diffusion of thorium ions; see also the diffusion plot for $ThO_2$, Fig. 13.6. These parameters correspond to the measurements of King (1971). They were chosen because the activation energy is close to that for power-law creep.

(n)    Diffusion of thorium ions. The parameters match the data of King (1968) and the activation energy matches certain creep-data—see text.

(o)    Based on data of Wolfe and Kaufman (1967), see Fig. 13.4; see also note (h).

(p,q)    Based on data of Ryshkewitch (1941, 1942).

stabilizes vacancies on the uranium sublattice, enhancing uranium-ion diffusion. They found a maximum activation energy of 561 kJ/mole at the composition $UO_{2\cdot00}$.

At lower stresses the oxides show a well-defined regime of diffusional flow, sometimes with a threshold stress. The creep is almost linear–viscous ($n \approx 1$), and strongly dependent on grain size. The

maps show that it is well described by the rate equation for diffusional flow (eqn. (2.29)).

Data for $UO_2$ are more extensive than for $ThO_2$ though there is a lack of good hardness measurements for both. Inspection of the maps for $UO_2$ shows that they give a good description of most of the data in all three fields. That for $ThO_2$ is less satisfactory. But the coincidence between the two

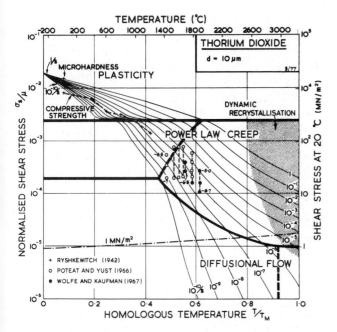

Fig. 13.3. A stress/temperature map for $ThO_{2.00}$ with a grain size of 10 $\mu$m. Data are labelled with $\log_{10} \dot{\gamma}$.

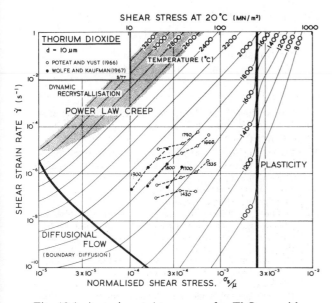

Fig. 13.4. A strain-rate/stress map for $ThO_{2.00}$ with a grain size of 10 $\mu$m. Data are labelled with the temperature in °C.

**References for Section 13.1**

Seltzer, M. S., Perrin, J. S., Clauer, A. H. and Wilcox, B. A. (1971) *Reactor Technol.* **14**, 99.

## 13.2  DATA FOR $UO_2$ AND $ThO_2$

The preferred data for the two oxides are listed in Table 13.1. The shear modulus was calculated from single-crystal data (referenced in the table) using:

$$\mu_0 = \sqrt{\tfrac{1}{2}c_{44}(c_{11} - c_{12})}$$

Its temperature dependence was obtained from polycrystal data.

The melting points of these oxides depend on stoi-

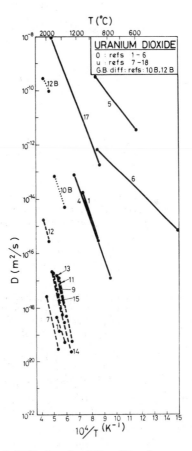

Fig. 13.5. Diffusion in $UO_2$. The data are from (1) Dornelas and Lacombe, 1967; (2) Contamin and Stefani, 1967; (3) Contamin and Slodzian, 1968; (4) Marin and Contamin, 1969; (5) Lay, 1970; (6) Contamin *et al.*, 1972; (7) Reinmann and Lundy, 1969; (8) *ibid*; 1968; (9) Auskern and Belle, 1961; (10) Alcock *et al.*, 1966; (11) Lindner and Schmitz, 1961; (12) Yajima *et al.*, 1966; (13) Hawkins and Alcock, 1968; (14) Nagels, 1966; (15) McNamara, 1963; (16) Carter and Lay, 1970; (17) Bittel *et al.*, 1969; (18) Iida, 1967; (19) Matzke, 1969; (20) Murch *et al.*, 1975.

is good enough that one can conclude with confidence that, when normalized, the mechanical behaviour of the two are closely similar. For this reason we think that oxides with the fluorite structure, in which the cation is larger than the anion, probably form a mechanically-similar class, and that $PuO_2$ and the rare-earth oxides typified by $CeO_2$ probably behave very much as $UO_2$ does.

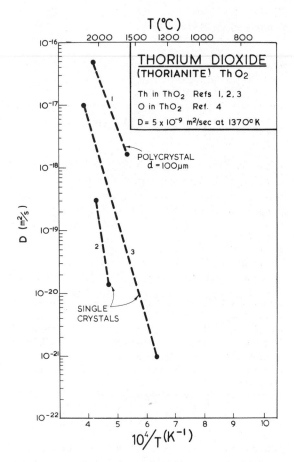

Fig. 13.6. Diffusion in ThO₂. The data are from (1) Hawkins and Alcock, 1968; (2) King, 1971; (3) Morgan and Poteat, 1968; (4) Roberts and Roberts, 1967.

chiometry. The values used here are from Shunk (1969) and refer to $UO_{2.00}$ and $ThO_{2.00}$. Both slip with a slip vector of a $\langle 1\bar{1}0\rangle$ on $\{100\}$, $\{110\}$ and $\{111\}$ planes (Edington and Klein, 1966; Gilbert, 1965).

The diffusion data for the two oxides are shown in Figs. 13.5 and 13.6. The metal (broken line) is the slower-diffusing ion in this class of oxide, and therefore the one likely to control creep and diffusional flow (see Chapter 12, MgO, for a detailed discussion of this point). Although there is considerable spread in quoted activation energies for metal ion diffusion, the figures show that the absolute values are in fairly good agreement. (Measurements of diffusion of oxygen show much greater scatter, partly because the measurements are difficult, and partly because deviations from stoichiometry may effect it.) We have selected a high value of activation energy for lattice diffusion, $Q_v$, of the metal ion, for each oxide: 452 kJ/mole for $UO_2$ and 625 kJ/mole for $ThO_2$ (references are given in the table and the figures). This choice comes closest to agreement with creep data discussed below.

Grain-boundary diffusion of the metal ion has

been measured in $UO_2$. We have used the data of Alcock et al. (1966) which are some three orders of magnitude slower than those of Yajima et al. (1966) at $0.5\ T_M$. When used in the diffusional creep equation, it comes closest to describing the creep observations of Poteat and Yust (1968), predicting a creep rate about one order of magnitude slower than that observed.

In deriving power-law creep parameters, we have used the activation energy for lattice self-diffusion of the metal ion given in Table 13.1 and chosen $n$ and $A$ to give the best fit to the power-law creep data. As Fig. 13.2 shows, our values for $n$ and $A$ for $UO_2$ match the data of Poteat and Yust (1968) at 1535°C, and give a tolerable fit to the rest of the high-stress data for stoichiometric $UO_2$ from 1400 to 2000°C. Those for $ThO_2$ (Fig. 13.4) match Wolfe and Kaufman's (1967) data between 1700 and 1900°C, and are a tolerable fit to the high-stress data of Poteat and Yust (1966).

At lower stresses and smaller grain-sizes, creep in $UO_2$ becomes Newtonian–viscous ($n = 1$) and strongly dependent on grain size (Poteat and Yust, 1966, 1968; Bohaboy et al., 1969; Seltzer et al., 1971; Burton and Reynolds, 1973). The data for both oxides are consistent with diffusional flow, usually involving grain-boundary transport, and sometimes with a threshold stress (Burton and Reynolds, 1973). An example is the study of the compressive creep of $ThO_2$, with a grain size of 10 $\mu$m, of Poteat and Yust (1966). They observed an activation energy of $470 \pm 30$ kJ/mole, about three-quarters of that for lattice self-diffusion of Th ions. The absolute magnitude of the creep rate is faster, by a factor of $10^4$, than that predicted by the Nabarro–Herring equation, but is broadly consistent with the Coble equation (Chapter 2, eqn. 2.29). We have adopted $Q_b = 0.6Q_v$ and $\delta D_{0b} = 10^{-14}$ m³/s, which adequately describes the diffusion measurements of King (1968) for low-angle boundaries, and have used this in evaluating Coble-creep in $ThO_2$, giving a tolerable match to the observed creep rates.

The creep data for both oxides were obtained from samples with grain sizes in the range 7 to 55 $\mu$m. Inspection of the figures, which are constructed for a grain size of 10 $\mu$m, suggests that the predicted position of the boundary between power-law creep and diffusional flow lies at stresses which are too low. It must be remembered that diffusion and creep in these oxides depends on stoichiometry and on purity, and because of this, a low activation energy for creep, like that observed by Poteat and Yust, may imply that diffusion is extrinsic.

Parameters describing the low-temperature plasticity of $UO_2$ (Table 13.1) are based on yield strength measurements of Radford and Terwilliger

(1975) and room-temperature hardness and compressive strength measurements of Rice (1971). Those for ThO$_2$ are based on the compressive strength data of Ryshkewitch (1941, 1942).

## References for UO$_2$ and ThO$_2$

Alcock, G. B., Hawkins, R. J., Hills, A. W. D. and McNamara, P. (1966) *Thermodynamics* (IAEA) Vienna, SM 66/36.

Auskern, A. B. and Belle, J. (1961) *J. Nucl. Mat.* **3**, 311.

Bechmann, R. and Hearmon, R. F. S. (1969) *Landolt–Bornstein Tables*, 3–1, p. 5.

Bittel, J. T., Sjodahl, L. H. and White, J. F. (1969) *J. Am. Ceram. Soc.* **52**, 446.

Bohaboy, P. E., Asamoto, R. R. and Conti, A. E. (1969) *USAEC Report*, GEAP. 10054, General Electric Co.

Burton, B. and Reynolds, G. L. (1973) *Acta Met.* **21**, 1641.

Carter, R. E. and Lay, K. W. (1970) *J. Nucl. Mat.* **36**, 77.

Contamin, P., Bacmann, J. J. and Marin, J. F. (1972) *J. Nucl. Mat.* **42**, 54.

Contamin, P. and Slodzian, G. (1968) *C.R. Acad. Sci., Paris, Ser. C* **267**, 805.

Contamin, P. and Stefani, R. (1967) *Commis. Energ. At. Rapp.* No. 3179.

Dornelas, W. and Lacombe, P. (1967) *C.R. Acad. Sci., Paris, Ser. C* **265**, 359.

Edington, J. W. and Klein, M. J. (1966) *J.A.P.* **37**, 3906.

Gilbert, A. (1965) *Phil. Mag.* **12**, 139.

Hawkins, R. J. and Alcock, C. B. (1968) *J. Nucl. Mat.* **26**, 112.

Iida, S. (1967) *Jap. J. Appl. Phys.* **6**, 77.

King, A. D. (1968) *J. Nucl. Mat.* **26**, 112.

King, A. D. (1971) *J. Nucl. Mat.* **38**, 347.

Lay, K. W. (1970) *J. Am. Ceram. Soc.* **53**, 369.

Lindner, R. and Schmitz, F. (1961) *Z. Naturforsch.* **16A**, 1373.

Marin, J. F. and Contamin, P. (1969) *J. Nucl. Mat.* **30**, 16.

Marlow, M. O. and Kaznoff, A. I. (1976) General Electric Co. Vallecites Nuclear Center, Pleasanton, Calif. 94506, private communication.

Matzke, Hj. (1969) *J. Nucl. Mat.* **30**, 26.

McNamara, P. (1963) Ph.D. thesis, quoted by Nagles, P. (1966) *Thermodynamics*. IAEA, Vienna, p. 311.

Morgan, C. S. and Poteat, C. E. (1968) ORNL Metals and Ceramics Divn. Ann. Progr. Rpt.

Murch, G. E., Bradhurst, D. H. and de Bruin, H. J. (1975) *Phil. Mag.* **32**, 1141.

Nagels, P. (1966) *Thermodynamics.* IAEA, Vienna, p. 311.

Poteat, L. E. and Yust, C. S. (1966) *J. Am. Ceram. Soc.* **49**, 410.

Poteat, L. E. and Yust, C. S. (1968) Grain Boundary Reactions during Deformation. In *Ceramic Microstructure*, (eds. Fulrath, R. M. and Pask, J. A.). Wiley, p. 646.

Radford, K. C. and Terwilliger, G. R. (1975) *J. Am. Ceram. Soc.* **58**, 274.

Reinmann, D. K. and Lundy, T. S. (1968) *J. Nucl. Mat.* **28**, 218.

Reinmann, D. K. and Lundy, T. S. (1969) *J. Am. Ceram. Soc.* **52**, 511.

Rice, R. W. (1971) The Compressive Strength of Ceramics. In *Ceramics in Severe Environments* (eds. Kriegel, W. W. and Palmour, H.). Plenum, p. 195.

Roberts, E. W. and Roberts, J. P. (1967) *Bull. Soc. Fr. Ceram.* **77**, 3.

Ryshkewitch, E. (1941) *Ber. Dtsch. Keram. Ges.* **22**, 54; (1942) *ibid.* **23**, 243.

Seltzer, M. S., Perrin, J. S., Clauer, A. H. and Wilcox, B. A. (1971) *Reactor Technology* **14**, 99.

Shunk, F. A. (1969) *Constitution of Binary Alloys*, second supplement. McGraw-Hill.

Spinner, S., Stone, L. and Kundsen, F. P. (1963) *J. Res. Natl. Bur. Std.* **676**, 93.

Wachtman, J. B. Jr., Wheat, M. L., Anderson, H. L. and Bates, J. L. (1965) *J. Nucl. Mat.* **16**, 39.

Wolfe, R. A. and Kaufman, S. I. (1967) "Mechanical Properties of Oxide Fuels". Westinghouse Report WAPD-TM-587.

Yajima, S., Fuynya, H. and Hirai, T. (1966) *J. Nucl. Mat.* **20**, 162.

# CHAPTER 14

# OXIDES WITH THE α-ALUMINA STRUCTURE:
## Al₂O₃, Cr₂O₃ AND Fe₂O₃

THREE common oxides have the α-alumina structure: corundum or sapphire ($Al_2O_3$), chromium sesqui-oxide ($Cr_2O_3$) and hematite ($Fe_2O_3$). The structure, if described by its smallest unit cell, is rhombo-hedral; but it is convenient to think of it in terms of a larger hexagonal unit cell. The oxygen ions are packed in a close-packed hexagonal arrangement with metal ions in two-thirds of the octahedral sites. The unoccupied octahedral sites are ordered within each close-packed layer, and alternate between layers, repeating every third layer. The hexagonal unit cell contains six oxygen layers.

These oxides are generally harder and more re-fractory than the rock-salt structured oxides, retain-ing their strength to higher temperatures. Alumina is used as a structural ceramic, as well as an abrasive and a coating for cutting tools. Chromium sesqui-oxide is perhaps most important as a surface layer on stainless steels and nickel-based alloys.

Maps for the three oxides are shown in Figs. 14.1 to 14.8. The parameters used to construct them are listed in Table 14.1.

## 14.1  GENERAL FEATURES OF THE MECHANICAL BEHAVIOUR OF α-ALUMINA STRUCTURED OXIDES

These oxides are generally stronger than those with the rock-salt structure. By comparing maps for α-alumina (Figs. 14.1–14.4) with those for magnesia (Figs. 12.1–12.4), one sees that the lattice-resistance persists to a much higher temperature ($0.5~T_M$ instead of $0.3~T_M$) and that creep, at the same values of $\sigma_s/\mu$ and $T/T_M$, is nearly ten times slower. Part of this difference is caused by the structure of α-alumina, which imposes restrictions on slip which do not exist in a simple hexagonal close-packing of ions. Because of the ordered filling of the octa-hedral sites (described above) certain shears create stacking faults; among these are the simple basal shear and simple twinning, both of which are made much more difficult. The easy slip systems are still those in the basal plane (with slip in the harder prismatic and pyramidal systems appearing at high

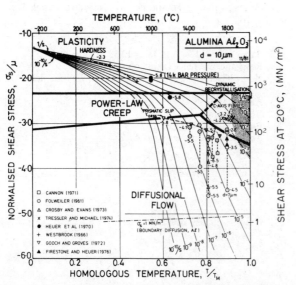

Fig. 14.1. A stress/temperature map for $Al_2O_3$ with a grain size of 10 $\mu$m. Data are labelled with $\log_{10}\dot{\gamma}$.

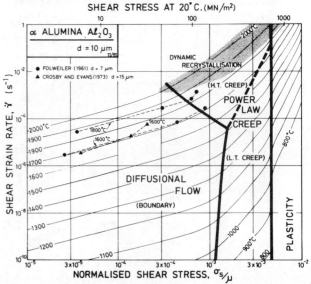

Fig. 14.2. A strain-rate/stress map for $Al_2O_3$ with a grain size of 10 $\mu$m. Data are labelled with the temperature in °C. Data for larger grain sizes are shown on Fig. 14.4.

**TABLE 14.1** Oxides with the α-alumina structure

| Material | $Al_2O_3$ | | $Cr_2O_3$ | | $Fe_2O_3$ | |
|---|---|---|---|---|---|---|
| *Crystallographic and thermal data* | | | | | | |
| Atomic volume, $\Omega$ (m³) | $4.25 \times 10^{-29}$ | | $4.81 \times 10^{-29}$ | (h) | $5.03 \times 10^{-29}$ | |
| Burger's vector, $b$ (m) | $4.76 \times 10^{-10}$ | | $4.96 \times 10^{-10}$ | (h) | $5.03 \times 10^{-10}$ | |
| Melting temperature, $T_M$ (K) | 2320 | | 2710 | | 1840 | (o) |
| *Modulus* | | | | | | |
| Shear modulus at 300 K, $\mu_0$ (MN/m²) | $1.55 \times 10^5$ | (a) | $1.30 \times 10^5$ | (i) | $8.82 \times 10^4$ | (p) |
| Temperature dependence of modulus, $\dfrac{T_M}{\mu_0}\dfrac{d\mu}{dT}$ | $-0.35$ | (a) | $-0.33$ | (j) | $-0.2$ | (q) |
| *Lattice diffusion: oxygen ion* | | | | | | |
| Pre-exponential, $D_{0v}$ (m²/s) | 0·19 | (b) | $1.59 \times 10^{-3}$ | (k) | $2.04 \times 10^{-4}$ | (r) |
| Activation energy, $Q_v$ (kJ/mole) | 636 | (b) | 423 | (k) | 326 | (r) |
| *Lattice diffusion: metal ion* | | | | | | |
| Pre-exponential, $D_{0v}$ (m²/s) | $2.8 \times 10^{-3}$ | (c) | — | | 40 | (s) |
| Activation energy, $Q_v$ (kJ/mole) | 477 | (c) | — | | 469 | (s) |
| *Boundary diffusion: oxygen ion* | | | | | | |
| Pre-exponential, $\delta D_{0b}$ (m³/s) | $10^{-8}$ | (d) | $10^{-15}$ | (l) | $4 \times 10^{-13}$ | (t) |
| Activation energy, $Q_b$ (kJ/mole) | 380 | (d) | 240 | (l) | 210 | (u) |
| *Boundary diffusion: metal ion* | | | | | | |
| Pre-exponential, $\delta D_{0b}$ (m³/s) | $8.6 \times 10^{-10}$ | (e) | — | | — | |
| Activation energy, $Q_b$ (kJ/mole) | 419 | (e) | — | | — | |
| *Core diffusion* | | | | | | |
| Pre-exponential, $a_c D_{0c}$ (m⁴/s) | $10^{-22}$ | (d) | $10^{-25}$ | (l) | $2.0 \times 10^{-25}$ | (l) |
| Activation energy, $Q_c$ (kJ/mole) | 380 | (d) | 240 | (l) | 210 | (l) |
| *Power-law creep* | | | | | | |
| Exponent, $n$ | 3·0 | (f) | 4·0 | (m) | 3·5 | (v) |
| Dorn constant, $A$ | 3·38 | (f) | $1 \times 10^3$ | (m) | 5·0 | (v) |
| Activation energy, $Q_{cr}$ (kJ/mole) | — | | — | | — | |
| *Obstacle-controlled glide* | | | | | | |
| 0 K flow stress, $\hat{\tau}/\mu_0$ | $5 \times 10^{-3}$ | | $5 \times 10^{-3}$ | | $5 \times 10^{-3}$ | |
| Pre-exponential, $\dot{\gamma}_0$ (s⁻¹) | $10^6$ | | $10^6$ | | $10^6$ | |
| Activation energy, $\Delta F/\mu_0 b^3$ | 0·5 | | 0·5 | | 0·5 | |
| *Lattice-resistance; controlled-glide* | | | | | | |
| 0 K flow stress, $\hat{\tau}_p/\mu_0$ | 0·05 | (g) | 0·05 | (n) | 0·05 | (n) |
| Pre-exponential, $\dot{\gamma}_p$ (s⁻¹) | $10^{11}$ | (g) | $10^{11}$ | | $10^{11}$ | |
| Activation energy, $\Delta F_p/\mu_0 b^3$ | 0·032 | (g) | 0·032 | (n) | 0·032 | (n) |

(a) Tefft (1966); $\mu_0 = (\frac{1}{2}c_{44}(c_{11} - c_{12}) - c_{14}^2)^{\frac{1}{3}}$,

$$\mu = \mu_0\left(1 + \frac{T-300}{T_M}\left(\frac{T_M}{\mu_0}\frac{d\mu}{dT}\right)\right).$$

(b) Oishi and Kingery (1960), intrinsic oxygen diffusion, see text; $D_v = D_{0v} \exp - (Q_v/RT)$.
(c) Paladino and Kingery (1962).
(d) Estimated using $Q_b = Q_c = 0.6Q_v$ (intrinsic); $\delta D_b = \delta D_{0b} \exp - (Q_b/RT)$; $a_c D_c = a_c D_{0c} \exp - (Q_c/RT)$.
(e) Estimate by Cannon and Coble (1974).
(f) Based on data of Cannon (1971), high stress data at 1650°C assuming $Q_{cr}$ to be that for intrinsic oxygen diffusion. The value of $A$ is for tensile creep; the maps are computed using $A_s = (\sqrt{3})^{n+1}A$.
(g) To match the hardness data of Westbrook (1966).

(h) Based on $a = 4.96$ Å and $c = 13.58$ Å.
(i) Rossi and Lawrence (1970); polycrystal data.
(j) Yevtushchenko and Levitin (1961); polycrystal data; see note (a).
(k) Hagel (1965).
(l) Estimated; see note (d).
(m) Burton (1975); see note (f).
(n) Assumed to be the same as $Al_2O_3$.
(o) See text.
(p) Chou and Sha (1971); see note (a).
(q) Anderson et al. (1968); polycrystal data; see note (a).
(r) Hagel (1966).
(s) Lindner et al. (1952) as corrected by Hagel (1966).
(t) To match creep data as described in text.
(u) Channing and Graham (1972).
(v) Crouch (1972); see note (f).

temperatures) but the overall strength is greater than that of the simpler oxides.

Neither α-alumina nor chromium sesquioxide show any appreciable deviations from exact stoichiometry. Both exist in other allotropic forms, stabilized by temperature and pressure, but no data

exist for these and we shall ignore them. Hematite, by contrast, is stable over a range of composition (depending on oxygen partial pressure) and decomposes to $Fe_3O_4$ and oxygen before it melts.

Certain problems exist in applying the simple rate equations of Chapter 2 to these oxides. First, there is

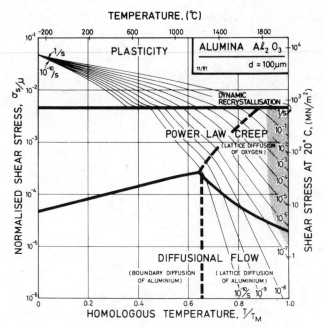

Fig. 14.3. A stress/temperature map for $Al_2O_3$ with a grain size of 100 $\mu$m. Data are shown on Fig. 14.1.

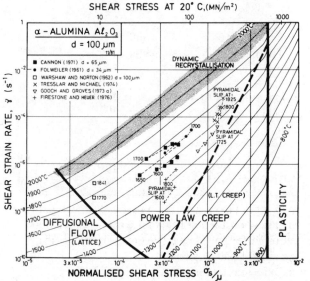

Fig. 14.4. A strain-rate/stress map for $Al_2O_3$ with a grain size of 100 $\mu$m. Data are labelled with the temperature in °C.

the problem of relating single crystal to polycrystal behaviour. Both at low temperatures and in creep, polycrystal plasticity is possible only if four or five independent slip systems operate: basal slip alone is insufficient. The polycrystal flow strength reflects some weighted average of the strengths of these systems, and is generally closer to that for the hard system (Hutchinson, 1977). Where necessary, we have assumed the two to be equal. Similarly, the

hardness is taken to be $3\sqrt{3}$ times the shear strength of the hard system.

The second problem is that of averaging diffusion coefficients of the components, discussed in Section 12.1 for rock-salt structured oxides. Mass transport requires the motion of metal and oxygen in the ratio 2:3. We therefore form the effective diffusion coefficient:

$$D_{\mathrm{eff}} = \frac{5D_{\mathrm{OX}}D_{\mathrm{MET}}}{3D_{\mathrm{MET}} + 2D_{\mathrm{OX}}}$$

and use this in place of the lattice diffusion coefficient. In diffusional flow there is a further problem in picking an appropriate molecular volume $\Omega$. When oxygen is the slower moving species, it is proper to use:

$$\Omega_{\mathrm{eff}} = \Omega/3$$

where $\Omega$ is the volume associated with the $Al_2O_3$ molecule (since work $\sigma\Omega/3$ is done every time one oxygen atom arrives at a surface). When the metal ion is slower, the proper effective volume becomes:

$$\Omega_{\mathrm{eff}} = \Omega/2$$

But diffusion may be extrinsic or intrinsic, and alternative diffusion paths (lattice and grain-boundary paths, for instance) may be available. As a general rule the creep rate is determined by the fastest path of the slowest species, and this may change with temperature and grain size as illustrated in Fig. 14.10, which helps understand the disposition of the fields in Figs. 14.3 and 14.4. A fur-

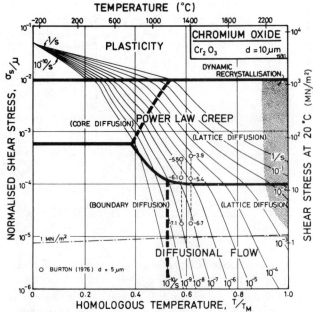

Fig. 14.5. A stress/temperature map for $Cr_2O_3$ with a grain size of 10 $\mu$m. Data are labelled with $\log_{10}\dot{\gamma}$.

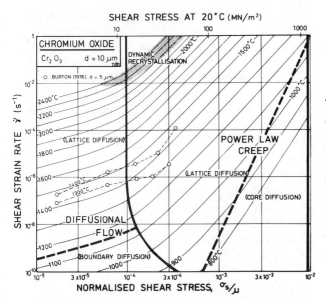

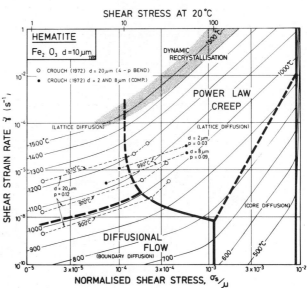

Fig. 14.6. A strain rate/stress map for $Cr_2O_3$ with a grain size of 10 $\mu$m. Data are labelled with the oxygen partial pressure in MN/m², the grain size in $\mu$m and the temperature in °C.

Fig. 14.8. A strain-rate/stress map for $Fe_2O_3$ with a grain size of 10 $\mu$m. Data are labelled with the temperature in °C.

## 14.2  DATA FOR Al₂O₃

The modulus $\mu_0$ was calculated from:

$$\mu_0 = (\tfrac{1}{2}c_{44}(c_{11} - c_{12}) - c_{14}^2)^{\frac{1}{2}}$$

using single-crystal data and their temperature dependence (Tefft, 1966). This modulus is the one determining the energy of a $\langle 11\bar{2}0 \rangle$ screw dislocation in a trigonal crystal (Chou and Sha, 1971).

Slip and twinning in alumina are discussed by Kronberg (1957). Because of the complicated packing of aluminium ions, the Burger's vector is large: 2·73 times larger than the oxygen ion spacing. Slip is easiest on the basal plane, but is observed (and required, for polycrystal plasticity) on the prism and pyramidal planes also.

Diffusion data for Al₂O₃ are shown in Fig. 14.9. Alumina does not deviate significantly from stoichiometry, but small levels of impurity can influence diffusion rates greatly. Cannon and Coble (1974), who have studied and reviewed creep of Al₂O₃, conclude that its diffusional flow can be understood if the boundary diffusion of oxygen is faster than that of aluminium either in the lattice or in the boundaries. If this is so, a regime will exist in which Al-diffusion controls the creep rate (see Fig. 14.10 and compare Figs. 14.1 and 14.2); data exist which support this view.

For the intrinsic lattice diffusion of oxygen, we have used the data of Oishi and Kingery (1960), shown in the diffusion plot of Fig. 14.9. Lattice

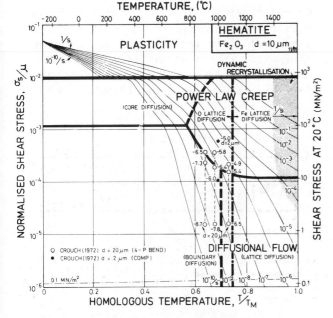

Fig. 14.7. A stress/temperature map for $Fe_2O_3$ with a grain size of 10 $\mu$m. Data are labelled with $\log_{10} \dot{\gamma}$.

ther discussion of this point is given in Section 12.1.

The maps have been constructed using these principles. It must be emphasized that the data are very limited. The maps are no better than the data used to construct them, and must be used with caution.

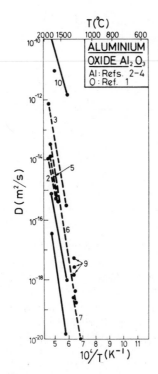

Fig. 14.9. Diffusion data for $Al_2O_3$. The data are from: (1) Oishi and Kingery (1960); (2) Paladino and Kingery (1962); (3) Folweiler (1961); (4) Roberts and Roberts (1967); (5) Warshaw and Norton (1962); (6) Vishnevskii et al. (1971); (7) Beauchamp et al. (1961); (8) Coble and Kingery (1956); (9) Stavrolakis and Norton (1950); (10) Jones et al. (1969).

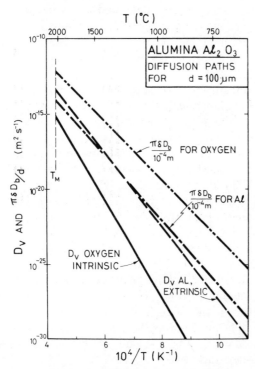

Fig. 14.10. Selected diffusion coefficients for $Al_2O_3$ with $d = 100\ \mu m$, illustrating the selection of the slowest species diffusing by the fastest path.

diffusion of $Al^{3+}$ appears to be extrinsic in all available creep experiments (Cannon and Coble, 1974). Its rate then depends on the concentration and nature of the impurities. We have used the data of Paladino and Kingery (1962), but when it limits the creep rate, the maps may be unreliable.

For boundary diffusion of $Al^{3+}$ we have used the estimate of Cannon and Coble (1974), though it leads to a value of $Q_b/RT_M$ which is exceptionally large. For boundary diffusion of oxygen we have estimated a value in line with data for other oxides. Its value is not critical since it is never rate-controlling. The core diffusion of oxygen is based on an estimate.

Power-law creep in $Al_2O_3$ is much slower than in the rock-salt-structured oxides at the same fraction of the melting point. The maps are based on the compression data of Cannon (1971) for large-grained (65 $\mu m$) material, taking a creep exponent $n$ of 3. Cannon reported an activation energy of 611 kJ/mole, large enough to be intrinsic lattice diffusion of oxygen (Table 14.1). These data are generally consistent with measurements of the critical resolved shear stress for pyramidal and for prismatic slip in single crystals.

The lattice-resistance controlled glide parameters are based on the hardness data of Westbrook (1966). The value of $\Delta F_p/\mu_0 b^3$ of 0·032 appears low only because the Burger's vector $b$ is large.

### References for $Al_2O_3$

Beauchamp, E. K., Baker, G. S., Gibbs, P. (1961) WADD Rept., Contract AF 33 (610)–6832 Project No. 0 (7–7350).

Cannon, R. M. (1971) Ph.D. thesis, Stanford.

Cannon, R. M. and Coble, R. L. (1974) in Deformation of Ceramic Materials (eds. Bradt, R. C. and Tressler, R. E.). Plenum, NY, p. 61.

Chou, Y. T. and Sha, G. T. (1971) Phys. Stat. Sol. (a) 6, 505.

Coble, R. L. and Kingery, W. D. (1956) J. Am. Ceram. Soc. 39, 377.

Crosby, A. and Evans, P. E. (1973) J. Mat. Sci. 8, 1573.

Firestone, R. F. and Heuer, A. H. (1976) J. Am. Ceram. Soc. 59, 24.

Folweiler, R. C. (1961) J. Appl. Phys. 32, 773.

Gooch, D. J. and Groves, G. W. (1972) J. Am. Ceram. Soc. 55, 105.

Gooch, D. J. and Groves, G. W. (1973a) Phil. Mag. 28, 623.

Gooch, D. J. and Groves, G. W. (1973b) J. Mat. Sci. 8, 1238.

Heuer, A. H., Firestone, R. F., Snow, J. D. and Tullis, J. D. (1970) Proc. 2nd Int. Conf. on The Strength of Metals and Alloys, Asilomar, Calif., ASM, p. 1165.

Hutchinson, J. W. (1977) Met. Trans. 8A, 1465.

Kronberg, M. L. (1957) Acta Met. 5, 507.

Oishi, Y. and Kingery, W. D. (1960) J. Chem. Phys. 33, 480.

Paladino, A. E. and Kingery, W. D. (1962) *J. Chem. Phys.* **37**, 957.

Roberts, E. W. and Roberts, J. P. (1967) *Bull. Soc. Fr. Ceram.* **77**, 3.

Stavrolakis, J. A. and Norton, F. H. (1950) *J. Am. Ceram. Soc.* **33**, 263.

Tefft, W. E. (1966) *J. Res. Natl. Bur. Std.* **70A**, 277.

Tressler, R. E. and Michael, D. J. (1974) In *Deformation of Ceramic Materials* (eds. Bradt, R. C. and Tressler, R. E.). Plenum, NY, p. 195.

Vishnevskii, I. I., Akselrod, E. I. and Talyanskaya, N. D. (1971) *Fiz. Tverd. Tela.* **13**, 3447.

Warshaw, S. I. and Norton, F. H. (1962) *J. Amer. Ceram. Soc.* **45**, 479.

Westbrook, J. H. (1966) *Rev. Hautes Temper. et Refract.* **3**, 47.

## 14.3 DATA FOR $Cr_2O_3$

Modulus measurements for $Cr_2O_3$ are confused by the antiferromagnetic to paramagnetic transition at 311 K, at which the shear modulus increases. No single-crystal constants are available, so we have used the polycrystal shear modulus of Rossi and Lawrence (1970) extrapolated to full density, and estimated a temperature dependence from the polycrystal data of Yevtushchenko and Levitin (1961).

Chromium sesquioxide, like alumina, exists only

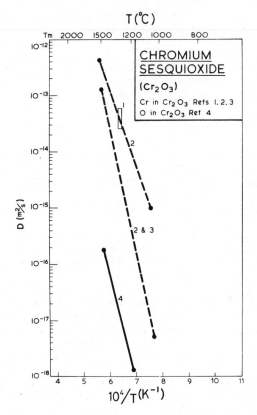

Fig. 14.11. Diffusion data for $Cr_2O_3$. The data are from (1) Walters and Grace (1965); (2) Hagel and Seybolt (1961); (3) Lindner and Åkerström (1956); (4) Hagel (1965).

close to stoichiometry. Its slip systems have not been studied and are assumed to be the same as those of alumina.

Lattice diffusion of oxygen is at least 100 times slower than that of chromium in $Cr_2O_3$ (Fig. 14.11), and should control the rate of creep. We have found no data for boundary or core diffusion for either species, and have made estimates based on data for other oxides. Because of the unequal number of $Cr^{3+}$ and $O^{2-}$ ions in a molecule, an effective diffusion coefficient $D_{eff} = \frac{5}{3}D_{oxygen}$ and $\Omega_{eff} = \Omega/3$ must be used in the diffusional flow equation, assuming that oxygen is always the slower-moving species.

The only creep data for $Cr_2O_3$ are those of Burton (1975) shown in Figs. 14.5 and 14.6. They are compatible with power-law behaviour at higher stresses, and with the transition to diffusional flow controlled by oxygen diffusion at lower stresses. There is an indication that a threshold stress may be involved in the diffusional flow field (not shown on the map).

We have found no hardness data for $Cr_2O_3$, and have therefore used glide parameters that give the same normalized plastic behaviour as $Al_2O_3$ at low temperatures.

### References for $Cr_2O_3$

Burton, B. (1975) Private communication.

Hagel, W. C. (1965) *J. Am. Ceram. Soc.* **48**, 70.

Hagel, W. C. and Seybolt, A. U. (1961) *J. Electrochem. Soc.* **108**, 1146.

Lindner, R. and Åkerström, Å. (1956) *Z. Phys. Chem. N.F.* **6**, 162.

Rossi, L. R. and Lawrence, W. G. (1970) *J. Am. Ceram. Soc.* **53**, 604.

Walters, L. C. and Grace, R. E. (1965) *J. Appl. Phys.* **36**, 2331.

Yevtushchenko, L. A. and Levitin, R. Z. (1961) *Fiz. Metal. Metalloved* **12**, 155.

## 14.4 DATA FOR $Fe_2O_3$

The modulus $\mu_0$ was calculated from the expression for the energy of a $\langle 11\bar{2}0 \rangle$ screw dislocation in the basal plane of a trigonal crystal:

$$\mu_0 = (\tfrac{1}{2}c_{44}(c_{11} - c_{12}) - c_{14}^2)^{\frac{1}{2}}$$

given by Chou and Sha (1971), and using the single-crystal constants cited by them. The temperature dependence is from the polycrystal measurements of Anderson *et al.* (1968).

The range of stability of $Fe_2O_3$ increases with oxygen partial pressure. In air it decomposes to $Fe_3O_4$ and $O_2$ at 1663 K; under one atmosphere of oxygen it is stable to 1732 K; and under 16 atmospheres of oxygen, Phillips and Mann (1960)

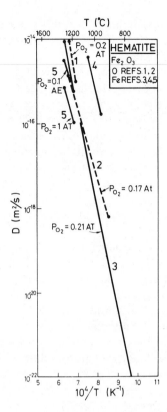

Fig. 14.12. Diffusion data for $Fe_2O_3$. The data are from (1) Kingery *et al.* (1960); (2) Hagel (1966); (3) Lindner *et al.* (1952); (4) Izvekov *et al.* (1962); (5) Chang and Wagner (1972).

found a eutectic between $Fe_3O_4$ and $Fe_2O_3$ at 1839 K. The *Handbook of Physics and Chemistry* (1972) gives the melting point as 1840 K. We have used this value.

Slip systems in $Fe_2O_3$ have not been studied. We assume them to be the same as $Al_2O_3$.

Lattice diffusion of both $Fe^{3+}$ and $O^{2-}$ in $Fe_2O_3$ have been measured (Fig. 14.12). They are of roughly equal magnitude, requiring the use of an effective diffusion coefficient:

$$D_{eff} = \frac{5D_{Ox}D_{Fe}}{3D_{Fe} + 2D_{Ox}}$$

and, for diffusional flow, an effective molecular volume $\Omega_{eff} = \Omega/2.5$. At high temperatures $Fe_2O_3$ can drift off stoichiometry, changing its diffusion coefficients. Chang and Wagner (1972), for example, found that the diffusion of iron decreased with increasing $PO_2$ (they found $D \propto PO_2^{-0.75}$) and Crouch (1972) observed a similar trend in the creep rate (he found $\dot{\varepsilon} \propto PO_2^{-0.14}$), although a quantitative explanation is lacking. We have used the lattice diffusion parameters of Hagel (1966) for oxygen and those of Lindner *et al.* (1952) for iron. The result is a switch from control by oxygen diffusion to control by iron diffusion at 1100°C. It is shown as a dash–dot line on Fig. 14.7.

We have derived creep parameters from the data of Crouch (1972) who observed power-law behaviour at higher stresses and linear behaviour at lower stresses (Figs. 14.7 and 14.8). A correction is required to compensate for the large porosity (12%) in these samples. Crouch (1972) reported that $\dot{\varepsilon} \propto$ exp 14 $\phi$, where $\phi$ is the porosity, so that specimens of zero porosity should creep more slowly, by a factor of 5, than the data plotted on Figs. 14.7 and 14.8. This has been taken into account in computing the maps.

Crouch (1972) interpreted his data in terms of Nabarro–Herring creep. But Hay *et al.* (1973), studying the low-stress creep of $Fe_2O_3$ found a grain size dependence which lay between $d^{-2}$ and $d^{-3}$, and a creep rate that appears to be too rapid for Nabarro–Herring creep, suggesting control by grain-boundary diffusion. Accordingly, we have calculated boundary diffusion coefficients from these data, and find that the value of $\delta D_b$ shows general agreement with those derived from sintering experiments by Johnson (1973); and that the activation energy is close to that for the oxidation of $Fe_3O_4$ to $Fe_2O_3$ (210 kJ/mole; Channing and Graham, 1972) which is believed to occur by boundary diffusion of oxygen. We have used this activation energy, and a pre-exponential which best matches Crouch's compression tests on fine-grained, low-porosity samples. This is why the fit to the rest of Crouch's data, which had a larger porosity, is poor.

We have found no hardness data for $Fe_2O_3$, and have therefore used for low-temperature plasticity the same normalized parameters as for alumina.

### References for $Fe_2O_3$

Anderson, O. L., Schreiber, E., Liebermann, R. C. and Soga, N. (1968) *Rev. Geophysics* **6**, 491.

Chang, R. H. and Wagner, J. B. Jr. (1972) *J. Am. Ceram. Soc.* **55**, 211.

Channing, D. A. and Graham, M. J. (1972) *Corrosion Sci.* **12**, 271.

Chou, Y. T. and Sha, G. T. (1971) *Phys. Stat. Sol. (a)* **6**, 505.

Crouch, A. G. (1972) *J. Am. Ceram. Soc.* **55**, 558.

Hagel, W. C. (1966) *Trans. AIME* **236**, 179.

*Handbook of Chemistry and Physics* (1972) 52nd edn. Chemical Rubber Co., Cleveland.

Hay, K. A., Crouch, A. G. and Pascoe, R. T. (1973) presented at the "International Conference on Physical Metallurgy of Reactor Fuel Elements".

Izvekov, V. I., Gobunov, N. S. and Babad-Zakhrapin, A. A. (1962) *Fiz. Met. Metall.* **14**, 195.

Johnson, D. L. (1973) quoted by Hay *et al.* (1973).

Kingery, W. D., Hill, D. C. and Nelson, R. P. (1960) *J. Am. Ceram. Soc.* **43**, 473.

Lindner, R., Anstrumdal, S. and Åkerström, Å. (1952) *Acta Chem. Scand.* **6**, 468.

Phillips, B. and Mann, A. (1960) *J. Phys. Chem.* **64**, 1452.

# CHAPTER 15

# OLIVINES AND SPINELS: $Mg_2SiO_4$ AND $MgAl_2O_4$

THE OLIVINES $(Mg, Fe)_2 SiO_4$ are orthorhombic. The simplest and most widely studied is forsterite, the magnesium end-member, to which natural olivines are close. It can be thought of as an almost close-packed hexagonal crystal of oxygen (the largest ion) containing silicon and magnesium in some of its tetrahedral and octahedral interstices. The $SiO_4$ tetrahedra are not linked (olivines are not network silicates) so that certain slips are possible without breaking Si—O bonds. For this reason, olivines have somewhat lower creep strengths than network silicates such as $SiO_2$. But examination of the structure shows that general plasticity without diffusion requires slip on additional planes on which Si—O bonds will be broken. Perhaps for this reason, the low temperature strength is high.

The spinels form a second large class of oxides, with the general formula $A B_2O_4$. They crystallize in a cubic structure which can be thought of as a combination of the rock-salt and the zinc blende structures. The oxygen ions are close packed in f.c.c. stacking, with the cations contained in the octahedral and tetrahedral interstices.

In the normal spinels, the $A^{2+}$ ions are contained in the tetrahedral holes, and the $B^{3+}$ in the octahedral holes: magnesium aluminate spinel, the spinel studied here, is of this type. In the inverse spinels, on the other hand, the $A^{2+}$ ions and half the $B^{3+}$ ions are in octahedral holes, and the remaining $B^{3+}$ ions are in tetrahedral holes. Among the normal aluminate spinels are $MgAl_2O_4$, and those obtained on replacing Mg by Fe, Co, Ni, Mn or Zn. Certain ferrites, too, have the normal spinel structure; among them are $Zn Fe_2O_4$ and $Cd Fe_2O_4$. The inverse spinel structure is adopted by many other ferrites, notably $Mg Fe_2O_4$ and those obtained on replacing Mg by Ti, Fe or Ni.

A number of compounds transform from the olivine to the spinel structure on increasing the pressure. Among these are forsterite and the magnesium-rich olivines.

Olivines form the dominant phase in the earth's upper mantle, in which shear must occur when the continental plates move. Much recent work has been motivated by the wish to understand and model this motion (Chapter 19, Section 19.8). Spinels have been studied for different reasons: they are used as refractories and their high strength and transparency makes them attractive for windows which must resist heat, impact or abrasion.

Maps for forsterite and for magnesium aluminate spinel are shown in Figs. 15.1 to 15.6. They are based on data plotted on the figures, and the parameters listed in Table 15.1.

## 15.1 GENERAL FEATURES OF THE DEFORMATION OF OLIVINES AND SPINELS

The mechanical behaviour of both classes of oxide closely resemble that of $\alpha$-alumina (Chapter 14). This is not surprising since their structures are closely related. All show a high lattice resistance, approaching roughly $\mu/20$ at 0 K, and all retain their strength to high homologous temperatures (0·5 $T_M$ and above). Diffusion rates in the spinels and olivines vary somewhat with composition, but are generally comparable with those in $Al_2O_3$. The rate-controlling species is often assumed to be oxygen, since this is the biggest ion in the structure. But evidence that diffusion of the metal ion can sometimes limit the creep rate in $Al_2O_3$ and $Fe_2O_3$ (Chapter 14) suggest that similar complications may arise in spinels and olivines also.

## 15.2 DATA FOR FORSTERITE, $Mg_2SiO_4$

The shear modulus and its temperature dependence for olivines have been measured by Graham and Barsch (1969) and Kamazama and Anderson (1969).

The slip systems have been determined by electron microscopy (Phakey et al., 1972) and optical examination (Raleigh, 1968; Carter and Ave' Lallemant, 1970; Young, 1969; Raleigh and Kirby, 1970). The easy slip system at low temperatures is that involving dislocations with $b = 5·98$ [001] Å gliding

**TABLE 15.1**    Olivine and spinel

| Material | $Mg_2SiO_4$ | | $MgAl_2O_4$ | |
|---|---|---|---|---|
| *Crystallographic and thermal data* | | | | |
|    Molecular volume, $\Omega$ (m³) | $4.92 \times 10^{-29}$ | (a) | $6.59 \times 10^{-29}$ | (h) |
|    Burger's vector, $b$ (m) | $6.0 \times 10^{-10}$ | (b) | $5.71 \times 10^{-10}$ | (i) |
|    Melting temperature, $T_M$ (K) | 2140 | | 2408 | (j) |
| *Modulus* | | | | |
|    Shear modulus at 300 K, (MN/m²) | $8.13 \times 10^4$ | (c) | $9.92 \times 10^4$ | (k) |
|    Temperature dependence of modulus, $\dfrac{T_M}{\mu_0}\dfrac{d\mu}{dT}$ | $-0.35$ | (c) | $-0.22$ | (k) |
| *Lattice diffusion: oxygen ion* | | | | |
|    Pre-exponential, $D_{0v}$ (m²/s) | 0.1 | (d) | $8.9 \times 10^{-5}$ | (l) |
|    Activation energy, $Q_v$ (kJ/mole) | 522 | (d) | 439 | (l) |
| *Boundary diffusion: oxygen ion* | | | | |
|    Pre-exponential, $\delta D_{0b}$ (m³/s) | $1 \times 10^{-10}$ | (e) | $1 \times 10^{-14}$ | (e) |
|    Activation energy, $Q_b$ (kJ/mole) | 350 | (e) | 264 | (e) |
| *Core diffusion: oxygen ion* | | | | |
|    Pre-exponential, $\delta D_{0c}$ (m⁴/s) | — | | $3 \times 10^{-25}$ | (e) |
|    Activation energy, $Q_c$ (kJ/mole) | — | | 264 | (e) |
| *Power-law creep* | | | | |
|    Exponent, $n$ | 3 | (f) | 2.7 | (m) |
|    Creep constant, $A$ | 0.45 | (f) | 0.16 | |
|    Activation energy, $Q_{cr}$ (kJ/mole) | 522 | (f) | — | |
| *Obstacle-controlled glide* | | | | |
|    0 K flow stress, $\hat{\tau}/\mu_0$ | $5 \times 10^{-3}$ | | $2 \times 10^{-2}$ | |
|    Pre-exponential, $\dot{\gamma}_0$ (s⁻¹) | $10^6$ | | $10^6$ | |
|    Activation energy, $\Delta F/\mu_0 b^3$ | 0.5 | | 0.5 | |
| *Lattice-resistance-controlled glide* | | | | |
|    0 K flow stress, $\hat{\tau}_p/\mu_0$ | $3.3 \times 10^{-2}$ | (g) | $8.5 \times 10^{-2}$ | (n) |
|    Pre-exponential, $\dot{\gamma}_p$ (s⁻¹) | $10^{11}$ | | $10^{11}$ | |
|    Activation energy, $\Delta F_p/\mu_0 b^3$ | 0.05 | (g) | 0.033 | (n) |

(a)   The volume of one $Mg_2SiO_4$ molecule.
(b)   See text.
(c)   Graham and Barsch (1960), and Kamazama and Anderson (1969)

$$\mu = \mu_0\left(1 + \frac{T-300}{T_M}\left(\frac{T_M}{\mu_0}\frac{d\mu}{dT}\right)\right)$$

(d)   See text and diffusion data plot, Fig. 15.7; $D_v = D_{0v}\exp -(Q_v/RT)$.
(e)   No data available. Obtained by taking $\delta D_{0b} = 10^{-9}D_{0v}$ and $Q_b = 0.67\,Q_v$; $\delta D_b = \delta D_{0b}\exp -(Q_b/RT)$, $a_c D_c = a_c D_{0c}\exp -(Q_c/RT)$.
(f)   Fitted to the creep data shown on Fig. 15.1. See Ashby and Verrall (1977). The value of $A$ is for tensile creep; the maps were computed using $A_s = (\sqrt{3})^{n+1}\,A$.

(g)   Based principally on hardness data of Evans, cited by Durham and Goetze (1977).
(h)   The volume of one $MgAl_2O_4$ molecule. Crystallographic data from Clark (1966).
(i)   Mitchell *et al.* (1976).
(j)   *Handbook of Chemistry and Physics* (1972).
(k)   Calculated from the single-crystal constant Lewis (1966);

$$\mu_0 = (\tfrac{1}{2}c_{44}(c_{11}-c_{12}))^{\frac{1}{2}}.$$

(l)   Ando and Oishi (1974) for intrinsic oxygen diffusion in $MgAl_2O_4$.
(m)   Fitted to data of Palmour (1966); see also note (f).
(n)   Fitted to data of Westbrook (1966), using the Marsh (1964) correction.

on (100) planes. If this slip system is suppressed by proper orientation of the crystal, then dislocations with Burger's vector $b = 4.76$ [100] Å gliding on (010) planes appear. At 1000°C these two systems operate with equal ease. Dislocations with a large Burger's vector ($b = 10.21$ [010] Å on (100)) appear above 800°C, particularly in crystals oriented so that the easy slip system is unstressed. We have used $b = 6 \times 10^{-10}$ m (Table 15.1) since this is broadly typical of the observations.

Data for oxygen transport in olivine are shown in Fig. 15.7. Borchardt and Schmaltzried (1972) reported that the oxygen-ion diffusivity at 1320°C was less than $10^{-17}$ m²/s; Goetze and Kohlstedt (1973) inferred diffusion coefficients from the kinetics of the annealing of prismatic dislocation loops; and Barnard (1975) measured oxygen ion diffusion by a proton activation method.

Studies of power-law creep in olivine, discussed below, are best described by an activation energy of $Q_{cr} = 522$ kJ/mol. If we assume that this creep is diffusion controlled, then $Q_{cr}$ can be identified (after minor corrections for the temperature dependence of the modulus) with the activation energy for mass transport in olivine—a process which is probably limited in its rate by oxygen-ion diffusion. We have, therefore, fitted a line with this slope to the data (full line on the figure) for which $D_0 = 0.1$ $m^2/s$ and $Q_v = 522$ kJ/mole. Also shown are plots of the diffusion equation proposed by Goetze and Kohlstedt (1973) (dash–dot line), and that by Stocker and Ashby (1973) (broken line). At high temperature they are all very close, and even at low temperatures the differences are small when compared with scatter in creep and other data.

This diffusion equation predicts a melting point diffusivity for $O^{2-}$ of $2 \times 10^{-14}$ $m^2/s$, compared to an average of around $10^{-14}$ $m^2/s$ for a number of other oxides, which, like olivine, have a close-packed oxygen sublattice, and in which oxygen is thought to diffuse as a single ion. Because the two numbers are similar we believe oxygen diffuses as a single ion, not as an $SiO_4^{4-}$ complex. The grain boundary diffusion coefficients are estimates, using $Q_b = 0.6$ $Q_v$ and $D_{0b} = 2b$ $D_{0v}$.

Published creep data (Fig. 15.1) for natural and synthetic olivines span the temperature range from 1000 to 1650°C (0.6–0.9 $T_M$) at strain-rates from $10^{-8}$ to $10^{-4}/s$, often under a confining pressure of up to $10^3$ $MN/m^2$. The results are complicated by the fact that talc was sometimes used as a pressure medium: above 800°C it releases water, which accelerates the creep of these (and most other) silicates.

Other experiments avoided these problems by using dry gas to apply pressure when it was required. We have given extra weight to these results in constructing the maps.

All investigators (Carter and Ave' Lallemant, 1970; Raleigh and Kirby, 1970; Goetze and Brace, 1972; Post and Griggs, 1973; Kirby and Raleigh, 1973; Kohlstedt and Goetze, 1974; Durham and Goetze, 1977; Durham et al., 1977) report creep rates which are consistent with a value of $n$ which increases from 3 at low stresses to 5 or more at high, with an activation energy of 522 kJ/mol. When all the data for dry olivine are normalized to 1400°C, using an activation energy of 522 kJ/mol, it shows power-law behaviour, with $n \approx 3$ at low stresses, followed by a long transition as stress is raised and glide contributes increasingly to the flow (Ashby and Verrall, 1977). The parameters of Table 15.1 were fitted to these data.

Three studies give information about low-temperature plasticity in olivines. First, the yield stress for a peridotite (60–70% olivine and 20–30% enstatite) of grain size 0.5 mm, was measured by Carter and Ave' Lallemant (1970) between 325 and 740°C. Secondly, Phakey et al. (1972) obtained stress-strain curves at 600 and 800°C for single-crystal forsterite. Four compression tests at 800°C with different orientations of the single crystals produced yield stresses ranging from 570 to 1300 $MN/m^2$. The highest value was obtained on the specimen orientated to produce no stress on the easy slip system; we have used this because it is the most representative of a polycrystalline material. Finally, hardness data for olivine exist (Evans,

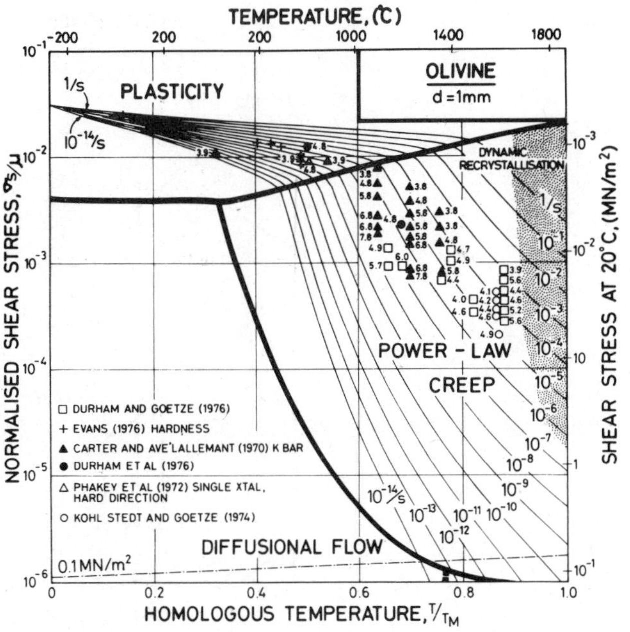

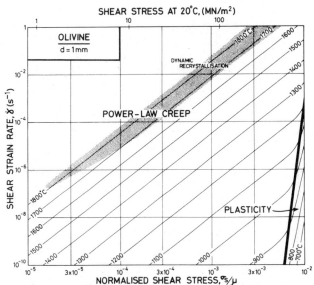

Fig. 15.1. A stress/temperature map for magnesium olivine with a grain size of 1 mm. Data are labelled with $\log_{10} \dot{\gamma}$.

Fig. 15.2. A strain-rate/stress map for magnesium olivine with a grain size of 1 mm. Data are shown on Fig. 15.1.

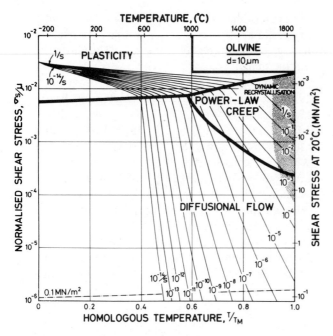

Fig. 15.3. A stress/temperature map for magnesium olivine with a grain size of 10 $\mu$m. Data are shown on Fig. 15.1.

cited by Durham and Goetze, 1977), and it is this which gives the most complete picture of the low-temperature strength. It and the yield strengths are plotted on Fig. 15.1.

### References for olivine

Ashby, M. F. and Verrall, R. A. (1977) *Phil. Trans. Roy. Soc. Lond.* **A288**, 59.

Barnard, R. S. (1975) Ph.D. thesis, Case-Western Reserve University, Department of Materials Science.

Borchardt, G. and Schmaltzried, H. (1972) *Ber. Dtsch. Keram. Ges.* **49**, 5.

Carter, N. L. and Ave' Lallemant, H. G. (1970) *Bull. Geol. Soc. Am.* **81**, 2181.

Durham, W. B. and Goetze, C. (1977) *J. Geophys. Res.* **36**, 5737

Durham, W. B., Goetze, C. and Blake, B. (1977) *J. Geophys. Res.* **36**, 5755

Goetze, C. and Brace, W. F. (1972) *Tectonophysics* **13**, 583.

Goetze, C. and Kohlstedt, D. L. (1973) *J. Geophys. Res.* **78**, 5961.

Graham, E. K. and Barsch, G. R. (1960) *J. Geophys. Res.* **64**, 5949.

Kamazama, M. and Anderson, O. L. (1969) *J. Geophys. Res.* **74**, 5961.

Kirby, S. H. and Raleigh, C. B. (1973) *J. Geophys. Res.* **78**, 5961.

Kohlstedt, D. L. and Goetze, C. (1974) *J. Geophys. Res.* **79**, 2045.

Paterson, M. S. (1974) *Proc. 3rd Int. Congress Soc. Rock. Mech.*, Denver, **1**, 521.

Phakey, P., Dollinger, G. and Christie, J. (1972) *Am. Geophys. Un., Geophys.* Monograph Series **17**, 117.

Post, R. L. and Griggs, D. T. (1973) *Science, N.Y.* **181**, 1242.

Raleigh, C. B. (1968) *J. Geophys. Res.* **73**, 5391.

Raleigh, C. B. and Kirby, S. H. (1970) *Mineral. Soc. Am. Spec. Paper* **3**, 113.

Stocker, R. A. and Ashby, M. F. (1973) *Rev. Geophys. Space Phys.* **11**, 391.

Young, C. (1969) *Am. J. Sci.* **267**, 841.

### 15.3 DATA FOR MAGNESIUM ALUMINATE SPINEL—MgAl$_2$O$_4$

The modulus and its temperature dependence were calculated from the single-crystal constant of Lewis (1966) using:

$$\mu = (\tfrac{1}{2} c_{44} (c_{11} - c_{12}))^{\frac{1}{2}}$$

Magnesium spinel, MgAl$_2$O$_4$, is cubic with a lattice parameter of 8·08 Å. The unit cell contains eight molecules, so that the molecular volume listed in Table 15.1 is one-eighth of the cell volume. Slip has been studied by Mitchell *et al.* (1976) who found easy slip with a Burger's vector $a/2 \langle 110 \rangle$, which dissociated into $a/4 \langle 110 \rangle$ partials. The value listed for $b$ is that for undissociated dislocations.

Oxygen is the slowest-diffusing species in spinel (Oishi and Ando, 1975). The coefficients measured by Ando and Oishi (1974) and Oishi and Ando (1975) are in fair agreement with those inferred from sintering studies by Bratton (1969, 1971). We have used the coefficient given by Ando and Oishi for intrinsic oxygen diffusion in MgAl$_2$O$_4$.

The most extensive studies of creep of spinel are those of Palmour and his co-workers. The maps

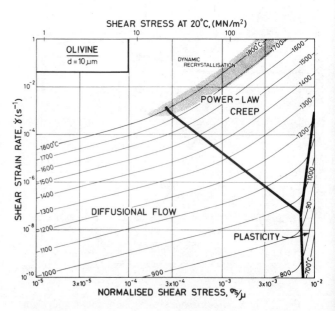

Fig. 15.4. A strain-rate/stress map for magnesium olivine with a grain size of 10 $\mu$m. Data are shown on Fig. 15.1.

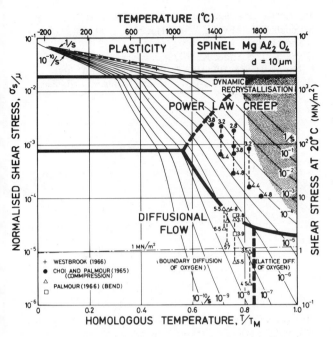

Fig. 15.5. A stress/temperature map for magnesium aluminate spinel with a grain size of 10 $\mu$m. Data are labelled with $\log_{10} \dot{\gamma}$.

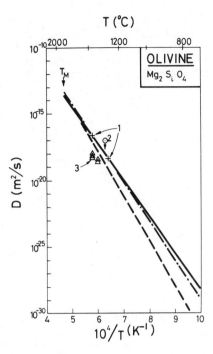

Fig. 15.7. Diffusion data for magnesium olivine. The data are from (1) Goetze and Kohlstedt (1973); (2) Borchardt and Schmatzried (1972); (3) Barnard (1975).

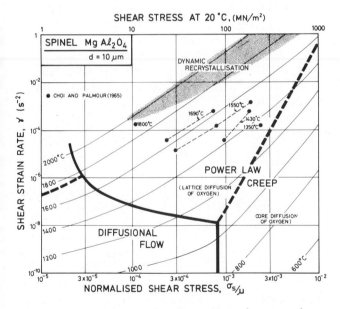

Fig. 15.6. A strain-rate/temperature map for magnesium aluminate spinel with a grain size of 10 $\mu$m. Data are labelled with the temperature in °C.

show data from the compression tests of Choi (1965) (see Choi and Palmour, 1965) and from the bend tests of Palmour (1966). This second set of tests gave data which lie at stresses which are too low, either because a steady state was not reached, or because of surface cracking. The parameters given in Table 15.1 are fitted to the compression data, and ignore the bend tests.

Diffusion flow was calculated by using the Ando and Oishi (1974) diffusion coefficient for intrinsic lattice diffusion of oxygen, with an effective molecular volume of:

$$\Omega_{eff} = \Omega/4$$

because there are four oxygen atoms per molecule. Enhanced boundary diffusion has been detected in spinels (Oishi and Ando, 1975) but it is not possible to extract a reliable coefficient from their results. The coefficients listed in Table 15.1 for both boundary and core diffusion are estimates.

The low-temperature strength of spinel is characterized by the hardness measurements of Westbrook (1966). There is evidence that cracking may have occurred during the test, making some of the data unreliable. The parameters in Table 15.1 are based on the higher hardnesses given by Westbrook, and have been corrected for elastic distortion using the Marsh (1964) correction. The results are then in general agreement with those for $Al_2O_3$ (Chapter 14) and $BeAl_2O_4$.

### References for spinel

Ando, K. and Oishi, Y. (1974) *J. Chem. Phys.* **61**, 625.
Bratton, R. J. (1969) *J. Am. Ceram. Soc.* **52**, 417.
Bratton, R. J. (1971) *J. Am. Ceram. Soc.* **54**, 141.
Choi, D. M. (1965) Ph.D. thesis, North Carolina State University at Raleigh (see Choi and Palmour, 1965).

Choi, D. M. and Palmour, H., III (1965) Technical Report No. 2, Army Research Office, Durham, USA.

Clark, S. P. Jr. (ed.) (1966) *Handbook of Physical Constants*. Memoir 97, The Geological Society of America.

*Handbook of Chemistry and Physics* (1972) 52nd edn. Chemical Rubber Co., Ohio.

Lewis, M. F. (1966) *J. Acoust. Soc. Am.* **40**, 728.

Marsh, D. M. (1964) *Proc. R. Soc.* **A279**, 420.

Mitchell, T. E., Hwang, L. and Heuer, A. H. (1976) *J. Mat. Sci.* **11**, 264.

Oishi, Y. and Ando, K. (1975) *J. Chem. Phys.* **63**, 376.

Palmour, H. III (1966) *Proc. Br. Ceram. Soc.* **6**, 209.

Westbrook, J. H. (1966) *Rev. Hautes Temper. et Refract.* **3**, 47.

# CHAPTER 16

# ICE, H₂O

Ice is a remarkably strong solid. When its normalized strength $(\sigma_s/\mu)$ is compared with that of other solids at the same fraction of their melting point $(T/T_M)$, ice is found to be among the strongest and hardest. In its mechanical behaviour it most closely resembles silicon and germanium (Chapter 9); but the unique character of its proton bonds (H-bonds) leads to forces opposing dislocation motion which are peculiar to ice. For this reason we compute maps for ice in two different ways: first, by treating the equations of dislocation glide and power-low creep of Chapter 2 as empirical, and fitting data for ice to them; and second, by using a set of equations for dislocation glide in ice which is based on a physical model of proton-rearrangement. The second approach is an example of glide-controlled creep and power-law breakdown (Section 2.4).

Data and maps for ice are shown in Figs. 16.1 to 16.6. They are based on data plotted in Figs. 16.1 and 16.2, and on the parameters listed in Table 16.1.

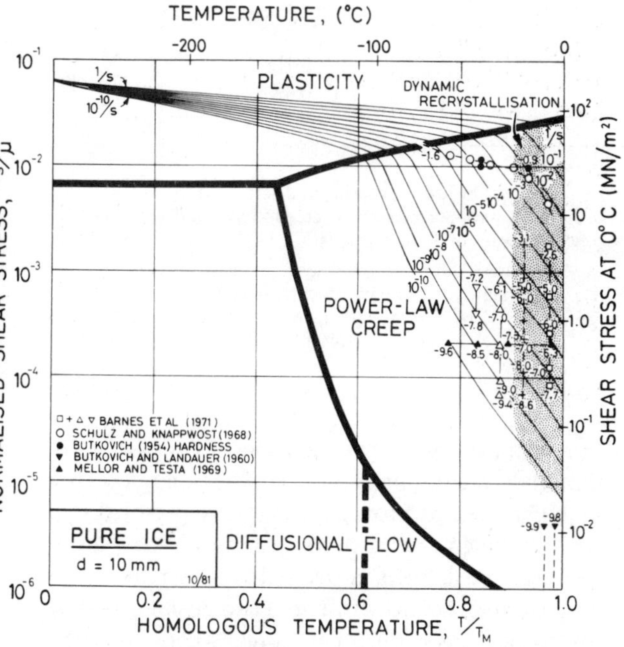

## 16.1 GENERAL FEATURES OF THE DEFORMATION OF ICE

At atmospheric pressure, ice exists in the hexagonal $l_h$ form. The motion of dislocations in it is made difficult by a mechanism peculiar to the ice structure (Glenn, 1968). The protons which form the bridge between the oxygen ions are asymmetrically placed (Fig. 16.7) occupying one of two possible positions on each bond. The protons are disordered, even at 0 K, so that the selection of the occupied site on a given bond is random. Shear, with the protons frozen in position on the bonds, creates defects: bonds with no proton on them ("L-type Bjerrum defects") and bonds with two protons ("D-type Bjerrum defects"). With the protons randomly arranged, a defect is, on average, created on every second bond; and the energy $F_f$ of such defects is high: 0·64 eV per defective bond (Hobbs, 1974).

The result is that the stress required to move a

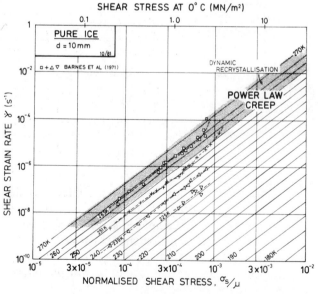

Fig. 16.1. A stress/temperature map for ice with a grain size of 10 mm, using the empirical creep equation (eqn. (16.4)). Data are labelled with $\log_{10} \dot{\gamma}$.

Fig. 16.2. A strain-rate/stress map for ice with a grain size of 10 mm using the empirical creep equation (eqn. (16.4)). Data are labelled with temperature in °C.

**TABLE 16.1  Ice**

| | Using eqns. of Ch. 2 | | Using eqns. of Ch. 16 | |
|---|---|---|---|---|
| *Crystallographic and thermal data* | | | | |
| Molecular volume, $\Omega$ (m$^3$) | $3{\cdot}27 \times 10^{-29}$ | (a) | $3{\cdot}27 \times 10^{-29}$ | (a) |
| Burger's vector, $b$ (m) | $4{\cdot}52 \times 10^{-10}$ | (a) | $4{\cdot}52 \times 10^{-10}$ | (a) |
| Melting temperature, $T_M$ (K) | $273{\cdot}15$ | | $273{\cdot}15$ | (a) |
| *Modulus* | | | | |
| Shear modulus at 273 K, $\mu_0$(MN/m$^2$) | $2{\cdot}91 \times 10^3$ | (b) | $2{\cdot}91 \times 10^3$ | (b) |
| Temperature dependence of modulus, $\dfrac{T_M}{\mu_0}\dfrac{d\mu}{dT}$ | $-0{\cdot}35$ | (b) | $-0{\cdot}35$ | (b) |
| *Lattice diffusion (normal)* | | | | |
| Pre-exponential, $D_{0v}$ (m$^2$/s) | $9{\cdot}1 \times 10^{-4}$ | (c) | $9{\cdot}1 \times 10^{-4}$ | (c) |
| Activation energy, $Q_v$ (kJ/mole) | $59{\cdot}4$ | | $59{\cdot}4$ | |
| *Boundary diffusion* | | | | |
| Pre-exponential, $\delta D_{0b}$ (m$^3$/s) | $8{\cdot}3 \times 10^{-13}$ | (d) | $8{\cdot}3 \times 10^{-13}$ | (d) |
| Activation energy, $Q_b$ (kJ/mole) | $38{\cdot}3$ | | $38{\cdot}3$ | |
| *Core diffusion* | | | | |
| Pre-exponential, $a_c D_{0c}$ (m$^4$/s) | $3{\cdot}7 \times 10^{-22}$ | | — | |
| Activation energy, $Q_c$ (kJ/mole) | $38{\cdot}3$ | (d) | — | |
| *Power-law creep* | | | | |
| Exponent, $n$ | $3{\cdot}0$ | | — | |
| Constant, $A$ (eqn. (16.4)) below $-8°$C (s$^{-1}$) | $4{\cdot}32 \times 10^{19}$ | (e) | — | |
| Constant, $A$ (eqn. (16.4)) above $-8°$C (s$^{-1}$) | $3{\cdot}31 \times 10^{27}$ | | — | |
| Activation energy below $-8°$C, $Q_{cr}$ (kJ/mole) | $80$ | | — | |
| Activation energy above $-8°$C, $Q_{cr}$ (kJ/mole) | $120$ | | — | |
| *Lattice-resistance-controlled glide* | | | | |
| 0 K flow stress, $\hat{\tau}_p/\mu_0$ | $0{\cdot}09$ | (f) | — | |
| Pre-exponential, $\dot{\gamma}_p/($s$^{-1})$ | $10^{11}$ | | — | |
| Activation energy, $\Delta F_p/\mu_0 b^3$ | $0{\cdot}5$ | | — | |
| *Proton rearrangement glide* (eqns. (16.2) and (16.3)) | | | | |
| Pre-exponential, $a_0$ | — | | $1{\cdot}2 \times 10^3$ | |
| Proton-rearrangement frequency, $v_0($s$^{-1})$ | — | | $1{\cdot}6 \times 10^{15}$ | |
| Reference stress $\tau_0/\mu_0$ | — | | $0{\cdot}09$ | |
| Defect creation stress, $\tau_G/\mu_0$ | — | | $0{\cdot}1$ | (g) |
| Activation energy for glide, $\Delta F_g/\mu_0 b^3$ | — | | $0{\cdot}48$ | |
| Activation energy for kink nucleation, $\Delta F_k/\mu_0 b^3$ | — | | $0{\cdot}15$ | |

(a) Fletcher (1970).
(b) Dantl (1968; $\mu_0 = (\tfrac{1}{2}c_{44}(c_{11} - c_{12}))^{\frac{1}{2}}$;

$$\mu = \mu_0\left(1 + \left(\frac{T-273}{273}\right)\left(\frac{T_M}{\mu_0}\frac{d\mu}{dT}\right)\right).$$

(c) Ramseier (1967); $D_v = D_{0v} \exp - (Q_v/RT)$.
(d) Estimated; $\delta D_b = \delta D_{0b} \exp - Q_b/RT$; $a_c D_c = a_c D_{0c} \exp - Q_c/RT$.

(e) Obtained by fitting experimental data to eqn. (16.4). Separate sets of parameters refer to creep above and below $-8°$C. The value of $A$ refers to tensile stress and strain rate. In computing the maps, $A_s = (\sqrt{3})^{n+1} A$ is used.
(f) Obtained by fitting experimental data to eqn. (2.12).
(g) Obtained by fitting experimental data to eqns. (16.2) and (16.3)—see Goodman et al., 1981.

dislocation in this way is very large (Glenn, 1968). If the energy absorbed, $F_f$ per defect, is equated to the work done by the applied stress, $2\sigma_s b^2 a$, when a length $a$ of dislocation moves forward by $2b$, one finds that the shear stress required to move a dislocation through the ice lattice with immobile protons is:

$$\sigma_G = \frac{F_f}{2b^2 a} \sim 0{\cdot}1\,\mu \qquad (16.1)$$

At and near absolute zero, it seems probable that dislocations can move only at this high stress—for

practical purposes equal to the ideal strength—and that they create many defects when they do. But at higher temperatures ice deforms plastically at stresses which are much lower than this and a dislocation must, then, move without creating such large numbers of defects. This it can do if it advances (by nucleating a kink pair) only where the local proton arrangement in its path is favourable—meaning that no defects (or an acceptably small number of defects) are created. The dislocation then drifts forward with a velocity which depends on the kinetics of proton rearrangement.

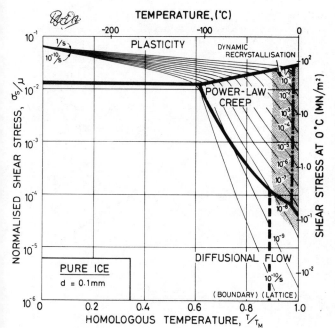

Fig. 16.3. A stress/temperature map for ice with a grain size of 0·1 mm using the empirical creep equation (eqn. (16.4)). Data are shown on Fig. 16.1.

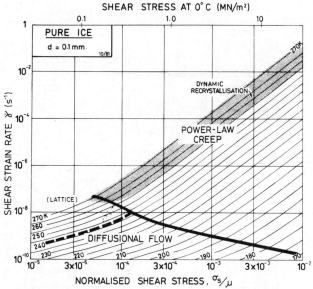

Fig. 16.4. A strain-rate/stress map for ice with a grain size of 0·1 mm using the empirical creep equation (eqn. (16.4)). Data are shown on Fig. 16.2.

This drift velocity can be calculated (Whitworth *et al.*, 1976; Frost *et al.*, 1976; Goodman *et al.*, 1981), and from it the strain-rate can be estimated. The results of such calculations suggest that, near the melting point and at low stresses, the strain-rate is determined by a linear–viscous drag caused by the sluggish rate of proton rearrangement, but that at higher stresses, kink nucleation becomes rate-controlling. At still higher stresses, dislocations create defects as they move, and are no longer limited in their velocity by the rate of proton rearrangement. Finally at 0 K the entire resistance to glide is caused by defect-creation. An equation for the dislocation velocity, which includes this compound resistance to flow, is developed by Goodman *et al.* (1981). The result is that flow follows the faster of two equations. The first is dominant at low stresses, has the form of a simple power-law, with $n = 3$:

$$\dot{\gamma}_{LS} = a_0 v_0 \left[\frac{\sigma_s}{\mu}\right]^3 \exp -\left[\frac{\Delta F_g}{kT}\right] \qquad (16.2)$$

The second is dominant at higher stresses, and includes the regime of power-law breakdown, and of low-temperature plasticity:

$$\dot{\gamma}_{HS} = \dot{\gamma}_{LS} \left(\frac{kT}{2\cdot5\sigma_s b^3(1 - \sigma_s/\tau_G)}\right)^{\frac{1}{2}}$$
$$\exp\left\{\frac{\Delta F_k}{kT}\left(\frac{\sigma_s}{\tau_0}\right)^{\frac{1}{2}} - 0\cdot06\left(\frac{\mu}{\sigma_s}\right)^{\frac{1}{2}}\right\} \qquad (16.3)$$

These equations, together with the standard equations of diffusional flow, were used to construct Figs. 16.5 and 16.6. Figs. 16.1 to 16.4 were constructed using the empirical, power-law description of creep. Both are fitted to the same sets of experimental data shown in Fig. 16.1. and 16.2. The parameters extracted from this data are listed in Table 16.1.

Observations of the creep of ice in the diffusional-flow regime are confused. Bromer and Kingery (1968) found behaviour consistent with Nabarro–Herring creep at $\sigma_s \approx 10^{-5} \mu$, but Butkovich and Landauer (1960) and Mellor and Testa (1969) who also measured creep rates at stresses near $10^{-5} \mu$, did not. Some of the confusion may be due to the contribution of transient creep (Weertman, 1969): unless the test is carried to strains exceeding 1%, transients associated with dislocation motion and grain boundary sliding may mask the true diffusional flow. And grains in ice are often large, so that rates of diffusional flow are small. But there seems no reason to suppose that diffusional flow will not occur in ice at the rate given approximately by eqn. (2.29), and that, for the appropriate regime of stress and temperature, it will appear as the dominant mechanism. This predicted regime of dominance is shown on the maps.

It is worth noting that mechanisms of flow which involve dislocation motion will create a fabric (a preferred orientation of the *c*-axis) in ice, but that diffusional flow will tend to destroy one. The presence or absence of fabric may—if the strains are large enough—help identify the dominant flow mechanism.

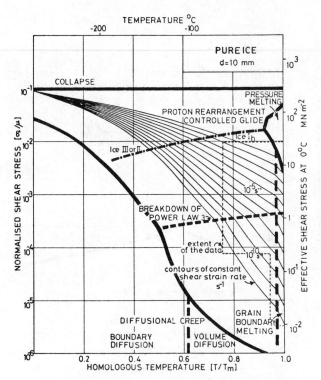

Fig. 16.5. A stress/temperature map for ice of grain size 10 mm, using the model-based equations (eqns. (16.2) and (16.3)) (Goodman *et al.*, 1981; courtesy *Phil. Mag.*).

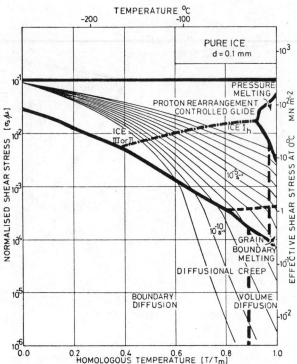

Fig. 16.6. A stress/temperature map for ice of grain size 0·1 mm, using the model-based equations (eqns. (16.2) and (16.3)) (Goodman *et al.*, 1981; courtesy *Phil. Mag.*).

Certain other mechanisms appear in ice. *Dynamic recrystallization* (Section 2.4) has been observed in ice after only a few per cent strain, both in compressive creep (Steinemann, 1958) and during indentation experiments (Barnes *et al.*, 1971), and almost certainly accelerates the rate of creep; the shading shows, roughly, where it occurs. *Grain-boundary melting*, too, accelerates creep. Barnes *et al.* (1971) observed that creep in polycrystals (but not in single crystals) above −8°C was faster than that predicted from creep data obtained below this temperature, and that its activation energy was larger. The onset of this regime is shown by a dash–dot line on the maps. Finally, it is quite feasible that *twinning* can occur in ice, although we know of no direct observations of it, and it is not shown on the maps.

## 16.2   DATA FOR ICE

The crystallographic and thermal data for ice are from standard sources (see, for example, Fletcher, 1970). The shear modulus and its temperature dependence were calculated from the single-crystal data of Dantl (1968) using:

$$\mu = (\tfrac{1}{2}c_{44}(c_{11} - c_{12}))^{\frac{1}{2}}$$

Ice l$_h$ is hexagonal, and slips most readily on the basal plane in the close-packed directions: (0001) $\langle 11\bar{2}0\rangle$. This basal slip provides only two independent slip systems; non-basal slip must occur when polycrystalline ice deforms to large strains. Etch-pit observations (Mugurma and Higashi, 1963) suggest that the most likely systems are the prismatic $\{10\bar{1}0\}$ $\langle1\bar{2}10\rangle$ and the pyramidal $\{11\bar{2}2\}$ $\langle\bar{1}\bar{1}23\rangle$ ones. Both require higher stresses to activate them than the basal system does, and so will control the flow strength.

Self-diffusion in ice has been extensively studied (for a review, see Hobbs, 1974). The measurements show that H$^2$, H$^3$ and O$^{18}$ diffuse at almost the same rate, indicating that the H$_2$O molecule tends to diffuse as a unit, probably by a vacancy mechanism (Ramseier, 1967). In calculating the rate of diffusional flow we have used the lattice-diffusion coefficients of Ramseier (1967) and have estimated the boundary-diffusion coefficients by setting:

$$\delta D_{0b} = 2b\, D_{0v}$$
$$Q_b = 0.6\, Q_v$$

though it must be recognized that this is merely a guess.

The secondary creep of polycrystalline ice has been thoroughly reviewed by Hobbs (1974), Weertman (1973) and Goodman *et al.* (1981). There exist, too, numerous studies of the creep of single crystals

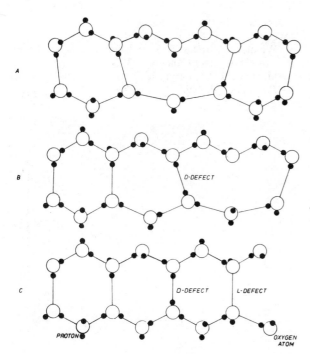

Fig. 16.7. The proton structure of ice, showing a dislocation and defects caused by its motion.

$$\dot{\varepsilon} = A\left(\frac{\sigma}{\mu}\right)^n \exp - (Q_{cr}/RT) \qquad (16.4)$$

or, equivalently,

$$\dot{\gamma} = \left(\sqrt{3}\right)^{n+1} A\left(\frac{\sigma_s}{\mu}\right)^n \exp - (Q_{cr}/RT)$$

(see also Chapter 7, MAR–M200). Here $\dot{\varepsilon}$ and $\sigma$ are the tensile strain-rate and stress, and $Q_{cr}$ is the activation energy for creep. Data for ice below $\sigma_s = 2 \times 10^{-4} \mu$ (1 MN/m²) fit this equation well, with $n = 3$, provided two activation energies and two values of $A$ are used, one pair to describe creep below $-8°C$, the other above. Above $\sigma_s = 2 \times 10^{-4} \mu$ the power-law breaks down, partly because, in simple compression, cracking starts to interfere with steady creep, and partly because of a transition to low-temperature plasticity (Barnes et al., 1971; Section 2.4).

We have fitted the creep data both to the proton-rearrangement controlled glide equations (eqns. (16.2) and (16.3)) which are based on a model of dislocation motion, and to the empirical equation (eqn. (16.4)), using separate creep laws below and above $-8°C$. The resulting parameters are given in the two columns of Table 16.1.

### References for ice

Atkins, A. G., Silvério, A. and Tabor, D. (1966) J. Inst. Met. **94**, 369.

Barnes, P., Tabor, D. and Walker, J. C. F. (1971) Proc. R. Soc. Lond. **A234**, 127.

Bromer, D. J. and Kingery, W. D. (1968) J. Appl. Phys. **39**, 1688.

Butkovich, T. R. (1954) Snow, Ice and Permafrost Research Establishment Research Report **20**, p. 1.

Butkovich, T. R. and Landauer, J. K. (1960) Snow, Ice and Permafrost Establishment Research Report **72**, p. 1.

Dantl, G. (1968) Phys. Kondens. Mat. **7**, 390.

Fletcher, N. H. (1970) The Chemical Physics of Ice. Cambridge University Press. (Cambridge Monographs on Physics).

Frost, H. J., Goodman, D. J. and Ashby, M. F. (1976) Phil. Mag. **33**, 951.

Glenn, J. W. (1955) Proc. R. Soc. **A228**, 519.

Glenn, J. W. (1968) Phys. Kondens. Mat. **7**, 43.

Goodman, D. J., Frost, H. J. and Ashby, M. F. (1981) Phil. Mag. **43**, 665.

Hawkes, I. and Mellor, M. (1972) J. Glaciol. **11**, 103.

Hobbs, P. V. (1974) Ice Physics. Oxford, Clarendon Press.

Kuon, L. G. and Jonas, J. J. (1973) Physics and Chemistry of Ice (eds. Whalley, E., Jones, S. J. and Gold, L. W.). Ottawa: Royal Society of Canada, p. 370.

Mellor, M. and Testa, R. (1969a) J. Glaciol. **8**, 131.

Mellor, M. and Testa, R. (1969b) J. Glaciol. **8**, 147.

Michel, B. (1978) Can. J. Civil. Eng. **5**, 286.

Mugurma, J. and Higashi, A. (1963) J. Phys. Soc. Japan **18**, 1261.

of ice (see Hobbs, 1974, for a review). They differ from those of polycrystalline material because easy slip occurs on the basal plane, and because large changes in the dislocation density usually take place near the beginning of the test, leading to a yield drop (in a tensile test) or a long inverse transient (in creep).

Selected data for ice are shown on Figs. 16.1 and 16.2. They serve to emphasize that creep has been studied only over a limited range of temperature, close to its melting point (0·8 to 1·0 $T/T_M$; $-50$ to $0°C$). This is in contrast with creep data for other materials, little of which is obtained at such high homologous temperatures, and it is one of the reasons that ice is sometimes thought of as a weak material. This view is misleading; as the maps show, ice is surprisingly creep-resistant.

Hardness data (Atkins et al., 1966; Tabor, 1970) have been included in Fig. 16.1. The indent size in a creeping solid depends on the loading time as well as the indentation pressure $p$, and and can be related to the creep properties of the material beneath the indenter. We assume (following Tabor, 1951) that the mean strain beneath the indenter is about 8%, and calculate the effective shear strain-rate as $0·08 \sqrt{3}/t$ (where $t$ is the loading time) and the effective shear stress as $p/3\sqrt{3}$.

The activation energy for the creep of ice differs from that for diffusion, so that the standard power-law creep equation (eqn. (2.21)) is inapplicable. When this is so, we use, instead, the equation:

Ramseier, R. O. (1967) *J. Appl. Phys.* **38**, 2553.

Schultz, H. H. and Knappwost, A. (1968) *Wear* **11**, 3.

Steinemann, Von S. (1958) *Beitrage zur Geologie der Schwiez, Hydrologie* Nr. 10, p. 1.

Tabor, D. (1951) *The Hardness of Metals.* Oxford University Press.

Tabor, D. (1970) *The Hardness of Solids*, Reviews in Physics and Technology **1**, (3), 145.

Weertman, J. (1969) *J. Glaciol.* **8**, 494.

Weertman, J. (1973) *Physics and Chemistry of Ice* (eds. Whalley, E., Jones, S. J. and Gold, L. W.). Ottawa: Royal Society of Canada, p. 362.

Whitworth, R. W., Paren, J. G. and Glenn, J. W. (1976) *Phil. Mag.* **33**, 409.

# CHAPTER 17

# FURTHER REFINEMENTS: TRANSIENT BEHAVIOUR; VERY HIGH AND VERY LOW STRAIN RATES; HIGH PRESSURE

THE MAPS shown so far in this book are based on simplified rate-equations which describe behaviour at steady structure or at steady state (Chapters 1 and 2), ignoring the effects of work-hardening and of transient creep. They ignore, too, deformation mechanisms which become important when strain-rates are very large (among them phonon drag and adiabatic heating) or very small (such as threshold effects associated with diffusional flow), and the influence of large hydrostatic pressures. This was done because the data for these mechanisms are so meagre that their rates, and even their placings on the maps, are often uncertain. But as a better understanding of them becomes available they can be included. In this chapter we discuss the present level of understanding, and show maps illustrating their characteristics.

## 17.1 TRANSIENT BEHAVIOUR AND TRANSIENT MAPS

When strains are small, as they are in service life of most engineering structures, the steady-state approximation is a poor one. At ambient temperatures, metals *work-harden*, so that the flow stress (at a given strain-rate) changes with strain. At higher temperatures most materials show *primary* or *transient creep* as well as steady-state flow; if small strains concern us we cannot neglect their contribution.

Flow at steady structure or steady state is described (Chapter 1) by an equation of the form:

$$\dot{\gamma} = f(\sigma_s, T) \tag{17.1}$$

The state variables $S_i$ (dislocation density and so forth) do not appear as independent variables because they are either fixed, or uniquely determined by $\sigma_s$ and $T$. But during non-steady flow the state variables change with time or strain:

$$S_i = S_i(t \text{ or } \gamma) \tag{17.2}$$

and a more elaborate constitutive law is needed, containing either time $t$ or a strain $\gamma$ as an additional variable:

$$\dot{\gamma} = f(\sigma_s, T, t) \tag{17.3}$$

If this law is integrated to give:

$$\gamma = F(\sigma_s, T, t) \tag{17.4}$$

we can construct maps, still using $\sigma_s/\mu$ and $T/T_M$ as axes, but with contours showing the *strain $\gamma$ accumulated during the time t*. Such maps can include work-hardening, and both transient and steady state creep (Ashby and Frost, 1976).

In doing this, we lose some of the generality of the steady-state maps. The *strain-rate* (which is used as the dependent variable in the steady-state maps) is a differential quantity which depends only on the current structure ($S_i$) of the material. The *strain* (which is the variable we use in the transient maps) is an integral quantity: it depends not on the current structure, but on its entire history. As a result, the constitutive laws we use are largely empirical, and the maps refer to monotonic loading at constant temperature only.

We start by listing the equations used to construct *transient maps* for a stainless steel, and then show three examples of them. They are computed from the data listed in Table 17.1. An application of such maps is described in Chapter 19, Section 19.3.

### Elastic deformation

A stress $\sigma_s$ produces an elastic strain:

$$\gamma = \frac{\sigma_s}{\mu} \tag{17.5}$$

Since we are now concerned with strain (not strain-rate), this elastic contribution must be added to the plastic strain to calculate the total strain.

### Low-temperature plasticity

Polycrystal stress–strain curves can, in general, be fitted to a work-hardening law, which for tensile straining, takes the form:

$$\sigma = \sigma_0 + K\varepsilon_p^m$$

117

TABLE 17.1   Data for Type 316 stainless steel

| Crystallographic and thermal data | | |
|---|---|---|
| Atomic volume, $\Omega$ (m$^3$) | $1 \cdot 21 \times 10^{-29}$ | |
| Burger's vector, $b$ (m) | $2 \cdot 58 \times 10^{-10}$ | (a,b) |
| Melting temperature, $T_M$ (K) | 1810 | |
| *Modulus* | | |
| Shear modulus at 300 K, | | |
| $\mu_0$ (MN/m$^2$) | $8 \cdot 1 \times 10^4$ | (a) |
| Temperature dependence of | | |
| modulus, $\dfrac{T_M}{\mu_0}\dfrac{d\mu}{dT}$ | $-0 \cdot 85$ | |
| *Lattice diffusion* | | |
| Pre-exponential, $D_{0v}$ (m$^2$/s) | $3 \cdot 7 \times 10^{-5}$ | (a) |
| Activation energy, $Q_v$ (kJ/mole) | 280 | |
| *Boundary diffusion* | | |
| Pre-exponential, $\delta D_{0b}$ (m$^3$/s) | $2 \times 10^{-13}$ | (a) |
| Activation energy, $Q_b$ (kJ/mole) | 167 | |
| *Power-law creep* | | |
| Exponent, $n$ | $7 \cdot 9$ | |
| Dorn constant, $A$ | $1 \cdot 0 \times 10^{10}$ | (a) |
| *Obstacle-controlled glide* | | |
| 0 K flow stress, $\hat{\tau}/\mu_0$ | $6 \cdot 5 \times 10^{-3}$ | |
| Pre-exponential, $\dot{\gamma}_0$ (s$^{-1}$) | $10^6$ | (a) |
| Activation energy, $\Delta F/\mu_0 b^3$ | $0 \cdot 5$ | |
| *Work-hardening* | | |
| Initial yield stress, $\sigma_{0s}/\mu_0$ | $7 \cdot 5 \times 10^{-4} - 2 \cdot 2 \times 10^{-7}\,T$ | |
| Hardening exponent, $m$ | $0 \cdot 31 + 6 \times 10^{-5}\,T$ | (c) |
| Hardening constant, $K_s/\mu_0$ | $2 \cdot 5 \times 10^{-3} - 5 \cdot 7 \times 10^{-7}\,T$ | |
| *Transient power-law creep* | | |
| Transient strain, $\gamma_t$ | $0 \cdot 087$ | (d) |
| Transient constant, $C_s$ | $46 \cdot 0$ | |

(a) Except where noted, the data are the same as those given in Table 8.1.

(b) All maps are normalized to 1810°C ($T_M$ for pure iron). This choice is arbitrary; one could use 1680 K (the solidus for 316 stainless steel) thereby expanding the abscissa slightly. The choice influences the computation only via the normalized temperature dependence of the modulus, $T_M/\mu_0$ $(d\mu/dT)$ in evaluating this we used $T_M = 1810$ K for consistency.

(c) Based on the data of Blackburn (1972). The temperature $T$ is in degrees centigrade.

(d) These data are based on an average of values given by Garofalo *et al.* (1963) and Blackburn (1972). A more precise description of the transient creep of 316 stainless steel requires two transient terms (Blackburn, 1972).

where $\varepsilon_p$ is the plastic tensile strain. Inverting, and converting from tensile to equivalent shear stress and strain, gives:

$$
\left.
\begin{aligned}
\gamma &= \left(\frac{\sigma_s - \sigma_{0s}}{K_s}\right)^{1/m} \\
\text{where} \qquad & \\
K_s &= \frac{K}{3^{\left(\frac{m+1}{2}\right)}}
\end{aligned}
\right\} \qquad (17.6)
$$

and $\sigma_{0s}$ is the initial shear strength. Data are available for many metals and alloys. Those for Type 316 stainless steel are listed in Table 17.1.

## Power-law creep

When an intrinsically soft material such as a metal is loaded, dislocations are generated and the material usually work-hardens. If the stress is now held constant these dislocations rearrange, finally attaining a steady structure, and the sample creeps at a steady state. During the rearrangement, the sample creeps faster than at steady state. This *normal transient* has been studied and modelled by Dorn and his co-workers (Amin *et al.*, 1970; Bird *et al.*, 1969; Webster *et al.*, 1969).

On loading an intrinsically hard material (Si, Ge, ice, probably most oxides, silicates, etc.) it appears that too few dislocations are immediately available to permit steady flow. As they move they multiply; during this process the creep-rate increases to that of the steady state. This *inverse transient* has been studied and modelled by Li (1963), Alexander and Haasen (1968), Gilman (1969) and others.

We shall restrict the discussion to that of normal transients. Many engineering texts and papers use a law:

$$
\varepsilon = A\sigma^n t^{1/q} \qquad (17.7)
$$

when $n$ and $q$ are positive and greater than unity and $\sigma$ and $\varepsilon$ are the tensile stress and creep strain, and $t$ is time. Differentiating and rearranging gives laws of the two forms:

$$
\left.
\begin{aligned}
\dot{\varepsilon} &\propto \frac{\sigma^n}{t^p} \quad \text{time hardening} \\[2ex]
\dot{\varepsilon} &\propto \frac{\sigma^n}{\varepsilon^{p'}} \quad \text{strain hardening}
\end{aligned}
\right\} \qquad (17.8)
$$

Though analytically convenient, these law are physically unsound. Both predict infinite creep-rates at zero time (or strain) and no steady state; and, like all integral formulations, they cannot predict the effect of changes of stress (see the discussion of Finnie and Heller, 1959).

Some of these difficulties are removed in the formulation of Dorn and his co-workers (Webster *et al.*, 1969; Amin *et al.*, 1970). They demonstrate remarkable agreement of creep data for Al, Mo, Ag, Fe, Cu, Ni, Nb and Pt with the creep law.

$$
\varepsilon = \varepsilon_t\{1 - \exp - (C\dot{\varepsilon}_{ss}t)\} + \dot{\varepsilon}_{ss}t \qquad (17.9)
$$

where $\dot{\varepsilon}_{ss}$ is the steady-state creep-rate, $\varepsilon_t$ is the total transient strain, and $C$ is a constant. We shall use this equation to construct maps, though it, too, is incapable of describing transient behaviour due to change of stress during a test.

Converted to shear stress and strain-rate, eqn. (17.9) becomes:

$$\gamma = \gamma^T \{1 - \exp - (C_s \dot{\gamma}_{ss} t)\} + \dot{\gamma}_{ss} t \quad (17.10)$$

where $\dot{\gamma}_{ss}$ is the steady-state strain-rate (and thus is identical with the rate used to construct the steady state maps, eqn. (2.21)), $C_s = C/\sqrt{3}$ and $\gamma_t = \sqrt{3}\varepsilon_t$. Data for Type 316 stainless steel are given in Table 17.1.

## Diffusional flow

In a pure, one-component system, there is a small transient associated with diffusional flow. On applying a stress, grain boundary sliding generates an internal stress distribution which decays with time, ultimately reaching the steady-state level. The transient strain must be of the same order as the elastic strain $\sigma_s/\mu$ (since it is associated with the redistribution of internal stresses). The time constant is determined by the relaxation process involved; in this case, diffusion over distances comparable with the grain size, giving a relaxation time of about $\sigma_s/\mu\dot{\gamma}_{ss}$, where $\dot{\gamma}_{ss}$ is the steady-state strain-rate by diffusional flow and thus is identical with the rate used to construct the steady-state maps, eqn. (2.29). The strain then becomes:

$$\gamma = \frac{\sigma_s}{\mu}\left(1 - \exp - \left(\frac{\dot{\gamma}_{ss} t}{\sigma_s/\mu}\right)\right) + \dot{\gamma}_{ss} t \quad (17.11)$$

This transient involves no new data. In alloys, larger transients with larger relaxation times, associated with the redistribution of solute by diffusion, appear. We shall not consider them here.

## Construction of transient maps

Figs. 17.1, 17.2 and 17.3 show transient maps for Type 316 stainless steel with a grain size of 100 $\mu$m. The first shows the areas of dominance of each mechanism after a time of $10^4$ s (about 3 hours);

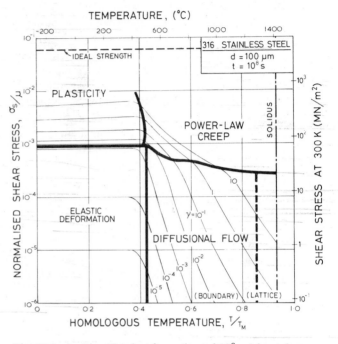

Fig. 17.2. As Fig. 17.1, but for a time of $10^8$ s (about 3 years).

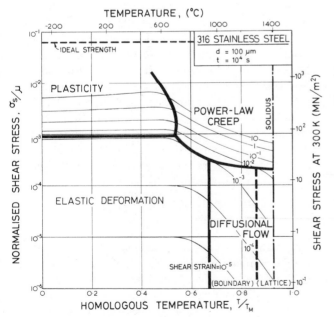

Fig. 17.1. A transient map for Type 316 stainless of grain size 100 $\mu$m, for a time of $10^4$ s (about 3 hours).

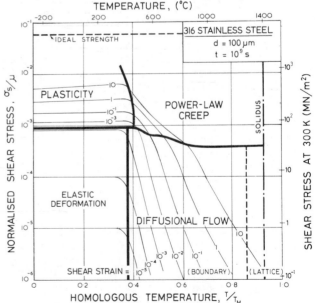

Fig. 17.3. As Fig. 17.1, but for a time of $10^9$ s (about 30 years).

DMM - 1

the second after a time of $10^8$ s (about 3 years), the third after $10^9$ s (30 years).

Within a field, one mechanism is dominant: it has contributed more *strain* to the total than any other. Superimposed on the fields are *contours of constant shear strain*: they show the total strain accumulated in the time to which the map refers: $10^4$, or $10^8$ or $10^9$ s in these examples.

The maps show an *elastic field*; within it, the elastic strain exceeds the total plastic strain (steady plus transient) due to all mechanisms. Above it lies the field of *low-temperature plasticity*; the spacing of the strain contours reflects work-hardening. The *power-law creep* and *diffusional flow* fields occupy their usual relative positions, but the boundaries separating them from each other and from the other mechanisms move as strain accumulates with time, because the various transients have different time constants.

An example of the use of these maps is given in Chapter 19, Section 19.3.

### References for Section 17.1

Alexander, H. and Haasen, P. (1968) *Solid State Physics* **22**, 27.

Amin, K. E., Mukherjee, A. K. and Dorn, J. E. (1970) *J. Mech. Phys. Solids* **18**, 413.

Ashby, M. F. and Frost, H. J. (1976) "The Construction of Transient Maps and Structure Maps". Cambridge University Engineering Department Report CUED/C/ MAPS/TR.26.

Bird, J. E., Mukherjee, A. K. and Dorn, J. E. (1969) *Quantitative Relation between Properties and Microstructure* (eds. Brandon, D. G. and Rosen, A.). Israel University Press, p. 255.

Blackburn, L. D. (1972) *The Generation of Isochronous Stress-Strain Curves*. ASME Winter Meeting, NY, published by ASME.

Finnie, I. and Heller, W. R. (1959) *Creep of Engineering Materials*. McGraw-Hill.

Garofalo, F., Richmond, C., Domis, W. F. and von Gemminger, F. (1963) in *Joint International Conference on Creep*. Inst. Mech. Eng., London, p. 1.

Gilman, J. J. (1969) *Micromechanics of Flow in Solids*. McGraw-Hill.

Li, J. C. M. (1963) *Acta Met.* **11**, 1269.

Webster, G. A., Cox, A. P. D. and Dorn, J. E. (1969) *Met. Sci. J.* **3**, 221.

## 17.2  HIGH STRAIN-RATES

Under impact conditions, and in many metal-working operations (Chapter 19, Section 19.3), strain-rates are high. They lie in the range 1/s to $10^6$/s, well above that covered by the maps shown so far. In this range, *phonon and electron drags* and

*relativistic effects* can limit dislocation velocities at low temperatures; and at high, the power law which describes creep breaks down completely. Further, if the material is deformed so fast that the heat generated by the deformation is unable to diffuse away, then it may lead to a localization of slip known as *adiabatic shear*.

Phonon and electron drags, and power-law breakdown, are easily incorporated into deformation maps by using the rate equations given in Chapter 2. The main problem is that of data: there are very few reliable measurements from which the drag coefficient $B$, and the power-law breakdown coefficient $\alpha'$, can be determined.

### Phonon and electron drag, and the relativistic limit

A moving dislocation interacts with, and scatters, phonons and electrons. If no other obstacles limit its velocity, a force $\sigma_s b$ per unit length causes it to move at a velocity:

$$v = \frac{\sigma_s b}{B} \qquad (17.12)$$

As the temperature increases, the phonon density rises, and the drag coefficient, $B$, increases. Experimental data (for review, see Klahn *et al.*, 1970 and Kocks *et al.*, 1975) show much scatter, but are generally consistent with a drag coefficient which increases linearly with temperature:

$$B = B_e + B_p\left(\frac{T}{300}\right) \qquad (17.13)$$

where $B_e$ is the electron drag coefficient, and $B_p$ is the phonon-drag coefficient at 300 K.

$B$ can be measured by direct observation of dislocation motion during a stress pulse, and can be inferred from measurements of internal friction, and from tensile or compression tests at very high strain-rates. The three techniques, properly applied, show broad agreement (Klahn *et al.*, 1970). For the metals and ionic crystals for which measurements exist, $B$ increases from about $10^{-5}$ Ns/m² at 4·2 K to about $10^{-4}$ Ns/m² at room temperature. Using the Orowan equation (eqn. (2.2)), we find:

$$\dot{\gamma}_D = \frac{\rho b^2 \mu / B_p}{B_e/B_p + T/300}\left(\frac{\sigma_s}{\mu}\right)$$

The high strain-rate experiments of Kumar *et al.* (1968), Kumar and Kumble (1969) and of Wulf (1979) allow the difficult term $\rho b^2 \mu / B_p$ to be evaluated; in all three sets of experiments the result is close to $5 \times 10^6$/s at room temperature. Combining these results gives an approximate rate-equation for phonon plus electron drag:

$$\dot{\gamma}_D = \frac{5 \times 10^6}{0\cdot1 + T/300}\left(\frac{\sigma_s}{\mu}\right) \qquad (17.14)$$

where $\dot{\gamma}$ is measured in units of $s^{-1}$. This equation has been used in constructing the maps described below.

As the dislocation velocity approaches that of sound, the stress required to move it increases more rapidly. This is in part due to the relativistic constriction of the strain field which causes the elastic energy to rise steeply, imposing a limiting velocity, roughly that of shear waves, on the moving dislocation. There is evidence (Kumar et al., 1968) that the mobile dislocation density, too, rises towards a limiting value, so that, from eqn. (2.2) an upper limiting strain-rate, $\dot{\gamma}_{LIM}$, exists which we take to be $10^6$ $s^{-1}$. Then the approach to this limit is described by the relativistic correction to the drag equation:

$$\dot{\gamma}_{RD} = \frac{\dot{\gamma}_D}{(1 + (\dot{\gamma}_D/\gamma_{LIM})^2)^{\frac{1}{2}}} \qquad (17.15)$$

These equations must be regarded as little more than first approximations, and they are fitted to minimal data. But they serve to show, roughly, the regimes on deformation maps in which the mechanisms have significant influence.

### Power-law breakdown

The transition from pure power-law creep to glide-controlled plasticity was described in Chapter 2, Section 2.4. An adequate empirical description is given by eqn. (2.26), which reduces, at low stresses, to the simple power-law of eqn. (2.21). The important new parameter is $\alpha'$, the reciprocal of the normalized stress at which breakdown occurs. Table 17.2 lists approximate values of $\alpha'$ derived from the data plots of previous chapters.

TABLE 17.2  **The power-law breakdown parameter**

| Materials and class | | $\alpha'$ |
|---|---|---|
| f.c.c. metals | (Cu, Al, Ni) | $10^3$ |
| b.c.c. metals | (W) | $2 \times 10^3$ |
| h.c.p. metals | (Ti) | $5 \times 10^2 \rightarrow 10^3$ |
| Alkali halides | (NaCl) | $2 \times 10^3$ |
| Oxides | (MgO, UO$_2$, Al$_2$O$_3$) | $10^3 \rightarrow 2 \times 10^3$ |
| Ice | | $2 \times 10^3$ |

### Adiabatic shear

Analyses of the onset of adiabatic shear vary in generality and complexity, but almost all are based on the same physical idea: that if the loss of strength due to heating exceeds the gain in strength due to the combined effects of strain hardening and of strain-rate hardening (which are locally higher if deformation becomes localized), then adiabatic shear will occur (Zener and Hollomon, 1944; Baron, 1956; Backofen, 1964; Culver, 1973; Argon, 1973; Staker, 1981).

Deformation generates heat, causing the flow strength $\sigma_y$ to fall. Work-hardening, or an increase in strain rate, raises $\sigma_y$. Treatments of diffuse necking (Considère, 1885, for example) assume that instability starts when the rate of softening first exceeds the rate of hardening. If the current flow strength is $\sigma_y$ and all work is converted into heat, then the heat input per unit volume per second is:

$$\dot{q} = \sigma_y \dot{\varepsilon} \qquad (17.16)$$

The flow strength $\sigma_y$ depends on strain, strain-rate and temperature:

$$\sigma_y = \sigma_y(\varepsilon, \dot{\varepsilon}, T)$$

Instability starts when $d\sigma_y = 0$, that is, when:

$$\left(\frac{\partial \sigma_y}{\partial \varepsilon}\right)_{T,\dot{\varepsilon}} d\varepsilon + \left(\frac{\partial \sigma_y}{\partial T}\right)_{\varepsilon,\dot{\varepsilon}} dT + \left(\frac{\partial \sigma_y}{\partial \dot{\varepsilon}}\right)_{\varepsilon,T} d\dot{\varepsilon} = 0 \quad (17.17)$$

This equation is the starting point of most treatments of adiabatic localization (see, for instance, Baron, 1956; Culver, 1973 or Staker, 1981).

Consider first the case when no heat is lost. (For this truly adiabatic approximation to hold, the strain-rate must be higher than the value $\dot{\varepsilon}_A$, calculated below.) At low temperatures we can assume (as Staker, 1981, does) that $(\partial \sigma_y/\partial \varepsilon) \gg (\partial \sigma_y/\partial \dot{\varepsilon})$, so that the instability condition simplifies to:

$$\left(\frac{\partial \sigma_y}{\partial \varepsilon}\right)_{T,\dot{\varepsilon}} = -\left(\frac{\partial \sigma_y}{\partial T}\right)_{\varepsilon,\dot{\varepsilon}} \frac{dT}{d\varepsilon} \qquad (17.18)$$

or, in words: work-hardening is just offset by the fall in strength caused by heating. If heating is uniform:

$$dq = C_p dT = \sigma_y d\varepsilon$$

or

$$\frac{dT}{d\varepsilon} = \frac{\sigma_y}{C_p} \qquad (17.19)$$

If work-hardening is described by a power-law:

$$\sigma_y = K\varepsilon^m \qquad (17.20)$$

we obtain the critical strain for localization under truly adiabatic conditions:

$$\left.\begin{aligned} \varepsilon_c &= \frac{-mC_p}{(\partial \sigma_y/\partial T)_{\varepsilon,\dot{\varepsilon}}} \\ &= \frac{-mC_p T_M}{\psi \sigma_y} \end{aligned}\right\} \qquad (17.21)$$

where $\qquad \psi = \dfrac{T_M}{\sigma_y}\left(\dfrac{\partial \sigma_y}{\partial T}\right)_{\varepsilon, \dot{\varepsilon}}$

The quantity $\psi$ is a dimensionless material property. Typically it lies in the range $-0.5$ to $-6$. The smaller number is appropriate if the yield stress varies with temperature only as the modulus does; the larger number is typical of a material with a strongly temperature-dependent yield strength, such as the b.c.c. metals below $0.1\ T_M$. For many engineering metals at room temperature, its value is about $-3$. Then the critical strain depends mainly on the current strength $\sigma_y$, the work-hardening exponent $m$, the specific heat $C_p$ and the melting point, $T_M$.

Eqn. (17.21) defines a sufficient condition for the onset of adiabatic shear provided no heat is lost from the sample. It is the basis of the approach used by Culver (1973) and Bai (1981), and by Staker (1981) who supports it with data on explosively deformed ($\dot{\varepsilon} \simeq 10^4$/s) AISI 4340 steel, heat-treated to give various combinations of $\sigma_y$ and $m$. But the assumption of no heat loss holds only when the rate of deformation is sufficiently large. So a second condition must also be met: that the strain-rate exceeds a critical value which we now calculate approximately.

Consider a uniform deformation (and thus heat input) but with heat loss to the surroundings at a rate (Carslaw and Jaeger, 1959, or Geiger and Poirier, 1973):

$$\dot{q} = \frac{\alpha k}{R}(T - T_s) \qquad (17.22)$$

Here $k$ is the thermal conductivity and $R$ a characteristic dimension of the sample (the radius of a cylindrical sample for example); $\alpha$ is a constant of order 2; $T$ is the temperature of the sample and $T_s$ is that of the heat sink.

The heat balance equation now becomes:

$$VC_p dT + A\dot{q}dt = V\sigma_y(\varepsilon)d\varepsilon \qquad (17.23)$$

where $V$ is the volume of the sample and $A$ its surface area (Estrin and Kubin, 1980, for example, base their analysis on this equation). Taking $A/V = 2/R$ we find:

$$\frac{dT}{dt} + \frac{2\alpha k}{C_p R^2}(T - T_s) = \sigma_y(\varepsilon)\frac{\dot{\varepsilon}}{C_p} \qquad (17.24)$$

Now the factor $(C_p R^2)/(2\alpha k) = \tau$ is the characteristic time (in seconds) for thermal diffusion to occur and is almost independent of temperature except near 0 K (it depends only on the temperature dependencies of $k$ and $C_p$). Eqn. (17.24) now becomes:

$$\frac{dT}{d\varepsilon} + \frac{1}{\tau\dot{\varepsilon}}(T - T_s) = \frac{\sigma_y(\varepsilon)}{C_p} \qquad (17.25)$$

If the critical strain for adiabatic shear is $\varepsilon_c$, we may write:

$$\frac{dT}{d\varepsilon} \simeq \frac{T - T_s}{\varepsilon_c}$$

Heat loss to the surroundings is significant only if the second term on the left-hand side of eqn. (17.25) becomes comparable to, or larger than, the first; adiabatic conditions therefore apply when:

$$\dot{\varepsilon} > \varepsilon_c/\tau$$

Using eqn. (17.21), we find the minimum strain-rate for adiabatic conditions to be, approximately:

$$\boxed{\dot{\varepsilon}_A = \frac{-2\alpha mk T_M}{\psi R^2 \sigma_y}} \qquad (17.26)$$

At an approximate level, then, adiabatic shear is expected when two conditions are met simultaneously: the strain must exceed the critical strain given by eqn. (17.21) and the strain-rate must exceed the critical strain-rate given by eqn. (17.26). In reality, shear localization can occur even when there is heat loss. Analyses which include it are possible (Estrin and Kubin, 1980) but are complicated, and, at the level of accuracy aimed at here, unnecessary.

### Deformation maps extended to high strain-rates

Using the equations and data developed above, the influence of drag (eqn. (17.14)), or of relativistic effects (eqn. (17.15)) and of adiabatic heating (eqn. (17.26)) can be incorporated into any one of the four types of map shown in Chapter 1. The first two are straightforward; the last requires further explanation.

The parameters $R$ and $\alpha$ which enter eqn. (17.26) are poorly known. But if in some standard state (say, room temperature) it is found that adiabatic localization occurs at a given strain rate $\dot{\varepsilon}_A^0$, then in some other state (say 4·2 K) it will occur at the strain-rate $\dot{\varepsilon}_A$ where:

$$\frac{\dot{\varepsilon}_A}{\dot{\varepsilon}_A^0} = \frac{mk}{\sigma_y \psi}\frac{\sigma_y^0 \psi^0}{m^0 k^0} \qquad (17.27)$$

where the superscript $^0$ refers to the standard state and the unsuperscripted parameters are the values in the other state. We have used the fact that commercially pure titanium at room temperature shows adiabatic localization at strain-rates above $10^2$/s (Winter, 1975; Wulf, 1979; and Timothy, 1982) to construct maps (Fig. 17.4) which show the field in which it will occur. The most useful is that with axes of strain-rate and temperature (Fig. 17.4); it displays most effectively the region in which high strain-rate effects are unimportant. The same

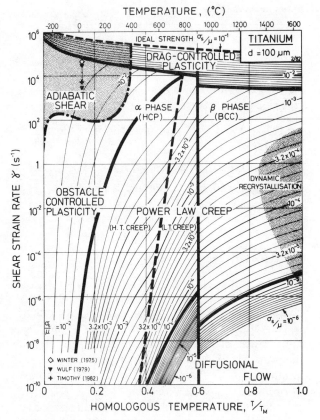

Fig. 17.4. A strain-rate/temperature map for titanium, showing the fields of drag-controlled plasticity and adiabatic shear. The relativistic limit is coincident with the top of the diagram.

information can, of course, be cross-plotted onto the others.

The maps are based on data described in Chapter 6, and on those listed in Table 17.3. In addition to the usual fields, they show a regime of drag-controlled plasticity (eqn. (17.14)), the relativistic limit (eqn. (17.15)) and the regime in which adiabatic heating can cause localization (eqn. (17.26)). Adiabatic localization, of course, can occur in compression or torsion, as well as in tension; but in tension the simple necking instability may obscure the adiabatic localization because it occurs first. Further details and examples are given by Sargent and Ashby (1983).

## References for Section 17.2

*American Institute of Physics Handbook* (1972), 34th edn., Table 4g-9.

Argon, A. S. (1973) in *The Homogeneity of Plastic Deformation.* A.S.M., Metals Park, Ohio, p. 161.

Backofen, W. A. (1964) in *Fracture on Engineering Materials.* A.S.M., Metals Park, Ohio, p. 107.

Bai, Y. (1981) Albuquerque Conference on High Strain Rates (to be published).

Baron, H. G. (1956) *J. Iron and Steel Inst.* **182**, 354.

Carslaw, H. S. and Jaeger, J. C. (1959) *Conduction of Heat in Solids.* Clarendon Press, Oxford.

Considère, M. (1885) *Annls. Ponts Chauss* **9**, 574.

Culver, R. S. (1973) *Metallurgical Effects at High Strain Rates* (eds. Rohde *et al.*). Plenum, p. 519.

Doner, M. and Conrad, H. (1973) *Met. Trans.* **4**, 2809.

Estrin, Y. and Kubin, L. P. (1980) *Scripta Met.* **14**, 1359.

Geiger, G. H. and Poirier, D. R. (1973) *Transport Processes in Metallurgy.* Addison Wesley, Ch. 7, p. 207 *et seq.*

Harding, J. (1975) *Archiwum Mechaniki Stosowanej* Warsaw. **27** (5–6).

Klahn, D., Mukherjee, A. K. and Dorn, J. E. (1970) *Proc. 2nd Int. Conf. on Strength of Metals and Alloys.* A.S.M., p. 951.

Kocks, U. F., Argon, A. S. and Ashby, M. F. (1975) *Prog. Mat. Sci.* **19**, Ch. 3.

Kumar, A., Hauser, F. E. and Dorn, J. E. (1968) *Acta Met.* **16**, 1189.

Kumar, A. and Kumble, R. C. (1969) *J. Appl. Phys.* **40**, 3475.

Sargent, P. M. and Ashby, M. F. (1983) to be published.

Staker, M. R. (1981) *Acta Met.* **29**, 683.

Tanaka, K., Ogowa, K. and Nojima, T. (1978) *IUTAM Symposium on High Strain Rate Deformation* (eds. Kawata, K. and Shiori, J.). Springer, NY.

Timothy, S. P. (1982) Personal communication.

Winter, R. E. (1975) *Phil. Mag.* **31**, 765.

Wulf, G. L. (1979). *Int. J. Mech. Sci.* **21**, 713.

Zener, C. and Holloman, J. H. (1944) *J. Appl. Phys.* **15**, 22.

TABLE 17.3 **Further material data for commercial-purity titanium**

| Property | Value | Reference |
|---|---|---|
| $\alpha'$ | $5 \times 10^2$ | Doner and Conrad (1973) |
| m | $0.11 - 8.6 \times 10^{-5} T$ | Harding (1975) |
| $k\ (Wm^{-1}\ K^{-1})$ | 5·8 (*at* 4·2 K); 33 (*at* 80 K); 20 (*at* 273 K) | *Am. Ins. Phys.* (1972) |
| $-\dfrac{\partial \sigma_y}{\partial T}$ (MPa K$^{-1}$) | $0.94 - (4.7 \times 10^{-4} T)$ for $T < 468$ K<br>$2.4 - (3.6 \times 10^{-3} T)$ for $468 < T < 664$ K<br>$0.0$ for $T > 664$ K | Tanaka *et al.* (1978) |

**TABLE 17.4  Apparent threshold stresses for creep in pure metals**

| Material | Grain size (a) ($\mu m$) | Temp. (K) | $\tau_{tr}$ ($MN/m^2$) (b) | $\tau_{tr}/\mu$    (c) | References |
|---|---|---|---|---|---|
| Cd | $80 \rightarrow 300$ | 300 | 0·2 | $7·5 \times 10^{-6}$ | Crossland (1974) |
| Mg | $25 \rightarrow 170$ | $425 \rightarrow 596$ | $0·88 \rightarrow 0·09$ | $6 \times 10^{-5} \rightarrow 6 \times 10^{-6}$ | Crossland and Jones (1977) |
| Ag | $40 \rightarrow 220$ | $473 \rightarrow 623$ | $1·0 \rightarrow 0·3$ | $4 \times 10^{-5} \rightarrow 1·3 \times 10^{-5}$ | Crossland (1975) |
| Cu | 35 | $523 \rightarrow 573$ | $0·6 \rightarrow 0·4$ | $1·5 \times 10^{-5} \rightarrow 1 \times 10^{-5}$ | Crossland (1975) |
| Ni | 130 | 1023 | 0·2 | $3·5 \times 10^{-6}$ | Towle (1975) |
| Al | $160 \rightarrow 500$ | 913 | 0·08 | $5 \times 10^{-6}$ | Burton (1972) |
| $\alpha$-Fe | $53 \rightarrow 89$ | $758 \rightarrow 1073$ | $0·3 \rightarrow 0·05$ | $6 \times 10^{-6} \rightarrow 1 \times 10^{-6}$ | Towle and Jones (1976) |
| $\beta$-Co | $35 \rightarrow 206$ | $773 \rightarrow 1113$ | $1·4 \rightarrow 0·6$ | $9 \times 10^{-6} \rightarrow 1·7 \times 10^{-5}$ | Sritharan and Jones (1979) |

(a)  Grain size = 1·65 × mean linear intercept.

(b)  Threshold stress in shear ($\tau_{tr} = \sigma_0/\sqrt{3}$ when the tensile threshold $\sigma_0$ is given).

(c)  Shear moduli at the test temperature calculated from data listed in Table 4.1, 5.1 and 6.1.

## 17.3  VERY LOW STRESSES

At very low stresses there is evidence that the simple rate equations for both power-law creep (eqn. (2.21)) and for diffusional flow (eqn. (2.29)) cease to be a good description of experiments. For pure metals the discrepancies are small, but for metallic alloys, and for some ceramics, they can be large. Most commonly, the strain-rate decreases steeply with stress at low stresses, suggesting the existence of a "threshold stress" below which creep ceases, or, more accurately, a stress below which the creep-rate falls beneath the limit of resolution of the experiment (typically $10^{-9}$/s).

Observed "threshold stresses", $\tau_{tr}$, for pure metals, alloys and one ceramic are listed in Tables 17.4 and 17.5. In pure metals, $\tau_{tr}$ increases with decreasing grain size and with decreasing temperature, and is of general order $5 \times 10^{-6}\mu$. In alloys containing a stable dispersion (of $ThO_2$, or of $Y_2O_3$, for example) it can be 10 to 100 times larger.

### Threshold stresses for power-law creep

In single crystals or large-grained polycrystals, power-law creep dominates for all interesting stresses. In pure metals the power law is well-behaved at low stresses. A fine dispersion of second phase introduces an apparent threshold stress, below which creep is too slow to measure with ordinary equipment. The most complete studies are those of Dorey (1968), Humphries et al. (1970) and Shewfelt and Brown (1974, 1977) who introduced up to 9% of $SiO_2$, $Al_2O_3$ and BeO into single crystals of copper. They found that, on introducing

the dispersion, the power $n$ rose from the value characteristic of pure copper (about 5) to a much larger value (10 or more) at low stresses, giving the appearance of a threshold on a $\log\dot\varepsilon$ vs. $\log\sigma$ diagram. The data in these tests extends down to a strain rate of $10^{-6}$/s, at which apparent thresholds of order $10^{-4}\mu$ were observed.

At low temperatures, plastic flow in a dispersion-hardened crystal requires a stress sufficient to bow dislocations between the dispersed particles; if their spacing is $\ell$, this *Orowan stress* is roughly:

$$\frac{\tau_{OR}}{\mu} = \frac{b}{\ell} \tag{17.28}$$

(Chapter 2 and Table 2.1). Shewfelt and Brown demonstrate that if dislocations can climb round particles instead of bowing between them, then flow is possible at a stress as low as one-third of this, so that a creep threshold of $\tau_{tr}/\mu = \frac{1}{3}(b/\ell)$ will appear (where $\mu$, of course, is the modulus at the test temperature). Their data, and those of Dorey (1968) and of Humphries et al. (1970) are consistent with this prediction.

The power-law behaviour of coarse-grained polycrystals is more complicated (Lund and Nix, 1976; Lin and Sherby, 1981). As with single crystals, a stable dispersion introduces an apparent threshold. In some instances it has the characteristics described above: the creep-rate becomes too small to measure below about $\frac{1}{3}\tau_{OR}$. But in others, perhaps because of the stress-concentrating effect of grain boundary sliding or the influence of a substructure created by previous working, the threshold is not quantitatively explained.

In fine-grained polycrystals there is a further aspect. Creep does not stop when power-law creep is suppressed because it is replaced by diffusional flow. This mechanism, too, is retarded by alloying,

**TABLE 17.5  Apparent threshold stresses for creep in alloys and ceramics**

| Material | Vol. fr (%) | Grain size (μm) | Temp. (K) | $\tau_{tr}$ (MN/m²) (a) | $\tau_{tr}/\mu$ (b) | Reference |
|---|---|---|---|---|---|---|
| N–ThO₂ | 2 | 1·3/25 | 1365 | 31 | $6 \times 10^{-4}$ | Whittenberger (1977) |
| Ni–Cr–ThO₂ | 2 | single | 1365 | $32 \to 40$ | $6 \times 10^{-4} \to 7 \times 10^{-4}$ | |
| Ni–Cr–Al–ThO₂ | 2 | 150/490 | 1365 | 30 | $5 \times 10^{-4}$ | |
| Ni–Cr–Al–ThO₂ | 2 | 120/1200 | 1365 | 24 | $4 \times 10^{-4}$ | |
| INCONAL–MA–754–Y₂O₃ | 0·6 | 115/130 | 1365 | $10 \to 42$ | $2 \times 10^{-4} \to 8 \times 10^{-4}$ | |
| Ni–Cr–Al–Y₂O₃ | 1 | 310/2500 | 1366 | 24 | $4 \times 10^{-4}$ | Whittenberger (1981) |
| INCONAL–MA–757–Y₂O₃ | 0·6 | 70/265 | $1144 \to 1477$ | $12 \to 52$ | $2 \times 10^{-4} \to 9 \times 10^{-4}$ | |
| Ni–Cr–ThO₂ | 2 | — | $1173 \to 1473$ | — | $9 \times 10^{-4}$ | Lund and Nix (1976) |
| Cu–SiO₂ | $2·3 \to 9$ | single | $1073 \to 1300$ | $4 \to 7$ | $1 \times 10^{-4} \to 1·8 \times 10^{-4}$ | Shewfelt and Brown (1974) |
| Cu–Al₂O₃ | $0·5 \to 1·5$ | 13 | 1173 | $0·3 \to 0·9$ | $1 \times 10^{-5} \to 3·4 \times 10^{-5}$ | Burton (1971) |
| Cu–SiO₂ | — | — | $816 \to 1118$ | $\sim 0·3$ | $\sim 10^{-5}$ | Clegg and Martin (1982) |
| Au–Al₂O₃ | 6·1 | — | $\sim 1000$ | 0·2 | $1·3 \times 10^{-5}$ | Sautter and Chen (1969) |
| Au–Al₂O₃ | 7·9 | — | $\sim 1000$ | 0·3 | $1·7 \times 10^{-5}$ | |
| Stainless steel–Nb (C,N) | 0·3 | 14 | $\sim 1000$ | $1 \to 2·6$ | $2 \times 10^{-5} \to 5 \times 10^{-5}$ | Crossland and Clay (1977) |
| UO₂–voids | 3 | 7 | $1523 \to 1723$ | $2 \to 8·5$ | $3 \times 10^{-5} \to 1 \times 10^{-4}$ | Burton and Reynolds (1973) |

(a) Threshold stress in shear ($\tau_{tr} = \sigma_0/\sqrt{3}$ when the tensile threshold $\sigma_0$ is given).
(b) Shear moduli at the test temperature calculated from data listed in Tables 4.1, 7.1, 8.1 and 13.1.

but to a lesser degree than power-law creep. For this reason, diffusional flow at low stresses is of particular interest to us here.

### Diffusional flow at low stresses

When a grain or phase boundary acts as a sink or source for a diffusive flux of atoms, the flux has its sources and sinks in *boundary dislocations* which move in a non-conservative way in the boundary plane. Electron microscopy reveals dislocations of the appropriate kind (Gleiter, 1969; Schober and Balluffi, 1970) which multiply by the action of sources, so that their density increases with the stress up to about $10^7$ m/m$^2$. Both theory and experiment show that their Burger's vectors $b_b$ are not lattice vectors (they are smaller) and therefore that they are constrained to remain in the boundary plane when they move.

The model for diffusional flow must now be modified in two ways. First, one must ask how its rate is changed by the presence of discrete sinks and sources. The answer (Arzt *et al.*, 1982) is that the change is negligible unless the density of boundary dislocation is exceedingly low; we shall ignore it here. But second, one must ask how the mobility of defects influences the creep-rate. The answer is that in pure metals of normal grain size, the mobility is high; and the creep-rate is unaffected; that is why much data for pure metals follow the simple diffusion-controlled rate-equation (2.29). Only at very low stresses does the self-stress of the boundary dislocations limit their mobility, giving a threshold. But in alloys or compounds, particularly those containing a fine, stable dispersion of a second phase, the mobility of boundary dislocations is much reduced. Then anomalously slow creep, and large apparent threshold stresses, are found. We now develop these ideas a little further.

### The threshold stress in pure materials

A boundary dislocation of the kind of interest here cannot end within a solid. It must either be continuous, or link, at nodes (at boundary triple-junctions, for instance) to one or more other dislocations such that the sum of the Burger's vectors flowing into each node is zero. If this line now tracks across the undulating surface making up the boundary between grains, its length fluctuates, and it may be pinned by the nodes. If the self-energy of the line is:

$$E_s \simeq \frac{Gb_b^2}{2} \qquad (17.28)$$

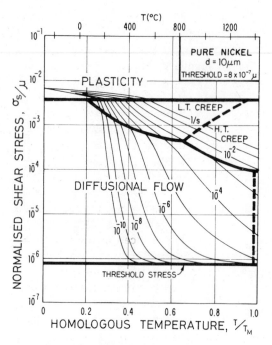

Fig. 17.5. A map for pure nickel with a threshold $\tau_{tr}$ of $8 \times 10^{-7}$ $\mu$. The self-energy of the boundary dislocations will introduce a threshold of general magnitude $\hat{\tau}/\mu = b_b/d$.

then the length fluctuation, or the pinning, will introduce a threshold for dislocation motion. Both are easily modelled (e.g. Burton, 1972; Arzt *et al.*, 1982) and lead in an obvious way to the result:

$$\frac{\tau_{tr}}{\mu} = C\frac{b_b}{d} \qquad (17.29)$$

where $d$ is the grain size and $C$ is a constant close to 1. Taking $b_b$ to be $10^{-10}$ m (about one-third of a lattice Burger's vector), we predict threshold stresses between $2 \times 10^{-7}$ $\mu$ (for a grain size of 500 $\mu$m) and $10^{-5}$ $\mu$ (for a grain size of 10 $\mu$m). They are a little smaller than those reported in Table 7.4, which probably reflect a combination of this with impurity drag, discussed next.

The influence of this threshold on the map for nickel is shown in Fig. 17.5. It was computed by replacing $\sigma_s$ in the diffusional flow equation (eqn. (2.29)) by $(\sigma_s - \tau_{tr})$. The map has been extended downwards by a decade in stress to allow the threshold to be seen. Over a wide range of stress, creep follows eqn. (2.29). Only near $\tau_{tr}$ is any change visible.

### Mobility-controlled diffusional flow in alloys and compounds

A solid solution, or dissolved impurities, can segregate to a boundary dislocation; then, when the dislocation moves, the segregant may diffuse along with it, exerting a *viscous drag* which can limit its mobility. Discrete particles of a second phase, too, interact with a boundary dislocation, pinning it, and introducing a *threshold stress* in much the same way that particles pin lattice dislocations and inhibit power-law creep. But because the average Burger's vector of the boundary dislocation is smaller (by a factor of perhaps 3) than that of a lattice dislocation, the extent of the segregation is less, and the pinning force is smaller. For this reason, strengthening mechanisms suppress power-law creep more effectively than diffusional flow, causing the latter to become dominant.

When the boundary-dislocation motion is impeded, part of the applied stress (or of the chemical-potential gradient it generates) is required to make the dislocations move, and only the remaining part is available to drive diffusion. The creep-rate is then slower than that given by eqn. (2.29), to an extent which we now calculate.

The velocity $v$ of a boundary dislocation is related to the force $F$ per unit length acting on it through a mobility equation (cf. eqn. (2.4)):

$$v = MF = M\sigma_s^i b_b \qquad (17.30)$$

Here $\sigma_s^i$ is the part of the applied stress required to move the boundary dislocations. (We assume that, in pure shear, the boundaries are subjected to normal tractions of $\pm\sigma_s$.) The strain-rate is related to $b_b$, $v$ and the density of boundary dislocations, $\rho_b$ (a length per unit area) by:

$$\dot{\gamma} = \frac{2\rho_b b_b v}{d}$$

(cf. eqn. (2.2)), where the factor 2 appears because the motion produces a normal, not a shear strain. Together these equations become:

$$\dot{\gamma} = \frac{2\rho_b b_b^2 M}{d}\sigma_s^i \qquad (17.31)$$

This strain-rate must match that produced by the transport of matter across the grain, by diffusion, driven by the remaining part of the stress, $\sigma_s - \sigma_s^i$. From eqn. (2.29), this is:

$$\dot{\gamma} = \frac{42 D_{\text{eff}}\Omega}{kTd^2}\left(\sigma_s - \sigma_s^i\right) \qquad (17.32)$$

Eliminating $\sigma_s^i$, and solving for $\dot{\gamma}$ gives:

$$\dot{\gamma} = \frac{\dfrac{42 D_{\text{eff}}\sigma_s\Omega}{kTd^2}}{\left(1 + \dfrac{21 D_{\text{eff}}\Omega}{kTd\rho_b b_b^2 M}\right)} \qquad (17.33)$$

This is the basic equation for diffusional flow when boundary dislocation mobility is limited (Ashby,

1969, 1972). Note that when $M$ is large, the equation reduces to the classical diffusional flow law (eqn. (2.29)); but when $M$ is small, it reduces to eqn. (17.31) with $\sigma_s^i = \sigma_s$. In particular, if boundary dislocations are pinned ($M = 0$), creep stops. To progress further we require explicit equations for $\rho_b$ and $M$.

The most reasonable assumption for the density of boundary dislocations, $\rho_b$, is that it increases linearly with stress (Burton, 1972; Ashby and Verrall, 1973):

$$\rho_b = \alpha\left(\frac{\sigma_s}{\mu b_b}\right) \qquad (17.34)$$

where $\alpha$ is a constant of order unity. This is simply the two-dimensional analog of eqn. (2.3) for the dislocation density in a crystal.

When a solute or impurity drag restricts dislocation motion, the drag will obviously increase as the atom fraction $C_0$, of solute or impurities, increases. Arzt *et al.* (1982) show that:

$$M = \frac{D_s\Omega}{\beta kTb_b^2 C_0} \qquad (17.35)$$

where $D_s$ is the diffusion coefficient for the solute or impurity, and $\beta$ is a constant (for a given impurity) of between 1 and 20. As an example of the influence of impurity drag, we set:

$$M = \frac{D_b\Omega^{1/3}}{kT}$$

(i.e. we set $D_s = D_b$ and $\beta C_0 b_b^2/\Omega^{2/3} = 1$) and compute maps for nickel using eqn. (17.33), and eqn.

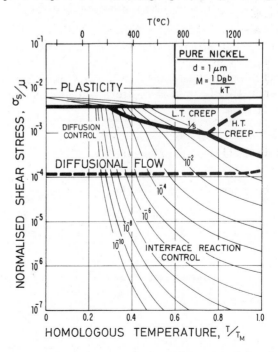

Fig. 17.6. A stress/temperature map for nickel of grain size 1 $\mu$m, with limited boundary-dislocation mobility.

(17.34) for diffusional flow. The results, for two grain sizes, are shown in Figs. 17.6 to 17.9. The figures illustrate how the field of diffusion-controlled flow, for which $\dot\gamma \propto \sigma_s$, is replaced at low stresses by one of mobility-controlled flow ("interface reaction control") for which $\dot\gamma \propto \sigma_s^2$. The size of this new field increases as the grain size decreases, and for a sufficiently large concentration $C_0$, or a sufficiently small diffusion coefficient $D_s$, it can completely replace diffusion-controlled flow.

This field rarely appears in pure metals. The small grain size necessary to observe it will not survive the tendency to grain growth at high temperatures;

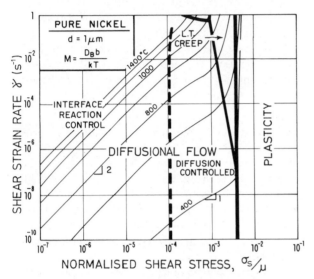

Fig. 17.7. A strain-rate/stress map for nickel of grain size 1 $\mu$m, with limited boundary-dislocation mobility.

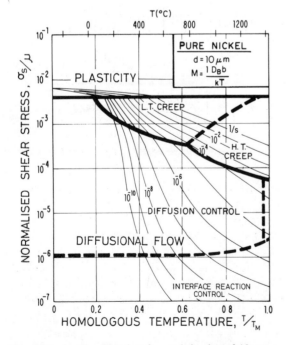

Fig. 17.8. As Fig. 17.6, but for a grain size of 10 $\mu$m.

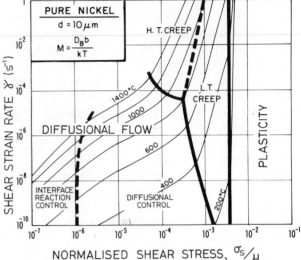

Fig. 17.9. As Fig. 17.7, but for a grain size of 10 $\mu$m.

and the mobility of boundary dislocations is too high. Impurities lower the mobility and suppress grain growth; larger alloying additions, leading to a precipitate or a duplex microstructure, do so even more effectively. For this reason the source and sink mobility may dictate the behaviour of fine-grained two-phase superplastic alloys, which commonly show a sigmoidal stress/strain-rate relation like that shown in Fig. 17.7.

A fine dispersion of a stable phase pins boundary dislocations, so that below a threshold stress their mobility is zero. The diffusional creep behaviour is then given by eqn. (17.33) with an appropriate (non-linear) expression for $M$. But at a useful level of approximation we can think of the threshold stress as subtracting from the applied stress. Then the creep-rate is given by the classical diffusional flow equation (eqn. (2.29)) with $\sigma_s$ replaced by $(\sigma_s - \tau_{tr})$; maps computed in this way are shown as Figs. 7.5 and 17.5. We anticipate that $\tau_{tr}$ is about one-third of the Orowan stress for boundary dislocations:

$$\frac{\tau_{tr}}{\mu} \approx \frac{1}{3}\frac{b_b}{\ell} \qquad (17.36)$$

for the reasons given earlier in this section. Note that, because the boundary Burger's vector $b_b$ is less than that of lattice, this threshold stress is lower, by a factor of perhaps 3, than that for power-law creep. Dispersion-hardened materials show apparent threshold stresses of general order $10^{-4}\,\mu$ (Table 17.5), implying an obstacle spacing of about 3000 $b_b$.

Apparent threshold stresses are often observed to be temperature-dependent. This can be understood (Arzt $et\ al.$, 1982) as a superposition of discrete obstacles and an impurity, or solute, drag: the drag causes the stress at which the strain-rate falls below the limit of resolution of the equipment to depend on temperature.

### References for Section 17.3

Arzt, E., Ashby, M. F. and Verrall, R. A. (1982) CUED Report CUED/t/MATS/TR.60, Cambridge University.
Ashby, M. F. (1969) *Scripta Met.* **3**, 837; (1972) *Surface Sci.* **31**, 498.
Ashby, M. F. and Verrall, R. A. (1973) *Acta Met.* **21**, 149.
Burton, B. (1971) *Met. Sci. J.* **5**, 11; (1972a) *Mat. Sci. Eng.* **10**, 9; (1972b) *Phil. Mag.* **25**, 645.
Burton, B. and Reynolds, G. L. (1973) *Acta Met.* **21**, 1073.
Clegg, W. J. and Martin, J. W. (1982) *Met. Sci.* **16**, 65.
Crossland, I. G. (1974) *Phys. Stat. Sol.* **a23**, 231; (1975) in *Physical Metallurgy of Reactor Fuel Elements* (eds. Harris, J. E. and Sykes, E. C.). Metals Society, London, p. 66.
Crossland, I. G. and Clay, B. D. (1977) *Acta Met.* **25**, 929.
Crossland, I. G. and Jones, R. B. (1977) *Mat. Sci.* **11**, 504.
Gleiter, H. (1969) *Acta Met.* **17**, 565.
Hayward, E. R. and Greenough, A. P. (1960) *J. Inst. Met.* **88**, 317.
Lin, J. and Sherby, O. D. (1981) *Res Mechanica* **2**, 251.
Lund, R. W. and Nix, W. D. (1976) *Acta Met.* **24**, 469.
Sautter, F. K. and Chen, E. S. (1969) Proc. 2nd Bolton Landing Conf. on *Oxide Dispersion Strengthening.* Gordon & Breach, p. 495.
Schober, T. and Balluffi, R. W. (1970) *Phil. Mag.* **21**, 109.
Shewfelt, R. S. W. and Brown, L. M. (1974) *Phil Mag.* **30**, 1135; (1977) *Phil. Mag.* **35**, 945.
Sritharan, T. and Jones, H. (1979) *Acta Met.* **27**, 1293.
Towle, D. J. (1975) Ph.D. thesis, University of Sheffield, England.
Towle, D. J. and Jones, H. (1976) *Acta Met.* **24**, 399.
Whittenberger, J. D. (1977) *Met. Trans.* **8A,** 1155; (1981) *Met. Trans.* **12A**, 193.

### 17.4  THE EFFECT OF PRESSURE ON PLASTIC FLOW

In engineering design it is normal to assume that it is the shearing, or *deviatoric*, part of the stress field which causes flow (see Chapter 1). There is some justification for this: neither low-temperature plasticity nor creep are measurably affected by pressures of less than $K/100$, where $K$ is the bulk modulus. But when the pressure exceeds this value, the flow strength is increased and the creep rate is slowed. It is still the deviatoric part of the stress field which causes flow, but the material properties (such as $\hat{\tau}$, $\Delta F$, $A$ and $Q$) have been altered by the pressure. Pressure must then be regarded, with temperature and shear stress, as an independent variable. In certain geophysical problems pressure is as important a variable as temperature: at a depth of 400 km below the earth's surface, for example, the pressure is about $K/10$, and its influence on material properties is considerable.

The influence of pressure on plastic flow has not been studied in anything like the same detail as that of temperature. But enough information exists to piece together a fairly complete picture. In this section we summarize the results required to incorporate pressure as an independent variable in computing deformation maps. Examples of their use are given in the final Case Study of this book (Section 19.8).

### Effect of pressure on the ionic volume, lattice parameter and moduli

In a linear-elastic solid of bulk modulus $K_0$, the atomic or ionic volume varies with pressure as:

$$\Omega = \Omega_0 \exp\{-(p - p_0)/K_0\}, \qquad (17.37)$$

and the lattice parameter $a$ (and the Burger's vector $b$) as:

$$a = a_0 \exp\{-(p - p_0)/3K_0\}, \qquad (17.38)$$

where $\Omega_0$, $K_0$ and $a_0$ are the values at atmospheric pressure, $p_0$. Values of $K_0$ for elements and compounds are tabulated by Huntington (1958) and Birch (1966).

To first-order, the moduli increases linearly with pressure and decrease linearly with temperature. We write this in the form:

$$\mu = \mu_0 \left\{ 1 + \left[ \frac{T_M}{\mu_0} \frac{d\mu}{dT} \right] \frac{(T - 300)}{T_M} \right\} + \left[ \frac{d\mu}{dp} \right] (p - p_0) \quad (17.39)$$

$$K = K_0 \left\{ 1 + \left[ \frac{T_M}{K_0} \frac{dK}{dT} \right] \frac{(T - 300)}{T_M} \right\} + \left[ \frac{dK}{dp} \right] (p - p_0) \quad (17.40)$$

where $p_0$ is atmospheric pressure, which, for almost all practical purposes we can ignore.

The coefficients in the square brackets are dimensionless. Table 17.6 lists means and standard deviations of the temperature and pressure coefficients for a number of cubic elements and compounds. Most are metals, though data for alkali halides and oxides are included. The coefficients are approximately constant; when no data are available for a specific material, these constant values may reasonably be used.

TABLE 17.6   Temperature and pressure coefficients of the
moduli and yield strength

| Coefficient | Mean and standard deviation | Source of data |
|---|---|---|
| $\left[\dfrac{T_M}{\mu_0}\dfrac{d\mu}{dT}\right]$ | $-0.52 \pm 0.1$ | This book, Tables 4.1, etc. |
| $\left[\dfrac{T_M}{K_0}\dfrac{dK}{dT}\right]$ | $-0.36 \pm 0.2$ | Huntington (1958) |
| $\left[\dfrac{d\mu}{dp}\right]$ | $1.8 \pm 0.7$ | Huntington (1958) |
| $\left[\dfrac{dK}{dp}\right]$ | $4.8 \pm 1$ | Birch (1966) |
| $\left[\dfrac{K_0}{\mu_0}\dfrac{d\mu}{dp}\right]$ | $5 \pm 3$ | |
| $\left[\dfrac{K_0}{\sigma_{s0}}\dfrac{d\sigma_s}{dp}\right]$ | $8 \pm 2$ | Richmond and Spitzig (1980) |

### Effect of pressure on low-temperature plasticity

Experiments on a number of steels (Spitzig *et al.*, 1975, 1976; Richmond and Spitzig, 1980) have characterized the effect of a hydrostatic pressure on the flow strength. To an adequate approximation it increases linearly with pressure.

$$\sigma_s = \sigma_{s0}\left(1 + \left[\frac{K_0}{\sigma_{s0}}\frac{d\sigma_s}{dp}\right]\frac{p}{K_0}\right) \qquad (17.41)$$

where (for pure iron and five steels) the dimensionless constant in square brackets has values in the range 6 to 11. The effect is far too large to be accounted for by the permanent volume expansion associated with plastic flow, and must be associated instead with a direct effect of pressure on the motion of dislocations.

This pressure-dependence can be accounted for almost entirely by considering the effect of pressure on the activation energies $\Delta F_p$ and $\Delta F$ and the strengths $\hat{\tau}_p$ and $\hat{\tau}$ which appear in eqns. (2.9) and (2.12). It is commonly found (see Kocks *et al.*, 1975, for a review) that the activation energy for both obstacle and lattice-resistance controlled glide scales as $\mu b^3$ and the strengths $\hat{\tau}$ scale as $\mu$. For the b.c.c. metals, for example, the activation energy is close to $0.07\,\mu b^3$ and the flow stress at absolute zero is close to $0.01\,\mu$ (Table 5.1, Chapter 5). As already described, both $\mu$ and $b$ depend on pressure, $\mu$ increasing more rapidly than $b^3$ decreases. Pressure, then, has the effect of raising both the activation energies ($\Delta F$) and the strengths ($\hat{\tau}$).

At absolute zero the shear stress required to cause flow is simply $\hat{\tau}$. Using eqn. (17.39) for the modulus, and neglecting $p_0$, we find by inserting the above proportionalities into eqn. (2.9) and inverting:

$$\sigma_s = \sigma_{s0}\left(1 + \left[\frac{K_0}{\mu_0}\frac{d\mu}{dp}\right]\frac{p}{K_0}\right) \qquad (17.42)$$

which has the form of eqn. (17.41), with:

$$\left[\frac{K_0}{\sigma_{s0}}\frac{d\sigma_s}{dp}\right] = \left[\frac{K_0}{\mu_0}\frac{d\mu}{dp}\right]. \qquad (17.43)$$

Values of both dimensionless quantities are listed in Table 17.6 for variety of materials. The calculated values range from 2 to 9 compared with the measured coefficient of 6 to 11, but the measurements, of course, were made at room temperature—about $0.2\,T_M$ for many of the listed materials. When a correction for this is made (Ashby and Verrall, 1977), closer agreement is obtained.

There are other contributions to the pressure-dependence of low-temperature plasticity, but they appear to be small. The presence of a dislocation expands a crystal lattice, partly because the core has a small expansion associated with it (about $0.5\,\Omega$ per atom length) and partly because the non-linearity of the moduli causes any elastic strain-field to produce an expansion (Seeger, 1955; Lomer, 1957; Friedel, 1964). It is this second effect which is, in general, the more important. The volume expansion per unit volume of uniformly strained material is roughly:

$$\Delta V = \tfrac{3}{2}\Delta E^{el}/\mu \qquad (17.44)$$

where $\Delta E^{el}$ is the elastic energy associated with the strain-field. If the activation energy which enters the rate-equations is largely elastic in origin (as it appears to be) then during activation there is a small temporary increase in volume, $\Delta V$. A pressure further increases the activation energy for glide by the amount $p\Delta V$. But when this contribution is compared with that caused by the change in moduli with pressure (eqn. (17.42)) it is found to be small. The influence of pressure on the low-temperature plasticity, then, is adequately described by eqn. (17.41). A pressure of $0.1\,K_0$ roughly doubles the flow strength.

### The effect of pressure on creep

There have been a limited number of creep tests in which pressure has been used as a variable; they have been reviewed by McCormick and Ruoff (1970). Typical of them are the observations of Chevalier *et al.* (1967), who studied the creep of indium under pressure. When the pressure was switched between two fixed values the creep-rate changed sharply but reversibly, returning to its earlier value when the additional pressure was removed.

When creep is glide-controlled, pressure should influence it in the way described in the last subsection. When, instead, it is diffusion-controlled, the main influence of pressure is through its influence on the rate of diffusion. Pressure slows diffusion because it increasess the energy required for an atom to jump from one site to another, and because it may cause the vacancy concentration in the solid to decrease. The subject has been extensively reviewed by Lazarus and Nachtrieb (1963), Girifalco (1964) and Peterson (1968); detailed calculations are given by Keyes (1963).

The application of kinetic theory to self-diffusion by a vacancy mechanism (see, for example, Shewmon, 1963) gives, for the diffusion coefficient:

$$D = \alpha a^2 n_v \Gamma, \tag{17.45}$$

where $\alpha$ is a geometric constant of the crystal structure, and $a$ is the lattice parameter (weakly dependent on pressure in the way described by eqn. (17.38)). The important pressure-dependencies are those of the atom fraction of vacancies, $n_v$, and the frequency factor, $\Gamma$. In a pure, one-component system, a certain atom fraction of vacancies is present in thermal equilibrium because the energy ($\Delta G_f$ per vacancy) associated with them is offset by the configurational entropy gained by dispersing them in the crystal. But in introducing a vacancy, the volume of the solid increases by $V_f$, and work $pV_f$ is done against any external pressure, $p$. A pressure thus increases the energy of forming a vacancy without changing the configurational entropy, and because of this the vacancy concentration in thermal equilibrium decreases. If we take:

$$\Delta G_f = \Delta G_{f0} + pV_f, \tag{17.46}$$

where the subscript "0" means "zero pressure", then:

$$n_v = \exp\{-(\Delta G_{f0} + pV_f)/kT\} \tag{17.47}$$

A linear increase in pressure causes an exponential decrease in vacancy concentration.

It is the nature of the metallic bond that the metal tends to maintain a fixed volume per free electron. If a vacancy is created by removing an ion from the interior and placing it on the surface, the number of free electrons is unchanged, and the metal contracts. For this reason, the experimentally measured values of $V_f$ for metals are small: about $\frac{1}{2}\Omega_0$ where $\Omega_0$ is the atomic volume. Strongly ionic solids can behave in the opposite way: the removal of an ion exposes the surrounding shell of ions to mutual repulsive forces. The vacancy becomes a centre of dilatation, and $V_f$ is large: up to $2\Omega_0$ where $\Omega_0$ is the volume of the ion removed. There are no data for oxides or silicates, but when the bonding is largely covalent one might expect the close-packed oxygen lattice which characterizes many of them to behave much like an array of hard sphere. Forming a vacancy then involves a volume expansion of $\Omega_0$, the volume associated with an oxygen atom in the structure.

There is a complicating factor. In a multi-component system vacancies may be stabilized for reasons other than those of entropy. Ionic compounds, for instance, when doped with ions of a different valency, adjust by creating vacancies of one species or interstitials of the other to maintain charge neutrality; pressure will not, then, change the vacancy concentration significantly. Oxides may not be stoichiometric, even when pure, and the deviation from stoichiometry is often achieved by creating vacancies on one of the sub-lattices. The concentration of these vacancies is influenced by the activity of oxygen in the surrounding atmosphere, so that the partial pressure of oxygen determines the rates of diffusion. For these reasons it is possible that the quantity $V_f$ in eqn. (17.46) could lie between 0 and $2\Omega_0$.

The jump frequency, too, depends on pressure. In diffusing, an ion passes through an activated state in which its free energy is increased by the energy of motion, $\Delta G_m$. The frequency of such jumps is:

$$\Gamma = v \exp(-\Delta G_m/kT), \tag{17.48}$$

where $v$ is the vibration frequency of the atom in the ground state (and is unlikely to depend on pressure). In moving, the ion distorts its surroundings, temporarily storing elastic energy. If all the activation energy of motion is elastic, then (by the argument leading to eqn. (17.44)) it is associated with volume expansion,

$$V_m = 3\Delta G_m/2\mu$$

per unit volume. Taking the activation energy for motion to be 0·4 of the activation energy of diffusion, we find, typically, $V_m = 0.2$–$0.4\,\Omega_0$, where $\Omega_0$ is the volume of the diffusing ion. Experimentally, $V_m$ is a little smaller than this, suggesting that the activation energy is not all elastic.

Assembling these results we find:

$$D = D^0(1 - 2p/3K)\exp(-pV^*/kT)$$
$$\approx D^0 \exp(-pV^*/kT) \tag{17.49}$$

where $D^0 = \alpha(a_0)^2 v \exp-(\Delta G_f + \Delta G_m)/kT$ is the diffusion coefficient under zero pressure, and:

$V^* = V_f + V_m$ for intrinsic diffusion;

$V^* = V_m$        for extrinsic diffusion.

Because experiments are difficult, there are few measurements of $V^*$, and these show much scatter. They have been reviewed by Lazarus and Nachtrieb (1963), Keyes (1963), Girifalco (1964), Goldstein *et al.* (1965), Brown and Ashby (1980) and Sammis and Smith (1981). Some results are summarized in Table 17.7: they lie between 0 and $2\Omega_0$, where $\Omega_0$ is the volume of the diffusing ion.

When creep is diffusion-controlled, the creep-rate should scale as the diffusion coefficient. The power-law creep equation (eqn. (2.21)) then becomes:

$$\dot{\gamma}_4 = \frac{A_2 D_{eff}^0 \mu b}{kT}\left(\frac{\sigma_s}{\mu}\right)^n \exp -\left(\frac{pV^*}{kT}\right) \quad (17.50)$$

where $\mu$ and $b$ are the values at the pressure $p$. The diffusional flow equation (eqn. (2.29)), similarly, becomes:

$$\dot{\gamma}_7 = \frac{42\sigma_s\Omega}{kTd^2}D_{eff}^0 \exp -\left(\frac{pV^*}{kT}\right) \quad (17.51)$$

where $\Omega$ is the atomic or ionic volume at the pressure $p$.

When pressures are large the creep-rate depends strongly on pressure: a pressure of $0.1\ K_0$ reduces the creep rate, typically, by a factor of 10. The principal contribution is that of the term involving $V^*$, so that the "activation volume" for creep, defined by:

$$V_{cr}^* = \frac{d(\ln\dot{\gamma})}{dp}$$

should be close to that for diffusion, $V^*$. Table 17.7 shows that this is so.

**TABLE 17.7 Activation volumes for diffusion and creep**

| Material | Structure | $V^*/\Omega_0$ for diffusion | $V_{cr}^*/\Omega_0$ for creep |
|---|---|---|---|
| Pb | f.c.c. | $0.8 \pm 0.1$ | 0.76 |
| Al |  | — | 1.35 |
| Na | b.c.c. | $0.4 \pm 0.2$ | 0.41 |
| K |  | — | 0.54 |
| In | h.c.p. | — | 0.76 |
| Zn |  | $0.55 \pm 0.2$ | 0.65 |
| Cd |  | — | 0.63 |
| AgBr | Rock salt | $1.9 \pm 0.5$ | 1.9 |
| Sn | Tetragonal | $0.3 \pm 0.1$ | 0.31 |
| P |  | $0.5 \pm 0.1$ | 0.44 |

Data from Lazarus and Nachtrieb (1963), Goldstein *et al.* (1965) and McCormick and Ruoff (1970).

## Incorporation of pressure dependence into deformation maps

The information summarized above allows the effect of pressure on material properties to be incorporated into deformation maps. The quantities $b$, $\Omega$, $\mu$ are calculated for the given pressure, using eqns. (17.37) to (17.39). Diffusion coefficients are corrected for pressure using eqn. (17.49), or (alternatively) the creep equations are modified to eqns. (17.50) and (17.51). Finally, the flow stress for low-temperature plasticity is replaced by that given in eqn. (17.41). The final Case Study (Section 19.8) shows maps computed in this way.

### References for Section 17.4

Ashby, M. F. and Verrall, R. A. (1977) *Phil. Trans. R. Soc. Lond.* **A288**, 59.

Birch, F. (1966) *Handbook of Physical Constants.* The Geological Society of America Memoir 97, Section 7.

Brown, A. M. and Ashby, M. F. (1980) *Acta Met.* **28**, 1085.

Chevalier, G. T., McCormick, R. and Ruoff, A. L. (1967) *J. Appl. Phys.* **38**, 3697.

Friedel, J. (1964) *Dislocations.* Pergamon, p. 25.

Girifalco, L. A. (1964) in *Metallurgy at High Pressures and High Temperatures* (eds. Gachneider, K. A., Hepworth, M. T. and Parlee, N. A. D.). Gordon & Breach, p. 260.

Goldstein, J. I., Hanneman, R. E. and Ogilvie, R. E. (1965) *Trans. AIME* **233**, 813.

Huntington, H. B. (1958) *Solid State Physics* 7, 213.

Keyes, R. W. (1963) *Solids under Pressure* (eds. Paul, W. and Warschauer, D. M.). McGraw-Hill, p. 71.

Kocks, U. F., Argon, A. S. and Ashby, M. F. (1975) *Prog. Mat. Sci.* **19**.

Lazarus, D. and Nachtrieb, N. H. (1963) in *Solids under Pressure* (eds. Paul, W. and Warschauer, D. M.). McGraw-Hill, p. 43.

Lomer, W. W. (1957) *Phil. Mag.* **2**, 1053.

McCormick, P. G. and Ruoff, A. L. (1970) in *Mechanical Behaviour of Materials under Pressure* (ed. Pugh, H. le D.). Elsevier.

Peterson, N. L. (1968) *Solid State Physics* **22**, 409.

Richmond, O. and Spitzig, W. A. (1980) *Proc. 15th Int. Congress on Theoretical and Applied Mechanics.* IUTAM, Toronto, p. 377.

Sammis, C. G. and Smith, J. C. (1981) *J. Geophys. Res.* **86**, 10707.

Seeger, A. (1955) *Phil. Mag.* **46**, 1194.

Shewmon, P. G. (1963) *Diffusion in Solids.* McGraw-Hill, p. 52.

Spitzig, W. A., Sober, R. J. and Richmond, O. (1975) *Acta Met.* **23**, 885; (1976) *Trans. Met. Soc. AIME* **7A**, 1703.

# CHAPTER 18

# SCALING LAWS AND ISOMECHANICAL GROUPS

An inescapable conclusion of Chapters 4 to 16 is that materials with the same crystal structure and similar bonding have similar deformation-mechanism diagrams. When plotted on the normalized axes $\sigma_s/\mu$ and $T/T_M$, the maps for all f.c.c. metals, each fitted to experimental data, have an almost identical disposition of fields and of strain-rate contours. Further, they are clearly differentiated from the b.c.c. metals, or the alkali halides, or the various classes of oxides, each of which has its own characteristic map.

Polycrystalline solids, then, can be classified into *isomechanical groups* by properly normalizing the data for them; the members of one group have similar mechanical properties, and are differentiated by the normalization from members of other groups. We now examine how best to normalize data to bring out the similarities and differences between materials. From this we derive the preliminary classification of materials into isomechanical groups given in Table 18.1.

## 18.1 THE NORMALIZATION OF MECHANICAL DATA

The maps of this book are plotted on normalized axes of $\sigma_s/\mu$ and $T/T_M$. This is a common choice for presenting mechanical data; but what are the

TABLE 18.1  Isomechanical groups

| Crystal system | Structure type | Point groups | Isomechanical group |
|---|---|---|---|
| Cubic | F.C.C. | m3m F | Al, Cu, Ag, Au, Pt, Pb, Ni, $\gamma$-Fe |
| | Diamond cubic | m3m F | C, Si, Ge, $\alpha$-Sn |
| | Rock salt: | m3m F | |
| | Alkali halides | | (K, Na, Li, Rb)(F, Cl, Br, I) AgCl, AgBr |
| | Simple oxides | | MgO, MnO, CaO, CdO, FeO, CoO |
| | Lead sulphide | | PbS, PbTe, etc. |
| | Metal carbides | | TiC, ZrC, UC, TaC, VC, NbC |
| | Fluorite | m3m F | $UO_2$, $ThO_2$, $CaF_2$, $BaF_2$, $CeO_2$ |
| | Spinel | m3m F | $MgAl_2O_4$, etc. |
| | B.C.C.: | m3m I | |
| | Alkali metals | | Li, Na, K, Cs, Rb |
| | Transition metals | | W, Ta, Mo, Nb, V, Cr, $\alpha$-Fe, $\delta$-Fe, $\beta$-Te, $\beta$-Ti, $\beta$-Ze, Eu, Er |
| | Rare earths | | $\gamma$-U, $\varepsilon$-Pu, $\delta$-Ce, $\gamma$-La, $\gamma$-Yb |
| | Caesium chloride | m3m P | CsCl, CsBr, LiTl. MgTl, TlI, AuZn, AuCd, $NH_4Cl$, $NH_4Br$ |
| | Zinc blende | $\bar{4}$ 3m F | $\alpha$-ZnS, InSb, $\beta$-SiC, AlAs, AlSb, GdSb, GaSb, BeS, HgS, AlP, BeTe |
| Tetragonal | $\beta$-Sn | 4/mmm I | $\beta$-Sn |
| | Rutile | 6/mmm P | $TiO_2$, $SnO_2$, $PbO_2$, $WO_2$, $MnO_2$, $VO_2$, $NbO_2$, $TaO_2$, $MgF_2$, etc. |
| Hexagonal | Zinc | 6/mmm P | Cd, Zn, Co, Mg, Re, Te, Be, Zr, Ti, Hf, Y, Gd, Dy, Ho, Er |
| | Wurtzite | 6/mmm | ZnO, BeO, $\beta$-ZnS, $\beta$-CdS, $\alpha$-SiC, AlN, InN |
| | Graphite | 6/mmm | C |
| | Ice | 6/mmm | $H_2O$ |
| Trigonal | $\alpha$-Alumina | $\bar{3}$m R | $\alpha$-$Al_2O_3$, $Cr_2O_3$, $Fe_2O_3$ |
| | Bismuth | 3 m R | Bi, Sb, As, Hg |

grounds for selecting it? There are two complementary ways to study the problem of revealing the intrinsic similarities and differences between groups of materials. First, a purely empirical approach: if the data for a group of materials, when normalized, are brought onto a single master curve, and separated in this way from data for other groups of materials, then the normalization is a successful one. We examine this approach in Section 18.2. Second, a method using the model-based constitutive laws: if a constitutive law is to be normalized such that data for a group of materials in the same structural state lie on a single master curve, then certain dimensionless combinations of material properties must be constant for that group. This approach, which gives a basis for dividing materials into isomechanical groups, is developed in Section 18.3.

Let the normalizing parameter for stress be called $\sigma^*$, that for temperature $T^*$, and that for strain-rate $\dot{\gamma}^*$. There are three obvious groups of normalizing stresses $\sigma^*$ (Table 18.2). The first are moduli, suggested by dislocation theory: shear moduli $\mu$, Young's modulus $E$, and the bulk modulus $K$ (because dislocation energies and line tensions depend on these). The second are based on energies and are suggested by microscopic models for the lattice resistance and for diffusion: the cohesive energy $\Delta H_c$, the energy of fusion $\Delta H_f$, and the energy of sublimation $\Delta H_{vap}$, each divided by the atomic or molecular volume, $\Omega$, so that $\sigma^* = \Delta H_c/\Omega$,

etc. Finally, there are normalizing parameters based on temperatures: $\sigma^* = kT_M/\Omega$ where $k$ is Boltzmann's constant and $\Omega$ is the atomic or molecular volume. The boiling point $T_B$ and the Debye temperatures $T_\theta$ can also be treated in this way.

Table 18.2 also lists normalizing parameters $T^*$ for temperature: they, too, can be based on the same three groups of material property used with stress, although, of course, the normalizing temperature must have the dimensions of temperature. First there are temperatures: $T_M$, $T_B$ and $T_\theta$. Then there are temperatures based on energy divided by the gas constant: $\Delta H_c/k$, for example. Finally there are temperatures based on moduli: $\mu\Omega/k$, for instance.

The normalization of the strain-rate is less obvious. In Section 18.3 a natural normalization appears: it is $\dot{\gamma}^* = D_{TM}/\Omega^{2/3}$ where $D_{TM}$ is the melting point diffusivity and $\Omega$ the atomic or molecular volume. It will be justified later.

The idea of normalizing mechanical data is not new. Dorn (1957), McLean and Hale (1961), Bird et al. (1969), Sherby and Burke (1967), Kocks et al. (1975), Sherby and Armstrong (1971) and Chin et al. (1972) advocate, variously, the use of $\mu$, $E$ and $c_{44}$ to normalize flow stress and hardness. The melting point, $T_M$, is widely used as a normalizing parameter for temperature (Sherby and Simnad, 1961; Sherby and Burke, 1967) although without convincing physical justification; by contrast, there are good physical grounds for using $\mu\Omega/k$ (Kocks et al., 1975).

#### TABLE 18.2 Normalizing parameters

| Normalizing stress $\sigma^*$ $(N/m^2)$ | Normalising temperature $T^*$ $(K)$ |
|---|---|
| **Modulus-based** | **Temperature-based** |
| $\mu = (\frac{1}{2}c_{44}(c_{11} - c_{12}))^{\frac{1}{2}}$ | $T_M$ = Melting point |
| $\mu' = $ Polycrystal shear modulus | $T_\theta$ = Debye temp. |
| $\mu'' = c_{44}$ | $T_B$ = Boiling point (or sublimation temperature) |
| $E$ = Young's modulus | |
| $K$ = Bulk modulus | |
| **Energy-based** | **Energy-based** |
| $\dfrac{\Delta H_c}{\Omega}$ = Cohesive energy/mol. vol. | $\dfrac{\Delta H_c}{k}$ = Cohesive energy/k |
| $\dfrac{\Delta H_{vap}}{\Omega}$ = Heat of vaporization/mol. vol. | $\dfrac{\Delta H_{vap}}{k}$ = Heat of vaporization/k |
| $\dfrac{\Delta H_f}{\Omega}$ = Heat of fusion/mol. vol. | $\dfrac{\Delta H_f}{k}$ = Heat of fusion/k |
| **Temperature-based** | **Modulus-based** |
| $\dfrac{kT_M}{\Omega}$ = $k \times$ Melting temp./mol. vol. | $\dfrac{\mu\Omega}{k}$ = Shear modulus $\times$ mol. vol./k |
| $\dfrac{kT_\theta}{\Omega}$ = $k \times$ Debye temp./mol. vol. | $\dfrac{E\Omega}{k}$ = Young's modulus $\times$ mol. vol./k |
| $\dfrac{kT_B}{\Omega}$ = $k \times$ Boiling temp./mol. vol. | $\dfrac{K\Omega}{k}$ = Bulk modulus $\times$ mol. vol./k |

$k$ = Boltzmann's constant
$\Omega$ = Atomic or molecular volume

The normalization $\dot{\gamma}/D$ (where $D$ is the lattice diffusion coefficient) has been suggested for strain-rate (Sherby and Burke, 1967), although this quantity is not dimensionless. Ashby and Frost (1975) use instead the dimensionless quantity $\dot{\gamma}\Omega^{2/3}/D_{T_M}$, as we do here.

There have, then, been numerous attempts at normalization. But only one systematic study has been undertaken: that reported by Brown (1980) and Ashby and Brown (1981). In the rest of this chapter we follow their approach, and examine the success or failure of applying the normalizing parameters of Table 18.2, in the two ways described earlier.

## 18.2 EMPIRICAL NORMALIZATION OF DATA

The maps of this book are based on experimental data, much of which are plotted on the maps themselves or on separate data plots. The striking similarity between the data plots for the b.c.c. transition metals of Chapter 5, for instance, confirms that the normalization $\sigma_s/\mu$ and $T/T_M$ has some merit in revealing intrinsic similarities and differences between materials. But the data on these plots derive from many different investigators each applying different techniques to samples of differing purity; the inevitable scatter makes it hard to distinguish between alternative normalizations.

To avoid this, Brown (1980) and Ashby and

Brown (1981) measured the Vickers hardness $H$ (MN/m$^2$) of single crystals of some 25 elements and compounds, forming seven distinct groups, over a wide range of temperature. The samples were, as nearly as possible, in a standard state of purity and history. Fig. 18.1 shows an example: in it, the hardnesses of nine alkali halides are plotted against temperature from 88 K to about 1000 K. At a given temperature, the hardness range is about a factor of 35.

Ashby and Brown then investigate the effect of normalizing these data in the various ways listed in Table 18.2. The data for each group of materials were replotted on normalized axes, a curve fitted to them, and the mean square deviation from the curve was used to measure the success of the normalization. Fig. 18.2 shows the data for the alkali halides after the normalization $H/\mu$ and $T/T_M$.

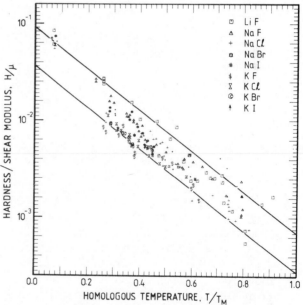

Fig. 18.2. The data of Fig. 18.1, replotted on the normalized axes $H/\mu$ and $T/T_M$. The spread of data is reduced from a factor of 25 to a factor of 2·5. A plot on axes of $H\Omega/\Delta H_c$ and $T/T_M$ is even more successful, but less convenient to use.

Some normalizations for the temperature are not successful. Normalizing by the Debye temperature $T_\theta$ (to form $T/T_\theta$), for example, increases the spread of the data. In every case the best correlation is given by normalizing temperature by the melting point, $T_M$, although the use of $E\Omega/k$ or $\mu\Omega/k$ (where $\Omega$ is the atomic or molecular volume) is almost as good. This is because the normalizing parameters are not independent. In particular, the moduli and the melting point are correlated for many elements and compounds. Fig. 18.3 shows the correlation which, to a good approximation, is described by:

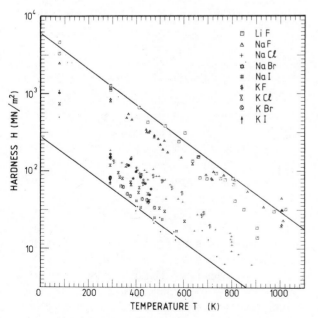

Fig. 18.1. The Vicker's hardness $H$ of nine alkali halides in, as nearly as possible, the same structural state, plotted against temperature.

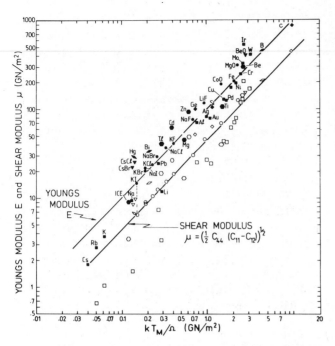

Fig. 18.3. The correlation between the melting temperatures and Young's moduli and the shear moduli of polycrystals.

$$E_0 = \frac{100 \; kT_M}{\Omega} \qquad (18.1)$$

and

$$\mu_0 = \frac{44 \; kT_M}{\Omega} \qquad (18.2)$$

(where $E_0$ and $\mu_0$ are the values of $E$ and $\mu$ at 300 K). This is important, since models for plasticity (Kocks *et al.*, 1975) suggest the normalization $kT/\mu\Omega$ rather than $T/T_M$ for temperature. The correlation makes $T_M$ an acceptable substitute for $\mu\Omega/k$, and an exceedingly convenient one, since the range for which the material is solid is then spanned by range $O < T/T_M < 1$.

The best normalization for the stress or the hardness is by the cohesive energy ($\Delta H_c/\Omega$). Applying it to the data for the alkali halides gives a scatter, as measured by a correlation coefficient, which is only slightly greater than that within data for a single alkali halide. Certain other normalizations are almost as good. For cubic materials, normalization by $\mu = (\frac{1}{2}c_{44}(c_{11} - c_{12}))^{\frac{1}{2}}$ works tolerably well. Normalization by a mean Young's modulus is better and has the advantage that only the bare minimum of data $(E, T_M)$ are necessary to perform it. The cohesive energy, despite its good performance, is not attractive for general use: it is hard to measure, and is not known for many compounds for which mechanical data exist; a modulus, though slightly less good, is much more convenient. We chose to

use $\mu$ because dislocation theory suggests that, in many cases, strength should scale as the line energy of a glide dislocation, and (for a screw) this is a proportional to $\mu$.

When data for groups of oxides and of elements are analysed in the same way, it is again found that the normalizations by $\mu$ and $T_M$ (or by $E$ and $T_M$) bring the hardnesses of members of one group onto a single master curve and distinguish them from those of other groups. This success suggests a less laborious way of identifying isomechanical groups. It is discussed in the next section.

## 18.3  NORMALIZATION OF CONSTITUTIVE LAWS

Chapter 2 described simple models for the mechanism of plastic flow. Each mechanism was described by a constitutive law of the form (Chapter 1, eqn. (1.1)):

$$\dot{\gamma} = f(\sigma_s, T, S_i, P_j) \qquad (18.3)$$

Here, it will be remembered, $S_i$ is a set of *state variables* which describe the microstructural state of the materials (its dislocation density, grain size, cell size, etc.) and $P_j$ is a set of *material properties* (bond energies, moduli, diffusion constants, and so forth) which are constant for a given material, but, of course, differ from one material to another.

If we choose a set of normalizing parameters, $\sigma^*, T^*, \dot{\gamma}^*$, with the dimensions of stress, temperature and strain rate, the constitutive law can be written:

$$\tilde{\dot{\gamma}} = f(\tilde{\sigma}_s, \tilde{T}, \tilde{S}_i, \tilde{P}_j)$$

where $\tilde{\sigma}_s = \sigma_s/\sigma^*$, $\tilde{T} = T/T^*$ and $\tilde{\dot{\gamma}} = \dot{\gamma}/\dot{\gamma}^*$   (18.5)

The quantities $\tilde{S}_i$ and $\tilde{P}_j$ are normalized, dimensionless state variables and material properties. It now follows that if the normalizing parameters, $\sigma_s^*, T^*$ and $\dot{\gamma}^*$ bring together data from a given group of materials, in an equivalent microstructural state ($\tilde{S}_i$ the same), thereby causing them to lie on a single master curve:

$$\tilde{\dot{\gamma}} = f(\tilde{\sigma}_s, \tilde{T}) \qquad (18.6)$$

*then the dimensionless material properties $\tilde{P}_j$ must be constants for the members of that group.*

The approach is most easily understood through an example. At high temperatures and low stresses, diffusional flow can become the dominant deformation mechanism (Chapter 2, Section 2.5). When lattice diffusion limits the creep rate, this rate (eqn. (2.29)) is given by:

$$\dot{\gamma} = \frac{42\sigma_s \Omega}{kTd^2} D_{0v} \exp\left\{-\frac{Q_v}{RT}\right\} \qquad (18.7)$$

This equation has the form of eqn. (18.3): it relates the macroscopic variables ($\dot{\gamma}$, $\sigma_s$, $T$) to the structure (the grain size, $d$) and to material properties (atomic volume, $\Omega$, and lattice diffusion constants $D_{0v}$ and $Q_v$). Normalizing by $\mu$ and $T_M$ gives:

$$\left(\frac{\dot{\gamma}\Omega^{2/3}}{D_{T_M}}\right) = 42\left(\frac{\Omega^{1/3}}{d}\right)^2 \left(\frac{\mu\Omega}{kT_M}\right)\left[\frac{\sigma_s}{\mu}\right]\left[\frac{T_M}{T}\right]$$

$$\exp-\left\{\left(\frac{Q_v}{RT_M}\right)\left(\left[\frac{T_M}{T}\right]-1\right)\right\} \qquad (18.8)$$

or $\qquad \tilde{\dot{\gamma}} = \dfrac{42\,\tilde{P}_1\,\tilde{\sigma}_s}{\tilde{S}_1^2}\,\dfrac{}{\tilde{T}} \exp-\left\{\tilde{P}_2\left(\dfrac{1}{\tilde{T}}-1\right)\right\} \qquad (18.9)$

where $\tilde{S}_1 = d/\Omega^{1/3}$, $\tilde{P}_1 = \mu\Omega/kT_M$ and $\tilde{P}_2 = Q_v/RT_M$. Here $D_{T_M}$ is the diffusion coefficient at the melting point and the strain-rate has been normalized by $D_{T_M}/\Omega^{2/3}$ to give $\tilde{\dot{\gamma}}$. The normalized stress is $\tilde{\sigma}_s = \sigma_s/\mu$ and normalized temperature is $\tilde{T} = T/T_M$.

This is a relationship between $\tilde{\sigma}_s$, and $T$ and $\tilde{\dot{\gamma}}$. Members of a group of materials, in the same microstructural state $\tilde{S}_1$, now have coincident diffusional creep behaviour if the properties $\tilde{P}_1$ and $\tilde{P}_2$ are constants for that group. So we obtain a second, quite separate, indicator of a good choice of normalizing parameter by examining how nearly constant the quantities $\tilde{P}_1$ and $\tilde{P}_2$ are.

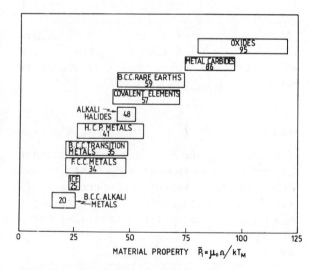

Fig. 18.4. The ranges of the dimensionless quantity $\mu_o \Omega/kT_M$ for classes of solids.

They are plotted in Figs. 18.4 and 18.5 (data from standard sources; see also Brown and Ashby, 1980a). The value of each, for a given group of materials, has a narrow range, and the values for different groups are distinct. In addition, the normalizing parameter $D_{T_M}/\Omega^{2/3}$ (which measured the atomic jump frequency at the melting point) is shown in

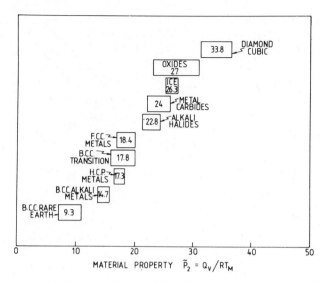

Fig. 18.5. The ranges of the dimensionless quantity $Q_v/RT_M$ for classes of solids.

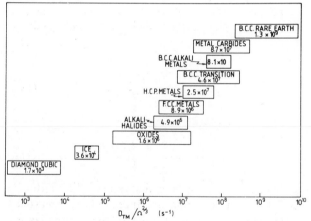

Fig. 18.6. The ranges of the quantity $D_{T_M}/\Omega^{2/3}$ ($s^{-1}$) for classes of solids.

Fig. 18.6. It, too, has a characteristic value for each group.

Diffusional flow may involve boundary diffusion also. If the full constitutive law (eqn. (2.29) with eqn. (2.30)) is normalized as before, two new material parameters appear:

$$\tilde{P}_3 = \frac{\delta D_{0b}}{\Omega^{1/3} D_{0v}} \quad \text{and} \quad \tilde{P}_4 = \frac{Q_b}{RT_M}$$

where $D_{0v}$ and $D_{0b}$ are the pre-exponentials for each diffusion path and $Q_b$ is the activation energy for boundary diffusion. Data are available for some 15 materials, but they are inadequate to construct plots like Figs. 18.4 and 18.5. Instead we tabulate the parameter $\tilde{P}_3$ and the ratio $\tilde{P}_4/\tilde{P}_2$ (Table 18.3; data from Brown and Ashby, 1980a). Both are roughly constant. As with $\tilde{P}_1$ and $\tilde{P}_2$, each group of materials

**TABLE 18.3  Approximate values of dimensionless material properties**

| | F.C.C. Metals | B.C.C. alkali metals | B.C.C. rare earths | B.C.C. transition | H.C.P. metals | Alkali halides | Metal carbides | Simple oxides | Covalent elements | Silicates | Ice |
|---|---|---|---|---|---|---|---|---|---|---|---|
| $\tilde{P}_1 = \dfrac{\mu_0\Omega}{kT_M}$ | $34 \pm 13$ | $20 \pm 5$ | $59 \pm 20$ | $35 \pm 14$ | $41 \pm 20$ | $48 \pm 4$ | $86 \pm 10$ | $95 \pm 50$ | $57 \pm 17$ | — | $25$ |
| $\tilde{P}_2 = Q_v/RT_M$ | $18{\cdot}4 \pm 1{\cdot}6$ | $14{\cdot}7 \pm 0{\cdot}5$ | $9{\cdot}3 \pm 1$ | $17{\cdot}8 \pm 2$ | $17{\cdot}3 \pm 1{\cdot}5$ | $22{\cdot}8 \pm 1{\cdot}5$ | $24{\cdot}0 \pm 2$ | $23{\cdot}4 \pm 1{\cdot}7$ | $33{\cdot}9 \pm 2{\cdot}3$ | $31 \pm 5$ | $26{\cdot}3$ |
| $\tilde{P}_3 = \dfrac{\delta D_{0b}}{\Omega^{1/3}D_{0v}}$ | $0{\cdot}5^*$ | — | — | $4^*$ | $0{\cdot}3^*$ | $0{\cdot}3^*$ | — | $0{\cdot}1^*$ | $1{\cdot}0^*$ | — | — |
| $\tilde{P}_4 = Q_b/RT_M$ | $10{\cdot}0 \pm 2$ | — | $15{\cdot}3 \pm 2$ | $11{\cdot}7 \pm 2$ | $10{\cdot}8 \pm 2$ | — | $12{\cdot}4 \pm 2$ | $14{\cdot}7 \pm 2$ | — | — | — |
| $\tilde{P}_5 = \sigma_0/\mu_0 \times 10^3$ | $1{\cdot}2$ | — | — | $0{\cdot}34\ddagger$ | $1{\cdot}0\ddagger$ | $0{\cdot}06\ddagger$ | — | $0{\cdot}06\ddagger$ | — | — | — |
| $\tilde{P}_6 = \Delta F_p/kT_M$ | $<10^{-5}$ | — | — | $3{\cdot}2 \pm {\cdot}7$ | $3 \pm 1$ | $7{\cdot}5 \pm 1$ | $6 \pm 1$ | $10 \pm 4$ | $35 \pm 4$ | $30 \pm 5$ | $36$ |
| $\tilde{P}_7 = \hat{\tau}/\mu_0 \times 10^2$ | $<10^{-3}$ | — | — | $1 \pm {\cdot}4$ | $0{\cdot}7 \pm {\cdot}2$ | $2{\cdot}7 \pm 0{\cdot}9$ | $4{\cdot}6 \pm 1$ | $3{\cdot}7 \pm 2$ | $7 \pm 1$ | $6 \pm 3$ | $9$ |
| $\tilde{P}_8 = \dfrac{\dot{\gamma}_0\Omega^{2/3}}{D_{T_M}}$ | $0{\cdot}3\dagger$ | — | — | $3 \times 10^3\dagger$ | $0{\cdot}1\dagger$ | $8 \times 10^4\dagger$ | — | $10^4\dagger$ | $7 \times 10^7\dagger$ | — | — |
| $\dot{\varepsilon}^* = D_{T_M}/\Omega^{2/3}$ (s$^{-1}$) | $9 \times 10^{6*}$ | $8 \times 10^{7*}$ | $1{\cdot}3 \times 10^{9*}$ | $5 \times 10^{7*}$ | $2{\cdot}5 \times 10^{7*}$ | $5 \times 10^{6*}$ | $9 \times 10^{7*}$ | $2 \times 10^{6*}$ | $2 \times 10^{3*}$ | — | $4 \times 10^{4*}$ |

The indicated range encompasses all data (it is the range, not the standard deviation).
* Range roughly a factor of $\times 10$.
† Range roughly a factor of $\times \div 5$.
‡ Range roughly a factor of $\times \div 3$.

138

has characteristic values of $\tilde{P}_3$ and $\tilde{P}_4$, which distinguish it from the other groups.

A similar procedure can be applied to power-law creep (Chapter 2, Section 2.4). At high temperatures, creep is frequently (though not always) limited by lattice diffusion. Then (eqn. (2.21)) its rate is:

$$\dot{\gamma} = \frac{A_2 D_v \mu b}{kT} \left(\frac{\sigma_s}{\mu}\right)^n \qquad (18.10)$$

This is the most difficult constitutive law with which to deal because, although the creep exponent $n$ is roughly constant for pure elements and compounds (generally $n = 4 \pm 1$), the creep "constant" $A$ varies by many orders of magnitude. This difficulty with the creep equation has been examined by Brown and Ashby (1980b) who rewrite the equation as:

$$\dot{\gamma} = A_3 \frac{D_v \sigma_0 b}{kT} \left(\frac{\sigma_s}{\sigma_0}\right)^n$$

where $\sigma_0$ is a material property. When $\sigma_0$ is properly defined it is found to be approximately constant for a given class of material; and with this value of $\sigma_0$, the dimensionless quantity $A_3$ is a constant also (its value is approximately $1\cdot6 \times 10^{-6}$).

Normalizing by $\sigma_0$ and $T_M$ gives:

$$\tilde{\dot{\gamma}} = A_3 \tilde{P}_1 \tilde{P}_5 \left(\frac{\tilde{\sigma}_s}{\tilde{P}_5}\right)^n \exp - \left\{\tilde{P}_2 \left(\frac{1}{\tilde{T}} - 1\right)\right\} \qquad (18.11)$$

where (as before) $\tilde{\sigma}_s = \sigma_s/\mu$, $\tilde{T} = T/T_M$ and $\tilde{\dot{\gamma}} = \dot{\gamma}\Omega^{2/3}/D_{T_M}$, and $\tilde{P}_1$ and $\tilde{P}_2$ were defined earlier. Although the equation contains three dimensionless material properties, only one is new: it is the quantity $\tilde{P}_5 = \sigma_0/\mu_0$. Reliable data for it are sparse; the best estimates we can make are given in Table 18.3 (Brown and Ashby, 1980b).

Finally, low-temperature plasticity (Chapter 2, Section 2.2) is described by a rate equation of the general form:

$$\dot{\gamma} = \dot{\gamma}_0 \exp - \left\{\frac{\Delta F}{kT} f\left(\frac{\sigma_s}{\hat{\tau}}\right)\right\} \qquad (18.12)$$

Here $\Delta F$ is the activation energy for glide, for either lattice-resistance or obstacle control. The pre-exponential $\dot{\gamma}_0$ has the dimensions of strain-rate, and $\hat{\tau}$, the flow strength at 0 K, has the dimensions of stress. The function $f$ is unimportant; it is commonly of the form:

$$f\left(\frac{\sigma_s}{\hat{\tau}}\right) = 1 - \frac{\sigma_s}{\hat{\tau}}$$

Normalizing by $(\mu, T_M)$ as before, leads to three new dimensionless properties:

$$\tilde{P}_6 = \Delta F/kT_M, \quad \tilde{P}_7 = \frac{\hat{\tau}}{\mu_0} \text{ and } \tilde{P}_8 = \frac{\dot{\gamma}_0 \Omega^{2/3}}{D_{T_M}}$$

Their approximate values for various material groups, for lattice-resistance controlled glide, are given in Table 18.3.

## 18.4 ISOMECHANICAL GROUPS

Both the investigation of scaling (Sections 18.2 and 18.3) suggest that materials can be classified into *isomechanical groups*. The members of a group are mechanically similar, and by proper scaling their mechanical properties can be brought, more or less, into coincidence. The same scaling separates and distinguishes this group from other groups.

The f.c.c. metals form an isomechanical group. So, too, do the diamond cubic elements. But it is not just the structure which determines mechanical strength. Rock-salt structured materials, for example, fall into four separate mechanical groups (Table 18.1). The ionically bonded alkali halides form one such group. But they are quite distinct from the transition-metal carbides and from the rock-salt structured oxides; and in ways that do not relate simply to the modulus and the melting point. The b.c.c. metals provide a second example. The data plotted in Figs. 18.4, 18.5 and 18.6 show a clear differentiation into three isomechanical groups: the alkali metals, the transition metals and the lanthanides.

It is necessary, then, to divide materials according to both structure and bonding if mechanical behaviour is to be grouped. But the groups are still large ones. A first attempt at this classification for elements and simple compounds is shown in Table 18.1.

## 18.5 SUMMARY AND APPLICATIONS

The systematic examination of scaling relations for mechanical behaviour summarized above shows that when stress, temperature and strain-rate are appropriately normalized, materials of the same structure and bonding, in the same microstructural state (grain size, purity, state of work-hardening and so forth) can be described by a single master-diagram. Successful normalizations bring data for members of a given isomechanical group together, and distinguish them from other groups.

This is because, in the broadest sense, mechanical behaviour is governed by a set of dimensionless material properties $\tilde{P}_j$. The values of $\tilde{P}_j$ are essentially constant for each isomechanical group—meaning solids with the same structure and similar bonding. The most successful normalization for $\sigma$,

**TABLE 18.4   Progression of material properties**

|  | F.C.C. | B.C.C./H.C.P. | Alkali halides | Simple oxides | Covalent solids |
|---|---|---|---|---|---|
| $\tilde{P}_7 = \hat{\tau}/\mu_0 \times 10^2$ | $< 10^{-3}$ | 1 | 2·5 | 4 | 7 |
| $\tilde{P}_6 = \Delta F_p/kT_M$ | $< 10^{-5}$ | 3 | 7·5 | 10 | 32 |
| $\tilde{P}_2 = Q_v/RT_M$ | 18 | 17 | 23 | 24 | 32 |

$T$ and $\dot{\gamma}$ is that described by $(\Delta H_c/\Omega, T_M, D_{T_M}/\Omega^{2/3})$, but the normalizations $(\mu, T_M, D_{T_M}/\Omega^{2/3})$ or $(E, T_M, D_{T_M}/\Omega^{2/3})$ are almost as good, and much more convenient.

The results have certain applications. First they allow one to see, at a very simple level, how classes of solids differ, and how the differences arise. There is a steady progression in the mechanical behaviour as the bonding becomes increasingly localized because this (i) generates an increasing lattice resistance, and (ii) raises the activation energy for both glide and for diffusion. Table 18.4 summarizes this progression. Note the increasing values of $\tilde{P}_7$, $\tilde{P}_6$ and $\tilde{P}_2$ as the bonding changes from metallic to ionic and covalent. It may be objected that material properties such as the stacking fault energy of f.c.c. crystals, or the c/a ratio of hexagonal crystals, or the stoichiometry of oxides, have been neglected in this treatment—and it is true that they have a considerable influence on a mechanical behaviour. But we are concerned here with the gross differences between broad classes of solids (the differences between b.c.c. metals and diamond cubic elements, for instance) and on this gross scale it is the quantities tabulated here which determine the characteristics of mechanical behaviour. Such quantities as stacking fault energy produce smaller variations about these characteristic or baseline properties.

The existence of such isomechanical groups also means that, at an approximate level, the mechanical behaviour can be predicted. Such a need appears in geology and geophysics, and (less often) in materials science and applied physics. Then data for one member of an isomechanical group can be scaled to give some idea of the behaviour of other, less well-characterized members of the group, provided its modulus and melting point are known. The maps shown in Chapters 4 to 16 can be thought of as typifying the group of solid to which they belong. Thus an approximate but useful description of the behaviour of (for instance) gold can be obtained from the map for copper or nickel, using the normalized axes (Chapter 4); an approximate description of NaI can be obtained from the map for NaCl (Chapter 10); and the mechanical behaviour of $PuO_2$ should be well approximated by that for $UO_2$ (Chapter 13).

### References for Chapter 18

Ashby, M. F. and Brown, A. M. (1981) *2nd Risø Conference on Materials Science* (eds. Hanson, N., Horsewell, A., Leffers, T. and Lilholt, H.), p. 1.

Ashby, M. F. and Frost, H. J. (1975) *Constitutive Relations in Plasticity* (ed. Argon, A.). MIT Press, p. 116.

Bird, J. E., Mukherjee, A. K. and Dorn, J. E. (1969) *Quantitative Relation between Properties and Microstructure* (eds. Brandon, D. G. and Rosen, A.). Haifa University Press, Israel, p. 255.

Brown, A. M. (1980) "The Temperature Dependence of the Vickers Hardness of Isostructural Compounds", Ph.D. thesis, University of Cambridge.

Brown, A. M. and Ashby, M. F. (1980a) *Acta Met.* **28**, 1085; (1980b) *Scripta Met.* **14**, 1297.

Burton, B. (1977), *Diffusional Creep in Polycrystalline Materials.* Trans. Tech. Publications.

Chin, G. Y., Van Uitert, L. G., Green, M. L. and Zydzik, G. (1972) *Scripta Met.* **6**, 475.

Dorn, J. E. (1957) *Creep and Recovery.* ASM, Ohio, p. 2255.

Kocks, U. F., Argon, A. S. and Ashby, M. F. (1975) *Prog. Mat. Sci.* **19**, 1.

McLean, D. and Hale, K. F. (1961) *Structural Processes in Creep.* Iron and Steel Institute, London, p. 86.

Mukherjee, A. K., Bird, J. E. and Dorn, J. E. (1969) *Trans. ASM* **62**, 155.

Poirier, J. P. (1978) *Acta Met.* **26**, 629.

Sherby, O. D. and Armstrong, P. E. (1971) *Met. Trans.* **2**, 3479.

Sherby, O. D. and Burke, P. M. (1967) *Prog. Mat. Sci.* **13**, p. 1.

Sherby, O. D. and Simnad, M. T. (1961) *Trans. ASM Q.* **54**, 227.

Sherby, O. D. and Weertman, J. (1979) *Acta Met.* **27**, 387.

# CHAPTER 19

# APPLICATIONS OF DEFORMATION-MECHANISM MAPS

## 19.1 USES OF THE MAPS

THE MAPS summarize, in a compact way, information about steady state and transient flow in a material. They can be used in the ways listed below. Each is illustrated by one of more of the case studies.

(a) They can be used to identify *the mechanism* by which a component or structure deforms in service, thereby identifying the *constitutive law or combination of laws* that should be used in design. Examples are given in the first three case studies: the lead pipe, the reactor components and the turbine blade. In particular, they can help in geophysical modelling as illustrated by the case studies of the polar ice cap and earth's upper mantle.

(b) They can be used to estimate approximately and quickly, the *strain-rate or the total strain* (both transient and steady state) of a component in service; the case studies of the reactor components and of the tungsten lamp filaments are examples.

(c) They can give *guidance in alloy design and selection*. Strengthening methods are selective: alloying, for instance, may suppress power-law creep but leave diffusional flow unchanged; increasing the grain size does the opposite. By identifying the dominant mechanism of flow, the maps help the alloy designer make a rational choice. The case studies of the lead pipes and of the turbine blades contain examples.

(d) They help in *designing experiments*. The ideal experimental conditions for studying a given mechanism can be read directly from even an approximate map—and the existence of iso-mechanical groups (Chapter 18) allows the construction of a rough map for almost any pure materials.

(e) Finally, they have *pedagogical value* in allowing a body of detailed information to be presented in a simple way. Each map is a summary of the mechanical behaviour of the material, and (because of the normalized axes) also gives an approximate description of other related

materials with similar bonding, crystal structure and purity (Chapter 18). The regime in which a structure operates, or a process takes place, can conveniently be illustrated on the maps: the case study of metal forming operations is an example, and the maps bring out the fundamental differences and similarities between the materials.

**Multiaxial stresses and strain rates**

Most practical problems in plasticity and creep involve multiaxial states of stress and of strain-rate. All the maps shown here have been plotted in a way which allows their use when this is so. The stress axis is that of *equivalent shear stress*:

$$\sigma_s = [\tfrac{1}{6}[(\sigma_1 - \sigma_2)^2 + (\sigma_2 - \sigma_3)^2 + (\sigma_3 - \sigma_1)^2]]^{\frac{1}{2}} = (\tfrac{1}{2}s_{ij}s_{ij})^{\frac{1}{2}} \quad (19.1)$$

and the contours are those of *equivalent strain-rate*.

$$\dot{\gamma} = [\tfrac{2}{3}[(\dot{\varepsilon}_1 - \dot{\varepsilon}_2)^2 + (\dot{\varepsilon}_2 - \dot{\varepsilon}_3)^2 + (\dot{\varepsilon}_3 - \dot{\varepsilon}_1)^2]]^{\frac{1}{2}} = (2\dot{\varepsilon}_{ij}\dot{\varepsilon}_{ij})^{\frac{1}{2}} \quad (19.2)$$

Here, $\sigma_1$, $\sigma_2$, $\sigma_3$, and $\dot{\varepsilon}_1$, $\dot{\varepsilon}_2$, $\dot{\varepsilon}_3$ are the principal stresses and strain rates and $s_{ij}$ is the deviatoric part of the stress tensor:

$$s_{ij} = \sigma_{ij} - \tfrac{1}{3}\delta_{ij}\sigma_{kk}$$

The maps show the relationship between $\sigma_s$, $\dot{\gamma}$ and $T$: they are a picture of the constitutive relations. The case study of 316 stainless steel contains examples of the use of eqns. (19.1) and (19.2), and shows how, if the two variables (in this case, $\sigma_s$ and $T$) are known, the map gives a third ($\dot{\gamma}$). The individual components of strain-rate are recovered by applying the *Levi Mises*, or *Associated Flow*, *Rule*, which for our purposes is best written as:

$$\frac{\dot{\varepsilon}_1}{\sigma_1 - \tfrac{1}{2}(\sigma_2 + \sigma_3)} = \frac{\dot{\varepsilon}_2}{\sigma_2 - \tfrac{1}{2}(\sigma_3 + \sigma_1)}$$

$$= \frac{\dot{\varepsilon}_3}{\sigma_3 - \tfrac{1}{2}(\sigma_1 + \sigma_2)}$$

$$= \frac{2}{9}\frac{\dot{\gamma}}{\sigma_s} \quad (19.3)$$

or, equivalently,

$$\dot{\varepsilon}_{ij} = \frac{\dot{\gamma}}{3\sigma_s} s_{ij}$$

## Procedure in applying the maps

The case studies follow a standard procedure that we have found to work well. It is as follows.

First, analyse the mechanics of the sample, component, or structure, tabulating the normalized stress ($\sigma_s/\mu$), homologous temperature ($T/T_M$) or strain-rate ($\dot{\gamma}$) to which it is subjected. In some instances all three are uniquely determined. More frequently they vary with position (as in the case study of the lead pipe and of the Polar ice cap) or with time (as with the turbine blade): then the ranges of stress, temperature and strain-rate should be tabulated.

Second, select or construct a map for the material of which the sample or part is made, with the appropriate grain size. Maps for some 40 materials are given in this book. Often, however, the material is one for which no map exists. Then, ideally, a map should be constructed, using the methods described in Chapter 3. But it is frequently adequate to start with the data for the pure base metal or ceramic—nickel for example—and modify those parts of it required to make the map fit experimental data for the alloy—Monel (Ni–30% Cu) for instance. To a first approximation, the lattice parameter, moduli and diffusion coefficients can be left unchanged, and the glide and creep data modified to fit experiment, though at a more precise level all would be modified. If little or no data are available (as is frequently the case for ceramics) an approximate map can sometimes be prepared by using the existence of iso-mechanical groups (Chapter 18), adjusting the map to fit limited data where they exist.

Third, plot the range of conditions of the sample or part onto the map. If the range of stress, temperature and strain-rate (or strain) are all known, an immediate *check for consistency* is possible. The lead-pipe study contains an example of this check.

Fourth, read off the mechanisms of flow, thus determining the appropriate constitutive law or combination of laws for design or modelling. If only two of the three variables ($\sigma_s$, $T$ and $\dot{\gamma}$) are known, use the map to determine the third. Maps for the same material with another grain size, or for a new material, can be used to examine how these changes might affect the deformation of the component, leading to the selection of an appropriate material.

At this stage a material has been chosen, the dominant deformation mechanism (and thus constitutive law) identified, and the approximate deformation rate determined. If design looks feasible, a detailed stress analysis of the sample or part should now be carried out, using the proper constitutive law fitted to data covering the range of operation of the part for the specific material, or batch, of which it is to be made. These may have to be requested from suppliers, or determined by specially commissioned tests.

The maps, then, should be used as the first phase of a design procedure only; but they allow this phase to be carried out very quickly and cheaply. Be cautious of attributing too much precision to the maps. They are only as good as the equations and data used to construct them—and both are often poor. An idea of the uncertainty can often be obtained from the data plots, of which many can be found in earlier chapters: the scatter in the data gives a measure of the batch-to-batch variation in material properties.

## 19.2  CASE STUDY: THE CREEP OF LEAD WATER PIPES

The first case study illustrates the use of deformation maps to identify the dominant mechanism of flow, and to prescribe a strengthening mechanism—in this instance, an increase in grain size—which will reduce its rate.

### Introduction

Since pre-Roman times lead has been used for water conduits and roof coverings. In England it has been common to use "soft lead" for these purposes: typically 99·5% lead containing a little Sb, Sn and lesser quantities of As, Cu, Ag, Fe and Zn, although elsewhere in Europe, particularly Germany, a "hard lead" containing 1 to 6% Sb was used. The corrosion resistance, ease of forming and welding, sound-absorbing properties and attractive appearance recommend lead for many exterior purposes—but it creeps at a rate which becomes alarming above 50°C. Even at room temperature (0·5 $T_M$) lead creeps (Fig. 19.1).

The earliest pipes and roofing sheet were cast. Even today, lead sheet is cast on a sand bed to a uniform thickness of about 3 mm for cathedral roofs; and lead down-spouting and piping is cast in sections for building use. This cast material has a large grain size (greater than 1 mm) and—for reasons which will become apparent—is much more

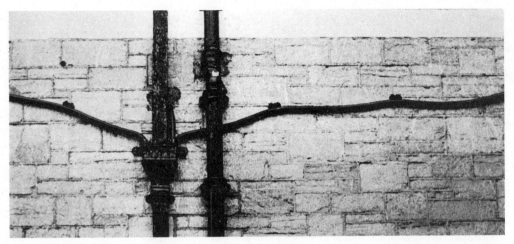

Fig. 19.1. Lead pipes on a 75-year-old building in southern England. The creep-induced curvature of these pipes is typical of Victorian lead water piping.

creep-resistant than the cheaper extruded pipe and rolled sheet introduced in Victorian times. The plastic deformation involved in these forming processes makes the lead recrystallize, giving a much smaller grain size (10–100 $\mu$) which, in commercial alloys, is stabilized by non-metallic inclusions and precipitates of an antimony-rich phase.

There is considerable evidence that a coarse grain size greatly reduces the creep rate. Pressurized lead pipe swells at the places where the grain size is finest (Hofmann, 1970, p. 435). Pure lead, and a number of its dilute alloys, show a strong inverse dependence of creep-rate on grain size at slow strain-rates (Hopkins and Thwaites, 1953); and these strain-rates are proportional to stress (Hofmann, 1970, p. 237), both observations suggesting diffusional flow.

## Analysis of sagging pipes

Given sufficient time, a Victorian lead pipe supported at discrete points along its length will sag under its own weight (Fig. 19.1). By what mechanism does this creep occur? And why is it that Roman lead ducting and piping, despite its greater age, does not appear to have sagged in this way?

We have examined two examples of creeping lead pipes: one an external drain pipe, 75 years old; the other, an interior hot-water pipe, aged about 70 years. Both had a grain size of about 50 $\mu$m. They are sketched in Fig. 19.2. The maximum strain is in the lower surface of the pipe; it can be calculated from the radius of curvature ($R$), the depth of the sag ($h$), the length ($l$) and the height ($b$) of the pipe. The average creep rate is given by dividing this by the life ($t$) of the pipe, and multiplying by $\sqrt{3}$ to convert to an equivalent shear strain-rate $\dot{\gamma}$. For

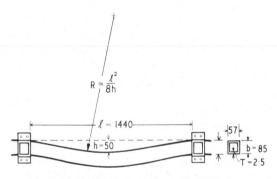

A. RECTANGULAR SECTION DRAIN PIPE (TRINITY HALL, CAMBRIDGE) AGE 75 YEARS

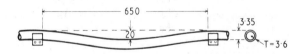

B. VICTORIAN HOT WATER PIPE (MONTGOMERY, WALES) AGE ABOUT 70 YEARS.

Fig. 19.2. The two examples of sagging lead pipes analysed in this case study. All dimensions are in mm.

### TABLE 19.1 Conditions to which the lead pipes are exposed

| Sample | Range of $\sigma_s/\mu$ | Range of $T/T_M$ | Maximum $\dot{\gamma}$ ($s^{-1}$) |
|---|---|---|---|
| External, drain, Cambridge, England, age: 75 years | $2\cdot5 \times 10^{-5}$ to $1\cdot0 \times 10^{-4}$ | $0\cdot44 \rightarrow 0\cdot5$ | $\sim 7 \times 10^{-12}$ |
| Hot water pipe, Montgomery, Wales, age: 70 years | $1 \times 10^{-5}$ to $4 \times 10^{-5}$ | $0\cdot46 \rightarrow 0\cdot53$ | $\sim 6 \times 10^{-12}$ |

small sags ($h < b$), the result is $\dot{\gamma} = 4\sqrt{3}bh/l^2t$. Applying this to the pipes of Fig. 19.2 gives the approximate strain-rates listed in Table 19.1. They are small—comparable with those permitted in engineering structures with a design life of 25 years.

The stresses are calculated from standard beam theory, treating the pipe as a thin-walled tube carrying a distributed load due to its own weight. The stress varies with position along the pipe; expressed as an equivalent shear stress it has a maximum value (which is an adequate measure for our purposes) of general magnitude $3 \times 10^{-5}\ \mu$; details are given in Table 19.1. These are low stresses: almost all the laboratory data for creep of lead have been obtained at stresses considerably higher than this.

### Use of the deformation map for antimonial lead

Maps for 0·5% antimonial lead (which typifies alloys used in England for pipes and roofing) are shown in Figs. 19.3 and 19.4. They are computed from the data listed for lead in Table 4.1, with three modifications. First the antimony lowers the melting point of lead to 595 K. Second, the creep constants have been modified to fit the data of Hofmann (1970, p. 237) for antimonial lead at 30°C; to do so we take $n = 4\cdot2$ and $A = 842$, and retain $Q = 109$ kJ/mole as before. Finally, the obstacle-controlled glide parameters have been changed to match the tensile strength of antimonial lead at

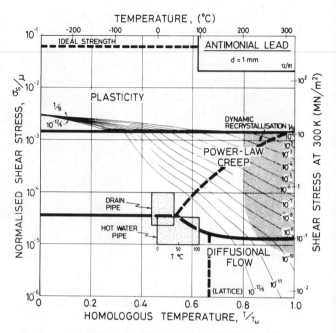

Fig. 19.4. Antimonial lead with a grain size of 1 mm. If the pipes had this grain size they would deform much more slowly than they do.

20°C and a strain rate of $10^{-2}$/s, given by Hofmann (1970, p. 94); to do so we take $\hat{\tau}/\mu_0 = 2\cdot9 \times 10^{-3}$.

The first map describes antimonial lead with a grain size of 50 $\mu$m, the grain size of the extruded pipes analysed here. On it are plotted, as shaded boxes, the ranges of stress and temperature to which the pipes are exposed. As a check on self-consistency, note that the strain rate contour of $10^{-12}$/s passes through both boxes: the map is in broad agreement with the observed mean strain rates (around $6 \times 10^{-12}$/s).

The second map shows the same lead, but with a grain size of 1 mm, roughly that found in cast lead pipe and sheet. Note that the predicted creep rates of the pipes are much smaller.

### Conclusions of the case study

The first map (Fig. 19.3) shows that the lead pipes deform by *diffusional flow*, of the kind controlled by *grain boundary diffusion*. This has certain consequences. The creep is linear–viscous, with little or no transient behaviour (this means that one can integrate over a history of stress and temperature). It is a very sensitive grain size—hence the advantage of cast sheet and pipes, a fact from which the Romans profited. An increase in grain size, to 1 mm (Fig. 19.4) slows the creep-rate by a factor of $10^3$. Cast lead has an even larger grain size than this—it is typically 5 mm—and creeps not only more slowly, but—for the loading conditions of our pipes

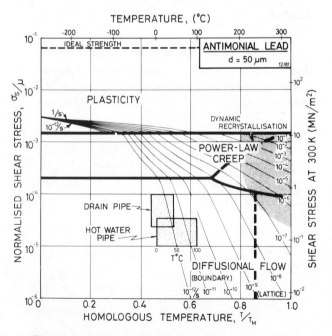

Fig. 19.3. A map for antimonial lead with a grain size of 50 $\mu$m, showing the conditions of operation of the pipes. Both deform by diffusional flow.

—by a different mechanism: suppressing diffusional flow causes the power-law creep field to expand until it includes both shaded boxes.

Though the creep of lead pipes is not a problem of pressing industrial importance, this case study has features which appear in reactor and turbine technology. The normalized stresses, homologous temperatures and strain-rates are similar to those involved in power-generating equipment (see the next two case studies). Diffusional creep, of the variety controlled by boundary diffusion, will frequently be the dominant mechanism of deformation in structures designed to last for 25 years at elevated temperatures. Yet it is often ignored in engineering design, which is normally based on an extrapolation of laboratory power-law creep data to the stresses encountered in the structure. The risk involved in doing this can be illustrated by the map of Fig. 19.3: if the power-law creep equation used to construct this map is extrapolated to predict the behaviour of the pipes, it gives the strain-rates slower by a factor of $10^3$ than those observed.

### References for Section 19.2

Hofmann, W. (1970) *Lead and Lead Alloys*, 2nd edn. Springer.

Hopkins, L. M. T. and Thwaites, C. J. (1953) *J. Inst. Met.* **82**, 181.

## 19.3  CASE STUDY: THE CREEP OF 316 STAINLESS STEEL IN A FAST NUCLEAR REACTOR

This case study illustrates the use of deformation maps to analyse components subjected to multiaxial stresses, and shows how they can help in selecting a constitutive law for design purposes. It further illustrates the use of transient (non-steady-state) maps to identify the dominant deformation mechanism, and the strains in the structure, when these are small.

### Introduction

The core of a fast reactor is quite small. To remove the heat, a coolant—either a liquid such as sodium, or a gas such as helium—must flow rapidly through it, and into a heat exchanger. In a liquid–metal-cooled reactor, the pressure differences needed to drive this flow, together with the weight of the structure and coolant itself, exert stresses on the structure supporting the core, on the pipework, and on the other components; in a gas-cooled reactor, the pressure differences are much smaller but the hydrostatic pressure needed to confine the gas imposes additional stresses.

Superimposed on these steady stresses are the thermal stresses which appear when the power output of the reactor changes. This is a problem of combined creep and low-cycle fatigue, and because they may not combine linearly, the rates of creep and the creep strains cannot be calculated safely from the equations of creep alone. The maps we present here should (if properly constructed) give a good description of the behaviour of reactor components under steady loading conditions, but they should not be used to give more than a qualitative picture of behaviour when loads change with time.

The temperature, and thus the efficiency of the reactor, is limited by the materials of which it is made. At present it appears likely that much of the internal structure of commercial fast reactors will be made of Type 316 stainless steel, a well-tried material for which long-term creep-rupture data are available. This choice limits, to a maximum of about 600°C, the temperature to which structural components of the reactor can be exposed. In the following hypothetical case study, we show how both steady-state and transient deformation maps may help a designer select the right constitutive law for his design calculations.

### Description of the Reactor and Approximate Stress Analysis

Fig. 19.5 shows, schematically and much simplified, a section through a hypothetical liquid–metal-cooled fast reactor. Liquid sodium contained in a pressure vessel is circulated through the core and

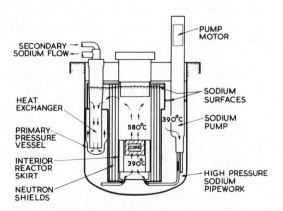

Fig. 19.5. Section through a hypothetical fast-breeder reactor.

heat-exchangers by pumps. We examine the creep of three components: the pressure vessel itself, the pipes leading from the pumps to the core, and the reactor skirt. We shall assume that all three are made of 316 stainless steel with a grain size of either 50 or 100 $\mu$m.

The *pressure vessel* operates at the input temperature of the sodium coolant: $390 \pm 30°C$. Since all the components are suspended from the roof of the reactor, the loading of the vessel is merely that due to the weight of the sodium it contains, to buoyancy forces, and to the small pressure of helium gas which covers the sodium. The stresses in the vessel are highest in the side wall near the bottom, about 8 m below the sodium surface, where the pressure due to the sodium alone is $0.074$ MN/m$^2$. To this we add the contribution of the overpressure $p_g$ of inert gas ($0.007$ MN/m$^2$), giving a total pressure $p$ of $0.08$ MN/m$^2$. These pressures, and the dimensions we assume below, broadly follow current designs. If, then, we take:

$a$ = wall thickness of pressure vessel = $0.0125$ m
$r$ = radius of pressure vessel = $6$ m
$M$ = mass of sodium contained in the vessel including buoyancy forces from displaced sodium $\simeq 10^6$ kg

the principal stresses in the vessel are:

$$\sigma_1 = \frac{pr}{2a} + \frac{Mg}{2\pi ra} = 22.5 \text{ MN/m}^2 \text{ (longitudinal)}$$

$$\sigma_2 = \frac{pr}{a} \qquad = 38.4 \text{ MN/m}^2 \text{ (circumferential)}$$

$$\sigma_3 = 0 \text{ (through wall thickness)}$$

The maximum equivalent shear stress (eqn. (19.1)) is $\sigma_s = 19.3$ MN/m$^2$. Near the top of the pressure vessel the equivalent shear stress in the wall is less ($p$ is reduced to $p_g$, and $\sigma_s$ to $12.5$ MN/m$^2$) and it is less in the hemispherical bottom because of its shape.

The *sodium input pipes*, too, operate at $390 \pm 30°C$. The stresses in them are due mainly to the pressure difference between the sodium inside and outside a pipe, though there is a small additional contribution (which we shall neglect) at a bend in the pipe due to inertial forces set up by the rapidly flowing sodium. If we use the following hypothetical dimensions:

$r_p$ = pipe radius $\quad = 0.125$ m
$a_p$ = pipe thickness $\quad = 0.005$ m
$p_i$ = internal pressure $= 0.83$ MN/m$^2$
$p_e$ = external pressure $= 0.03$ MN/m$^2$
$\quad$ (corresponding to a depth of 3 m of sodium)

then the principal stresses are:

$$\sigma_1 = \frac{(p_i - p_e)r_p}{2a_p} \simeq 10 \text{ MN/m}^2 \text{ (axial)}$$

$$\sigma_2 = \frac{(p_i - p_e)r_p}{a_p} \simeq 20 \text{ MN/m}^2 \text{ (circumferential)}$$

and the equivalent shear stress, $\sigma_s$, is 10 MN/m$^2$.

The *interior reactor skirt* is a cylinder, about 6 m in diameter, containing the core and associated structure. At least part of it is exposed to the hot sodium leaving the core at a temperature of $580 \pm 30°C$. It is stressed because of the difference of 2 m in the sodium level inside and outside the skirt, this difference driving the flow through the heat exchangers; the consequent pressure difference is about $0.02$ MN/m$^2$. If, as before, we take hypothetical values for the dimensions:

$r_s^*$ = skirt radius $\qquad = 3$ m
$a_s$ = wall thickness of skirt = $0.005$ m

we can calculate the stress state in the skirt. The circumferential stress, $\sigma_2$, is 11 MN/m$^2$; and since it is supported from below, the longitudinal stress $\sigma_1$ is negligible. The equivalent shear stress, $\sigma_s$, is $6.3$ MN/m$^2$.

This information, normalized by a (temperature-corrected) modulus, and by the melting temperature of iron (1810 K) is summarized in Table 19.2. We have allowed a $\pm 20\%$ variation in stress and a $\pm 30°C$ variation in temperature about the values given in the text.

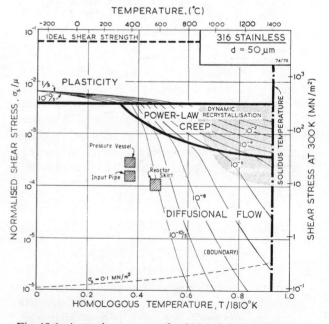

Fig. 19.6. A steady-state map for 316 stainless steel with a grain size of 50 $\mu$m, showing the operating conditions of the three reactor components, and the dominant steady-state flow mechanisms.

TABLE 19.2   Summary of conditions to which the reactor components are exposed

| Component | Range of $\sigma_s/\mu$ | Range of $T/T_M$ |
|---|---|---|
| Pressure vessel | $2{\cdot}3 \times 10^{-4} \rightarrow 3{\cdot}5 \times 10^{-4}$ | $0{\cdot}35 \rightarrow 0{\cdot}38$ |
| Sodium input pipes | $1{\cdot}2 \times 10^{-4} \rightarrow 1{\cdot}8 \times 10^{-4}$ | $0{\cdot}35 \rightarrow 0{\cdot}38$ |
| Reactor skirt | $8{\cdot}5 \times 10^{-5} \rightarrow 1{\cdot}3 \times 10^{-4}$ | $0{\cdot}45 \rightarrow 0{\cdot}49$ |

## Use of deformation maps to analyse reactor components

Deformation maps showing both the steady-state and transient flow of 316 stainless steel are described in Chapters 8 and 17; those relevant to this case study are shown in Figs. 19.6, 19.7 and 18.8. The data used to construct them are given in Tables 8.1 and 17.1.

Fig. 19.6 shows the service conditions of the reactor components on a steady-state deformation map with a grain size of 50 $\mu$m. All three lie well inside the *diffusional flow* field, in the regime in which *boundary diffusion* is dominant. The strain rates are small: $10^{-10}$/s or less, so that (for the conditions we have assumed here) the total strain which appears in the structure during—say—3 years ($10^8$ s) is 1% or less.

When strains are as small as this, the elastic and transient creep strains cannot be neglected. Though their construction requires more data, it makes sense to apply non-steady-state ("transient" maps, Chapter 17) to a problem such as this one. This is done in Figs. 19.7 and 18.8. They show maps for

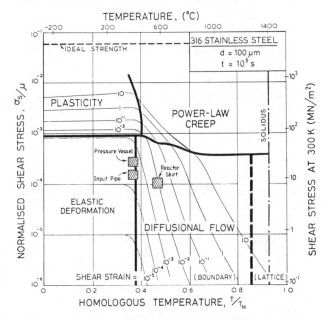

Fig. 19.8.  A transient map like that of Fig. 19.7, but showing the strains which have accumulated in $10^9$ s (roughly the life of the reactor). All three have suffered some creep strain; that in the reactor skirt may exceed 1%. Deformation is dominated by elasticity and diffusional flow.

316 stainless steel with a grain size of 100 $\mu$m. Transient maps, it will be remembered, show the strain reached in a given time, and the mechanism principally responsible for that strain. In a time $10^4$ s (about 3 hours, Fig. 19.7) the strain in all three components is predominantly elastic, and of order $2 \times 10^{-4}$.

As time elapses, the creep strains grow, so that after 30 years ($10^9$ s, Fig. 19.8) the creep strain in the reactor skirt is of order 1%; diffusional flow is the dominant mechanism. The other two components lie on the boundary between elastic deformation and diffusional flow, indicating that the contributions from the two are about equal ($\approx 2 \times 10^{-4}$).

These components are not exposed to a heavy neutron flux. If they were, a map incorporating radiation-induced, or radiation-enhanced, creep would be required. Such maps can be constructed (Ashby, 1971, unpublished), but are relevant only to the fuel, the fuel cans and the components immediately adjacent to the fuel pins.

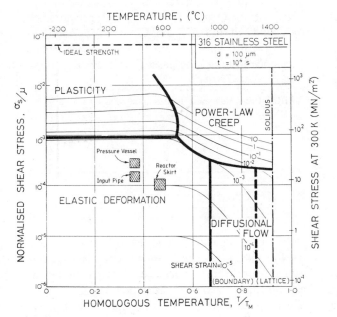

Fig. 19.7.  A transient map for 316 stainless steel with a grain size of 100 $\mu$m, showing the strains which appear in $10^4$ s. All three components deform elastically.

## Conclusions of the case study

Under the steady loads considered here, the distortion in the reactor components is elastic and by creep, the latter caused predominantly by diffusional flow. For much of the life of the reactor, the two contributions are of comparable magnitude. Transient contributions to diffusional flow and to power-law creep should not be neglected because their contribution is comparable with the elastic strain.

In certain reactor designs, the creep strain permitted in the life of the reactor—about $10^9$ s—is limited to 1%. Fig. 19.8, which is constructed for this time span, shows that the pressure vessel and the sodium input pipe are well inside this limit. The inner reactor skirt, however, would exceed this permitted strain if our assumptions about dimensions and pressures are correct, even though the stresses in it are lower than for the other two components. Reducing the stress by a factor of two would not remedy this. An acceptable creep rate in the skirt could be obtained by selecting a steel with the same composition but a larger grain size: an increase to 200 $\mu$m, for instance, lowers the creep strain by a factor of 8.

A designer, concerned about the distortion of these reactor components under steady load and wishing to analyse them in more detail, should use a constitutive law which combines the elastic and creep deformation, including the transient contribution. The elastic distortion (in tensor notation) is:

$$\varepsilon_{ij}^{EL} = \frac{1+v}{E}\left(\sigma_{ij} - \frac{v}{1+v}(\delta_{ij}\sigma_{kk})\right) \quad (19.4)$$

The dominant flow mechanism is Coble creep. Transient contributions to diffusional flow are discussed in Chapter 17; the relevant constitutive law is:

$$\dot{\varepsilon}_{ij} = \frac{s_{ij}}{2\eta}(1 + \exp - t/\tau_t)$$

where

$$s_{ij} = \sigma_{ij} - \tfrac{1}{3}\delta_{ij}\delta_{kk} \quad (19.5)$$

$$\eta = \frac{kTd^3}{42\ \delta D_b\Omega}$$

$$\tau_t = \eta/\Omega$$

and $t$ is time. The data necessary to evaluate this equation are given in Table 17.1. If the designer is concerned with occasional overloads, he must first add to this the contribution from power-law creep and (if the overloads are large enough) from yielding; and second, take into account the possibility of fatigue failure and of interaction between creep

and fatigue; these last are beyond the scope of this case study.

The case study leads to a further conclusion. Diffusional flow is poorly studied experimentally; the boundary-diffusion coefficients required to calculate its rate are not well documented, and the influence of alloying on it (Chapter 17) is only partly understood. There is a need for a major scientific study of diffusional flow in stainless steels.

## 19.4 CASE STUDY: CREEP OF A SUPERALLOY TURBINE BLADE

This case study is an example of the application of deformation maps to a component in which the stress and temperature vary with position. It illustrates the fact that strengthening methods are selective; a given method does not slow down all flow mechanisms equally.

### Introduction

Throughout the history of its development, the gas turbine has been limited in power and efficiency by the availability of materials that can withstand high stress at high temperatures. Where conditions are at their most extreme, in the first stage of the engine, nickel and cobalt-based superalloys are currently used because of their unique combination of high-temperature strength, adequate ductility and oxidation resistance. Typical of these is MAR–M200, an alloy based on nickel, strengthened by a solid solution of W and Co and by precipitates of $Ni_3(Ti,Al)$, and containing Cr to improve its resistance to attack by gases (Table 19.3).

TABLE 19.3 Nominal composition of MAR–M200 in wt. %

| Al | Ti | W | Cr | Nb |
|----|----|----|----|----|
| 5·0 | 2·0 | 12·5 | 9·0 | 1·0 |

| Co | C | B | Zr | Ni |
|----|---|---|----|----|
| 10·0 | 0·15 | ·015 | ·05 | Balance |

This particular alloy, and descendants of it, are used in the as-cast state—they cannot be rolled or forged—for turbine blades and vanes in commercial gas turbines. In aircraft use, start-up, shut-down, and frequent changes of power mean that steady-state conditions very seldom apply. One could, however, envisage a steady state being reached in other applications: for example, in the use of such

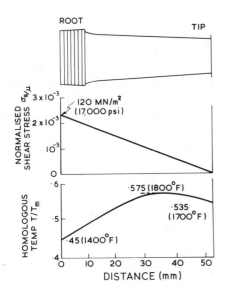

Fig. 19.9. The approximate distribution of axial stress and temperature along a turbine blade operating in the first sage of a typical turbine of the 1960s.

turbines for power generation, or in the use of MAR–M200 as a structure material in reactors or chemical plants. Here we shall use only steady-state maps for MAR–M200—insufficient data are available at present to construct transient maps for it. (An example of the use of transient maps for stainless steel is given in the case study preceding this one.)

### Stress and temperature of a turbine blade

When a turbine is running at a steady speed, centrifugal forces subject each rotor blade to an axial tension. If the blade has a constant cross-section, the tensile stress rises linearly from zero at its tip to a maximum at its root. As an example, a rotor of radius $r$ of 0·3 m rotating at an angular velocity $\theta$ of 1000 radians/s (11,000 r.p.m.) induces an axial stress $(r\theta^2\rho l)$ of order $10^{-3}\mu$, where $\rho$ is the density of the alloy and $l$ the distance from the tip. This stress and a typical temperature profile, for a blade in an engine in use in the 1960s, are shown in Fig. 19.9. The range of conditions is summarized in Table 19.4.

These temperatures and stresses are averages over the cross-section of the blade, and are useful in

**TABLE 19.4**
**Summary of average steady running conditions on blade**

| Range of temperature | Range of stress $\dfrac{\sigma_s}{\mu}$ | Maximum acceptable strain rate |
|---|---|---|
| 0·45–0·58 $T_M$ | $0 \rightarrow 2\cdot3 \times 10^{-3}$ | $\sim 10^{-8}$/s |

giving a general idea of the mechanism of creep, and its approximate rate, under steady service conditions. When the power changes, the surface temperature of the blade rises or falls sharply, creating thermal stresses which are superimposed on those caused by centrifugal forces. They can be large, sometimes large enough to stress the skin of the blade above its yield stress. But the average stress in the blade remains in the range given by Table 19.4.

### Deformation maps for nickel and MAR–M200

Figs. 19.10, 19.11 and 19.12 show deformation maps for pure nickel and for MAR–M200 of two grain sizes. The data on which they are based are listed in Tables 4.1 and 7.1, and discussed in Chapters 4 and 7.

### Conclusions of the case study

The steady running conditions given in Table 19.4 are plotted as a shaded box onto the maps. If made of pure nickel (Fig. 19.10) the blade would deform by *power-law creep*, at a totally unacceptable rate. By applying the strengthening methods used in MAR–M200 (Fig. 19.11) the rate of power-law creep has been reduced by a factor of $10^5$, and the dominant mechanism of flow has changed— from power-law creep to diffusional flow. Further

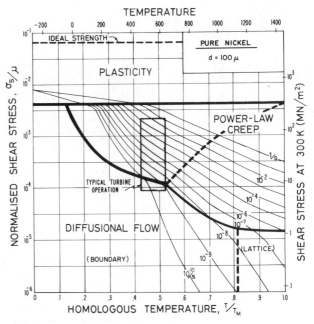

Fig. 19.10. A map for pure nickel with a grain size of 100 $\mu$m, showing the conditions of operation of the blade described by Fig. 19.9.

solution-strengthening or precipitation-hardening is now ineffective unless it slows this mechanism. A new strengthening method is needed: the obvious one is to increase the grain size. The result of doing this is shown in Fig. 19.12. This new strengthening method slows diffusional flow while leaving the

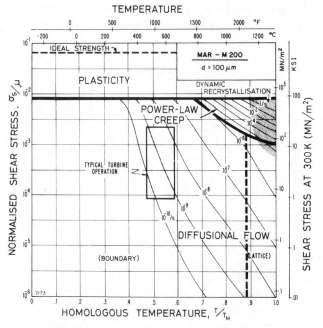

Fig. 19.11. A map for MAR–M200, with the same grain size as that for the nickel of Fig. 19.10 (100 μm). The shaded box shows the conditions of operation of the blade.

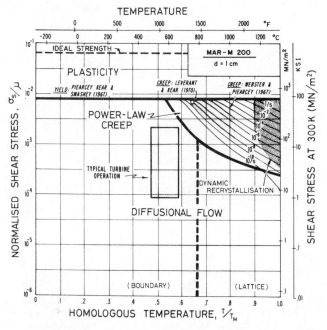

Fig. 19.12. A map for MAR–M200 with a large grain size (10 mm) approximating the creep behaviour of directionally solidified or single crystal blades. The shaded box shows the conditions of operation of the blade.

other flow mechanisms unchanged. The power-law creep field expands, and the rate of creep of the turbine blade falls to a negligible level.

The point to remember is that plastic flow has contributions from several distinct mechanisms; the one that is dominant depends on the stress and temperature applied to the material. If one is suppressed, another will take its place. Strengthening methods are selective: a method that works well in one range of stress and temperature may be ineffective in another. A strengthening method should be regarded as a way of attacking a particular flow mechanism. Materials with good creep resistance combine several strengthening mechanisms; of this, MAR–M200 is a good example.

## 19.5   CASE STUDY: THE CREEP OF TUNGSTEN LAMP FILAMENTS

Deformation maps allow the performance of components to be compared, and suggest ways in which design might be changed to give better performance or longer life. In this case study we examine the performance of two tungsten lamp filaments.

### Strain-rate and stress on filaments in service

The filament of a 25 or 40 Watt lamp is a single-coiled wire of doped tungsten. Typical dimensions of a 40 Watt lamp are given in Fig. 19.13. Low-Wattage lamps like these burn at a temperature of 2250 to 2500°C, with an average life-time of about 1000 hours. (Higher-powered lamps run at higher temperatures: up to 2765°C for ordinary lamps and up to 3160°C for photo-flood bulbs—but their life is shorter: as short as 3 hours.)

A lamp may fail in one of several ways. Most fail

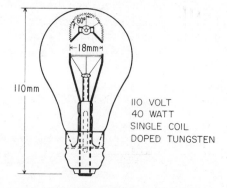

Fig. 19.13. Typical dimensions of a 40 Watt, 110 Volt, tungsten filament lamp. The filament is a simple coil of doped tungsten wire.

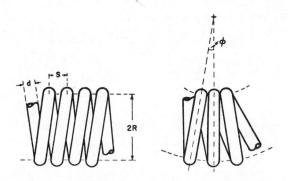

Fig. 19.14. The creep-failure of a tungsten filament. Torsional creep causes the windings to touch, causing overheating or shorting.

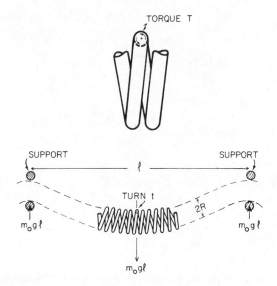

Fig. 19.15. The statics of a horizontal lamp filament.

because evaporation from the filament surface, or the formation of a bubble or void within it, locally reduces the cross-section, producing a hot spot which accelerates evaporation and finally causes melting. The reason that most fail in this way, and not by creep, is that design against creep failure is adequate. What factors enter this design problem? To answer this question, we must first consider failure by creep.

The most probable mechanism of failure by creep is illustrated in Fig. 19.14. An undistorted coil is shown on the left; its dimensions for two sizes of lamp are listed in Table 19.5. Lamps normally burn with the filament horizontal. Then sag by torsional creep of the wire leads to overheating, and ultimately to shorting between turns as shown on the right. Suppose elements of wire on the upper side of the coil suffer a torsional creep strain resulting in a twist of $\theta$ per unit length; those on the bottom suffer a similar twist in the opposite sense. Then the change of angle between the turns, $\phi$, is approximately $\phi = 2R\theta$ where $R$ is the coil radius. Contact occurs when:

$$\phi R = S - d$$

where $S$ is the turn spacing and $d$ the wire diameter

(Fig. 19.14). The shear strain at the surface of an element of the wire, $\gamma_{max}$, is related to the twist per unit length by $\gamma_{max} = \theta d/2$. The maximum permissible strain before shorting occurs is therefore:

$$\gamma_{max} \simeq \frac{d(S - d)}{4R^2}$$

If the lamp is to survive its rated lifetime, $t$, then the maximum permissible steady creep rate is:

$$\dot{\gamma}_{max} \simeq \frac{d(S - d)}{4R^2 t}$$

Inserting data from Table 19.5 gives:

$$\dot{\gamma}_{max} = 3 \cdot 0 \times 10^{-9}/s \text{ 25 watt}$$
$$\dot{\gamma}_{max} = 3 \cdot 6 \times 10^{-9}/s \text{ 40 watt}$$

To allow a margin of safety against this sort of creep failure, the lamp should be designed so that the maximum strain-rate in the wire is less than $10^{-9}/s$.

The shear stress on the filament is calculated to a sufficient approximation as follows. Consider the equilibrium of a section of the coil of length $l$ between the two supports, as shown in Fig. 19.15. If the mass per unit length of the coil is $m_0$, then the length $l$ between a pair of supports requires a force $m_0 g l$ to support it. The upper side of each turn of the coil is then subjected to a torque $T$, where:

$$T \simeq 1/2 \, m_0 g l^2$$

The coil sags as the wire deforms plastically under this torque. The mean shear stress in the wire, $\sigma_s$, is defined by:

$$\sigma_s \int_0^{d/2} 2\pi r^2 dr = T$$

TABLE 19.5  Typical light bulb specifications

| 110 volt single-coiled lamps | 25 Watt | 40 Watt |
|---|---|---|
| Burning temperature (°C) | 2250–2350 | 2400–2500 |
| Design life (s) | $3 \cdot 6 \times 10^6$ | $3 \cdot 6 \times 10^6$ |
| Turns/metre (m⁻¹) | $2 \cdot 6 \times 10^4$ | $2 \cdot 4 \times 10^4$ |
| Spacing of turns, $S$ (mm) | 0·038 | 0·043 |
| Wire diameter, $d$ (mm) | 0·030 | 0·036 |
| Coil diameter, $2R$ (mm) | 0·15 | 0·14 |
| Total length of wire (mm) | 660 | 430 |
| Total length of coil (mm) | 41 | 41 |
| Total mass of coil (mg) | 9·0 | 7·2 |
| Number of intermediate supports | 3 | 3 |

**TABLE 19.6  Summary of conditions under which filaments operate**

| Power of lamp | Range of $\sigma_s/\mu$ | Range of $T/T_M$ | Maximum $\dot\gamma$ $(s^{-1})$ |
|---|---|---|---|
| 25 Watt | $0 \to 1\cdot0 \times 10^{-4}$ | $0\cdot68 \to 0\cdot71$ | $3\cdot0 \times 10^{-9}$ |
| 40 Watt | $0 \to 6\cdot2 \times 10^{-5}$ | $0\cdot72 \to 0\cdot75$ | $3\cdot6 \times 10^{-9}$ |

from which 
$$\frac{\sigma_s}{\mu} = \frac{6m_0 g l^2}{\pi\mu d^3}$$

The resulting normalized stress for the coils, together with the homologous temperatures and maximum strain-rates, are shown in Table 19.6. The stresses are low, but the temperatures are high—up to three-quarters of the melting point.

### The use of deformation maps to analyse the filaments

Most filaments are made from *doped* tungsten: tungsten made from powder containing a little $Al_2O_3$, $SiO_2$ and $K_2O$ (or mixed oxides of these three elements) which gives the wire added creep strength. Microstructurally, doping introduces a fine dispersion of bubbles which stabilizes an elongated, highly interlocked grain structure, of a sort which cannot be obtained in pure tungsten. Partly because these elongated grains impart good creep resistance (they make grain boundary sliding difficult) and partly because individual bubbles pin dislocations, doped tungsten has superior creep strength. The effect of doping is obvious when

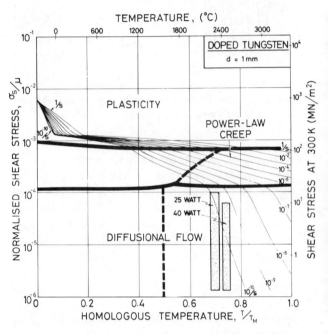

Fig. 19.17. Doped tungsten. In wire made of doped tungsten the grains are elongated; their long dimension can be as great as 1 mm. For that reason a grain size of 1 mm has been used here. The shaded boxes show the conditions of operation of the two lamps.

deformation maps for pure and doped tungsten are compared (Figs. 19.16 and 19.17).

The map for pure tungsten is constructed from the data listed in Table 5.1, and discussed in Chapter 5. That for doped tungsten is based on the measurements of creep in doped tungsten of Moon and Stickler (1970). Doping does not change the yield stress much; but it greatly reduces the rate of power-law creep, causing this field to shrink in size. We have fitted the map to Moon and Stickler's (1970) measurements by altering the creep constants to $n = 8\cdot43$ and $A = 8\cdot4 \times 10^{14}$, leaving all the other parameters unchanged.

Although the grains in doped tungsten are small in two dimensions, they are large in the third, being about 1 mm long in the direction of the wire axis. For diffusional flow, it is this long dimension which counts, so the map is constructed for grains of this size. The map for pure tungsten, too, is for 1 mm grains, partly to make comparison easy, and partly because pure tungsten readily recrystallizes to give large grains if heated above about 1600°C.

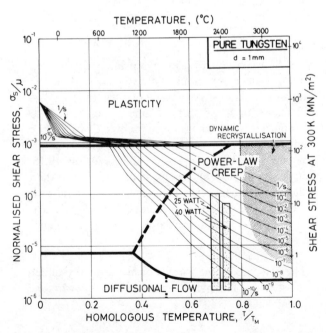

Fig. 19.16. Pure tungsten with a large grain size (1 mm). The shaded boxes show the conditions of operation of the two lamp filaments.

## Conclusions of the case study

The information summarized in Table 19.6 is shown on the maps as shaded boxes, each box defined by the range of temperature and stress to which a filament is subjected. Suppose, first, that the filaments were made from pure, large-grained tungsten (Fig. 19.16). The maximum creep rate (top of the box) would be about $10^{-4}$/s: the filament would fail by creep in about 30 s. The mechanism leading to this rate is *power-law creep*, so (unlike the examples of the lead pipe and turbine blade) increasing the grain size, if that were possible, would do no good. We require a strengthening mechanism which suppresses power-law creep without accelerating diffusional flow.

Doping successfully does this (Fig. 19.17). The boxes show that the maximum creep rate is now about $4 \times 10^{-10}$/s, comfortably below the limit required for adequate life. Further, the dominant creep mechanism may have been changed by doping: the tops of the boxes now lie close to the boundary between power-law creep and diffusional flow.

Fig. 19.17 can be used as a guide for change in filament design. If the temperature is raised, the maximum stress on the filament must be reduced —by providing more intermediate supports, for instance—so that the top of the box remains below the contour of $10^{-9}$/s. If it does not, failure by creep will occur in less than 1000 hours, and the life of the lamp will be reduced.

### References for Section 19.5

Moon, D. and Stickler, R. (1970) *Proc. Second Int. Conf. on the Strength of Metals and Alloys*. Asilomar, A.S.M.

## 19.6   CASE STUDY: METAL-FORMING AND SHAPING

The various regimes of metal-forming and machining can conveniently be presented on a deformation map. The result puts the regimes of metal-forming into perspective, shows the deformation mechanisms underlying each, and gives a rough idea of the way in which the forming forces change if the rate of the process, or the temperature at which it is performed, are altered.

### Introduction

Strain rates between $10^{-12}$/s and $10^5$/s—a range of 17 decades—are encountered in engineering practice. Those associated with metal-forming and shaping lie at the upper end of this range, between $0.1$/s for slow extrusion and $10^5$/s for fast machining. The temperatures, too, cover a wide range. Cold working is carried out at room temperature. Warm working involves temperatures in the creep regime but below those at which recrystallization will occur. Hot working requires temperatures above the recrystallization temperature. Extrusion involves yet higher temperatures.

TABLE 19.7   The conditions of metal-forming

| | True strain range | Velocity range (m/s) | Strain rate range ($s^{-1}$) | Temperature range | Reference |
|---|---|---|---|---|---|
| Cold working (rolling, forging, etc.) | $0.1 \rightarrow 0.5$ | $0.1 \rightarrow 100$ | $1 \rightarrow 2 \times 10^3$ | Room temperature (slight adiabatic heating) | |
| Wire drawing (sheet, tube drawing) | $0.05 \rightarrow 0.5$ | $0.1 \rightarrow 100$ | $1 \rightarrow 2 \times 10^4$ | Room temperature to $0.3\ T_M$ (adiabatic heating) | |
| Explosive forming | $0.05 \rightarrow 0.2$ | $10 \rightarrow 100$ | $10 \rightarrow 10^3$ | Room temperature | Bittans and Whitton (1972) |
| Machining | $\simeq 1$ | $0.1 \rightarrow 100$ | $10^2 \rightarrow 10^5$ | Room temperature to $0.4\ T_M$ (adiabatic heating) | Boothroyd (1965) |
| Warm working (rolling, forging, etc.) | $0.1 \rightarrow 0.5$ | $0.1 \rightarrow 30$ | $1 \rightarrow 10^3$ | $0.35 \rightarrow 0.5\ T_M$ | |
| Hot working (rolling, forging, etc.) | $0.1 \rightarrow 0.5$ | $0.1 \rightarrow 30$ | $1 \rightarrow 10^3$ | $0.55 \rightarrow 0.85\ T_M$ | Higgins (1970) |
| Extrusion | $\simeq 1$ | $0.1 \rightarrow 1$ | $0.1 \rightarrow 10^2$ | $0.7 \rightarrow 0.95\ T_M$ | Feltham (1956) |

The plastic strains involved in metal-working are often large—of order 1; by comparison the elastic strains are negligible, and the material can be thought of as reaching its steady-state flow stress. For this reason, we plot the regimes of metal-working onto a steady-state map (we have used copper as an example) based on data for the ultimate tensile strength.

In the following sections, the characteristics of a number of metal-forming operations are discussed. The data are summarized in Table 19.7.

## Rolling and forging

Rolling and forging are among the simplest of forming operations (Fig. 19.18). They can be performed *cold* at room temperature (though adiabatic and frictional heating may raise the temperature a little); *warm*, in the temperature range 0·3 to 0·5 $T_M$, too low for recrystallization, but high enough to reduce the forming forces; or *hot*, in the temperature range 0·6 to 0·9 $T_M$, the regime of dynamic recrystallization, when the forming forces are lower still.

Cold working gives greater precision, and much better surface finish than hot or warm forming. The velocities of rolling or forging can be high (as high as 100 m/s) but the reductions are generally small, giving strain-rates in the range of 0·1 to $10^3$/s (Table 19.7).

Hot working allows larger reductions and lower roll or forge pressures. Typical of hot working (Higgins, 1970) is the rolling of steel ingots. They are preheated to between 800 and 1250°C (0·65–0·85 $T_M$) and run through rolls which impose a strain of 0·1 to 0·5 per pass, at velocities of up to 1 m/s. Forging is carried out at comparable temperatures and speeds. For normal billet dimensions, the strain-rates are between 1 and $10^3$/s. Table 19.8 summarizes forging and extrusion temperatures for copper, aluminium and lead-based alloys, and for steels. It illustrates that hot rolling and forging are typically carried out at 0·6–0·8 $T_M$ while extrusion requires higher temperatures: 0·7–0·9 $T_M$.

Although the forces are higher during warm working, problems of oxidation and scaling are reduced, and the texture introduced by the forming operation is retained. Velocities and strain-rates are listed in Table 19.7; they are much the same as those for hot working.

Almost all the work done during a forming operation appears as heat. The temperature rise in the workpiece depends on many factors (Johnson and Mellor, 1973), but it can be large: sufficient to cause a temperature rise of 0·3 $T_M$.

## Wire, sheet and tube drawing

Fig. 19.19 illustrates a typical drawing operation. The wire or sheet is drawn through a lubricated die at speeds of 0·1 to 100 m/s. A strain of (typically) 0·05 to 0·5 is imposed as the metal passes through the die, which is in contact with the wire over a distance of 1–10 mm, giving strain rates which can exceed $10^4$/s—rather higher than those involved in cold rolling.

Wire is usually drawn cold (though adiabatic and frictional heating can raise the temperature to

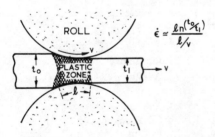

Fig. 19.18. Rolling. The average strain-rate depends on the reduction in section $t_0/t_1$, on the length of contact with the rolls, $l$, and the rolling velocity $v$. It is given approximately by the expression on the figure.

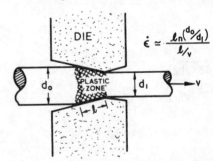

Fig. 19.19. Wire drawing. The average strain-rate depends on the reduction in diameter, $d_1/d_0$, the contact length $l$ and the velocity $v$.

TABLE 19.8  Temperature of hot working and extrusion

| Material | Forging or hot rolling | Extrusion |
|---|---|---|
| Al alloys | 320 → 450°C (0·65 → 0·8 $T_M$) | 450 → 500°C (0·7 → 0·85 $T_M$) |
| Cu alloys | 450 → 900°C (0·55 → 0·85 $T_M$) | 650 → 1050°C (0·7 → 0·95 $T_M$) |
| Pb alloys | 100 → 200°C (0·6 → 0·8 $T_M$) | 200 → 250°C (0·8 → 0·9 $T_M$) |
| Steels | 800 → 1250°C (0·6 → 0·85 $T_M$) | 1100 → 1300°C (0·75 → 0·85 $T_M$) |

$0.3\ T_M$). Refractory metals are the exception: to avoid the ductile-to-brittle transition, which can lie above room temperature, materials such as tungsten are preheated to around 600°C ($0.25\ T_M$) for wire drawing.

## High-velocity or explosive forming

Explosive forming is one of the more recent developments in metal-working. It lends itself to the making of one, or a few components—often large (such as the end caps for large pressure vessels) or of difficult materials (high-strength steels, for example). Typically an explosive charge is detonated, or an electrical discharge triggered, below the surface of a liquid which acts as a transfer medium (Fig. 19.20). The shock wave presses the material into a former, at a velocity $v$, which depends on the transfer of energy to the workpiece; it is typically 50 m/s. The time taken for it to do this determines the strain rate, which for "stand off" charges like that shown in Fig. 19.20, is relatively small: between 10 and $10^2$/s. The resulting temperature rise is slight, perhaps 20°C.

Sometimes the charge is applied to the surface of the workpiece itself. The strain-rates then depend on the rate at which the explosive burns, and can be higher: up to $10^4$/s (Bitans and Whitton, 1972).

## Machining

The highest strain-rates are encountered during machining. Fig. 19.21 illustrates the process, which typifies turning, drilling, milling or shaping. The workpiece and the cutting tool move with a relative velocity $v$, removing a chip of width $a$. The fundamental process is one of shear. In forming the chip, the material (if it does not fracture) suffers a shear of order 1 in traversing a distance of order $a$. The strain-rate is then:

$$\dot{\gamma} \simeq \frac{v}{a}$$

It can be very large: surface velocities during machining range from 0.1 to 100 m/s and the chip thickness is typically of order 1 mm, giving strain-rates in the range $10^2$ to $10^5$/s.

At these high strain-rates, deformation is purely adiabatic (Chapter 17, Section 17.2). Then the temperature rise is large—as much as $0.4\ T_M$—and localization of flow, leading to serrated machining chips and a poor surface finish, can result.

## Extrusion

Working temperatures are highest in extrusion (Fig. 19.22). Typically, a ram forces material to flow through a shaped die, at velocities of between 0.1 and 1 m/s. The strains involved are large—of order 1 giving strain-rates of up to $10^2$/s. The temperatures (Table 19.8) are in the range of 0.7 to 0.95 $T_M$.

## The use of deformation maps to display metal-forming conditions

Fig. 19.23 shows the regime of each forming operation (Table 19.7) plotted onto a map for

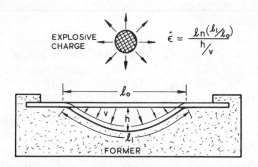

Fig. 19.20. Explosive forming. The average strain-rate is the strain ln ($l_1/l_0$), divided by the time $h/v$ required to achieve it. These strain rates are not as large as those often encountered in machining or wire drawing.

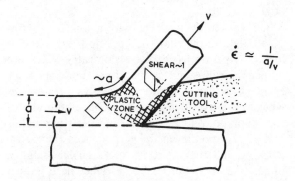

Fig. 19.21. Machining. The highest strain-rates are encountered during machining.

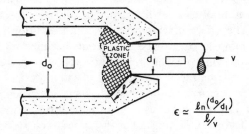

Fig. 19.22. Extrusion. The highest homologous temperatures are encountered during extrusion.

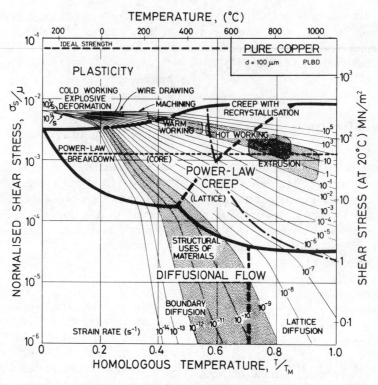

Fig. 19.23. A steady-state map for copper, based on data for the tensile strength and steady-state creep, and including power-law breakdown. The regimes of metal forming are shown, and are contrasted with the regime of the engineering application of alloys.

copper. This map is based on the data listed and discussed in Chapter 4, and includes power-law breakdown. The low temperature side of the map uses data for the (true) tensile strength since this approximates most closely the resistance to plastic flow reached in the large strains of metal-forming $(\hat{\tau}/\mu_0 = 6\cdot4 \times 10^{-3})$.

### Conclusions of the case study

The figure illustrates in a compact way the regimes of stress, strain rate and homologous temperature associated with each operation. Although the map is for copper, it broadly describes the forming of any f.c.c. metal if the normalized axes $\sigma_s/\mu$ and $T/T_M$ are used. The data for Table 19.7 can be replotted onto other maps in this book to illustrate forming of b.c.c. or h.c.p. metals, or their alloys.

The effect of changed working conditions on forming forces can be seen from the diagram. In going from cold working to hot working at the same strain-rate, the shear strength of the material falls by a factor of more than 4. If the coefficient of friction remains constant, then the forming forces and work of forming fall by the same factor. Increasing the rate of forming does not change the forces required for cold work much, but can have a large

effect in hot work (a factor of 100 in rate can double the forces).

As well as the regimes of metal-working, the figure shows the regime in which engineering alloys are generally used. Most structures are designed to last for some years; a strain-rate in excess of 1% per 1000 hours $(3 \times 10^{-8}/s)$ would certainly be considered excessive. Structures such as nuclear reactors are designed to much more rigid specifications, sometimes requiring as little as 1% strain in 20 years $(10^{-11}/s)$. The regime bounded by these limits is shown as a shaded band on the figure, which includes contours of strain-rate down to $10^{-14}/s$, a regime of interest to geologists and geophysicists, and to those concerned with the safe storage of nuclear waste.

This case study illustrates why an ability to extrapolate laboratory data is so important. Metal-forming operations involve strain-rates which are higher than those usually used in laboratory tests. Most structural operations involve strain-rates which are lower. The great body of careful laboratory data, not only for copper but for other metals and alloys, too, lies in the gap between the metal-working and the structural uses. To be useful in understanding the processes described here, it must be extrapolated, often through many decades of strain-rate.

**References for Section 19.6**

Bitans, K. and Whitton, P. W. (1972) *Int. Met. Rev.* **17**, 66.

Boothroyd, G. (1965) *Fundamentals of Metal Machining.* Arnold, Ch. 2 and 3.

Feltham, P. (1956) *Metal Treatment* **23**, 440.

Higgins, R. A. (1970) *Engineering Metallurgy.* Part II: *Metallurgical Process Technology.* English Universities Press.

Johnson, W. and Mellor, P. B. (1973) *Engineering Plasticity.* Von Nostrand Reinhold.

## 19.7  CASE STUDY:
## CREEP IN THE SOUTH POLAR ICE CAP

This is a simple, geophysical application of the maps. Phenomena such as glacial flow, the creep deformation of rocks, or the formation of salt domes might be treated in a similar way. It illustrates how the maps can be used to identify deformation mechanisms, even when the stress and temperature vary with position. A proper identification is, of course, essential if the phenomenon is to be modelled.

### Introduction

Of all the crystalline materials to be found on the earth's surface in a more or less pure form, ice is by far the commonest. The South Polar Ice Cap alone contains some $10^{15}$ cubic metres of pure ice; a large glacier might contain $10^{11}$ cubic metres. Acted on by gravity, much of it is creeping.

Studies are at present in progress with the ultimate goal of solving the coupled heat and mass-flow equations which would describe completely the accumulation, flow and attrition of ice bodies such as the South Polar Ice Cap (e.g. Budd *et al.*, 1971). The mechanical movement is complicated: part of the surface displacement (which can be measured) is due to creep within the ice; part may be due to sliding at the surface where the ice contacts bedrock. To model the creep contribution, the proper constitutive law must be chosen. This choice is the topic of this case study.

### Stresses and temperatures in the Antarctic Ice Cap

The surface temperature in Antarctica varies with the seasons, reaching a minimum of about $-58°C$. A few tens of metres below the surface, the temperature ceases to fluctuate with the seasons, and rises

slowly to $0°C$ at bedrock where the geothermal heat flux is thought to be sufficient to maintain a film of water. The elevation of the ice surface has a maximum value of 4000 m near the centre of the continent, where the rate of snow accumulation is highest; from this high point, the ice sheet slopes gently outwards for 1000 km or more, before falling steeply to the open sea. The ice sheet overlies bedrock. Where radar-echo surveying techniques have been applied, the shape of the underlying terrain is known to be irregular, with ridges up to 2000 m high separated by broad, deep valleys. This complicates the flow of the sheet, which is channelled by these valleys, at least near its edges.

Reduced to its simplest, the stress analysis for

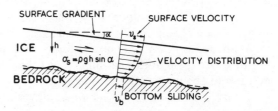

Fig. 19.24. The shear stress in an ice body with a surface gradient $\alpha$ of $\sigma_s$. This leads to the velocity distribution shown. There may be a velocity discontinuity due to sliding at bedrock.

gravity-driven flow in the ice sheet is shown in Fig. 19.24. The gradient in surface height generates a shear stress which increases linearly with depth, and has the magnitude:

$$\sigma_s = \rho gh \sin \alpha$$

where $\rho$ is the density of the ice, $g$ the acceleration due to gravity, $h$ the distance below the surface and

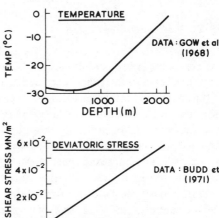

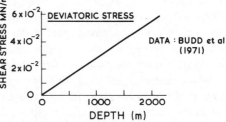

Fig. 19.25. The temperature and stress distribution in the ice sheet at Byrd Station, where the total depth of ice is 2·2 km.

α the surface gradient. The gradient at Byrd Station
($α = 2·5 \times 10^{-3}$) is broadly typical of the Antarctic
plateau; and at this point, the temperature profile
with depth has been measured by making borings.
These measurements, and the associated stresses,
are shown in Fig. 19.25. The pressure, even under
3 km of ice, is too small to change the mechanical properties significantly—unlike the situation
described in the next case study.

### Deformation maps for ice

Ice is remarkably well studied in the temperature
range of interest here: $-40°C$ to $0°C$. The data and
maps are described in Chapter 16. The grain size
in Antarctic polar ice is known (Gow et al., 1968):
cores drilled at Byrd Station show that near the
surface, where the porosity is considerable, the
grains are about 1 mm in diameter; deeper; (below
200 m) where the porosity has almost vanished, it
increases to about 10 mm. This is the grain size we
need, since almost all the deformation occurs below
this depth.

At these greater depths the ice has a texture or
fabric. (This may be evidence of flow by a mechanism involving dislocation motion, and supports the
conclusion we reach here, that power-law creep is
the dominant deformation mechanism; diffusional
flow is not expected to produce a texture, and may
even destroy one). Given sufficient data, one could
produce deformation maps for plastically anisotropic materials, though the method is clumsy and
we shall not attempt it: it requires a new map for
each direction of shear and each level of texture.
Instead, we make the considerable approximation
of treating the ice as isotropic.

### Conclusions of the case study

The stress/temperature profile of the ice cap at
Byrd Station, plotted onto a map for ice with a grain
size of 10 mm, is shown in Fig. 19.26. At bedrock,
both the stress and temperature are at their highest.
The creep strain rate is highest here—about $10^{-9}/s$
according to the map—and falls as shown with
decreasing depth. The important part of the stress–
temperature profile lies in the power-law creep field:
this is the dominant mechanism of plasticity. But
one must be careful: if (because, say, of impurities
or included dust particles) the ice elsewhere had a
smaller grain size—0·1 mm for example—then the
entire Antarctic ice cap would deform by diffusional
flow (Figs. 16.3 and 16.6).

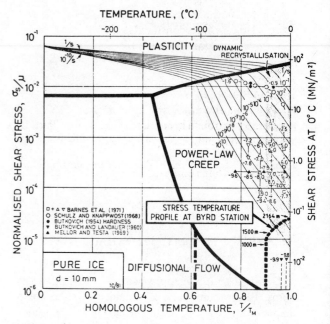

Fig. 19.26. A map for ice with a grain size of 10 mm. The
stress–temperature profile of the ice at Byrd Station is
plotted onto it. Most of the ice body deforms by power-
law creep.

### References for Section 19.7

Budd, W. F., Jensen, D. and Radok, U. (1971) ANARE
   Report 120, Commonwealth of Australia, Department
   of Supply.
Gow, J., Ueda, H. T. and Garfield, D. E. (1968) Science
   **161**, 1101.

## 19.8  CASE STUDY:
## THE RHEOLOGY OF THE UPPER MANTLE

This geophysical application of deformation
maps illustrates how large pressures influence the
mechanisms of plastic flow.

### Introduction

The continents float on, and drift in, a thick shell
of matter known as the upper mantle. Acoustic and
petrologic measurements suggest that the most
abundant phase in this shell is olivine, a silicate with
the approximate composition $(Mg_{0·9}Fe_{0·1})_2SiO_4$.
Samples ejected on the earth's surface show pale
green, transparent grains with a size of about 5 mm.
If, as is thought, olivine is the connected phase in
the mantle (though it contains other phases as well),
then to a first approximation flow in the upper
mantle might be treated as flow in a layer of pure
olivine.

There is considerable interest in modelling the process by which the continents slowly drift, moving about 1 cm a year. As with the ice-flow problem of the last case study, the stresses and temperatures vary with position, so a complete analysis requires the simultaneous solution of the differential equations of heat and mass-flow. But it is more difficult than the ice problem because the material properties are less well documented, and because the drift itself must be driven by convection in the mantle—a phenomenon hard to model. Any attempt to do so requires a flow-law describing the mechanical behaviour of olivine under the conditions found in the mantle. A deformation map for olivine gives insight into the proper choice for this law (Stocker and Ashby, 1973; Ashby and Verrall, 1977), and allows the effects of pressure to be investigated.

## Pressure, temperature, stress and strain-rate in the upper mantle

Values of the material properties and microstructural variables in the upper mantle are uncertain, and, at best, can only be assigned upper and lower limits. The one exception is the pressure: the rock densities and the gravitational constants needed to calculate it as a function of depth are known. To a sufficient approximation, it is given by:

$$p = 10^5 + 3.2 \times 10^4 \, \Delta \qquad (19.6)$$

where $p$ is the pressure in $N/m^2$ and $\Delta$ is the depth in metres. (This identifies the pressure as that due to the atmosphere plus a height $\Delta$ of rock of average density $3.25 \times 10^3 \, kg/m^3$.)

The temperature distribution is less certain. An acceptable approximation is arrived at by combining information about the temperature gradient at the earth's surface (about 12 K/km) and the melting temperature of certain phases in the mantle, with seismic data indicating the depths at which phase changes in the upper mantle and lower mantle occur (indicating a temperature of 1850 K at a depth of 500 km). The result is the temperature profile, or geotherm:

$$T = 300 + 1580(1 - \exp - (7.6 \times 10^{-6} \, \Delta)) \qquad (19.7)$$

when $T$ is the temperature (K).

The magnitude of the shear stress causing flow in the mantle can be inferred from the external gravitational potential, and from stress drops during earthquakes (see, for example, McKenzie, 1968). Stresses between 0.1 and 50 $MN/m^2$ encompass all values derived from these two methods although a narrower range—0.1 to 20 $MN/m^2$—is more likely.

The local stresses, of course, may be greater than this; but while they are important in the triggering of earthquakes, they can have little effect on flow on a global scale.

Finally, a limit can be placed on the strain-rates involved. Continents are known to have drifted distances of the order of 5000 km in times of order $2 \times 10^8$ years. If the shear involved occurs in a layer of the upper mantle 200 km or less in thickness (as is generally supposed) then the strain-rates must be greater than $10^{-15}/s$; and direct measurement of current rates of drift indicate that they are certainly less than $10^{-12}/s$.

The grain size in the upper mantle is not known. Where olivine is found on the earth's surface, the grains within it have a size between 0.1 and 10 mm, though the process by which it reached the surface may have altered its structure.

## Deformation maps for olivine, including the effects of pressure

The mechanical properties of olivine at atmospheric pressure are reviewed in Chapter 15. In the upper mantle the pressure is large enough to change most of them seriously: the lattice is compressed, increasing the moduli and the low-temperature strength; and (most important) the activation energies for diffusion and for creep are increased, reducing their rates. These changes are discussed in Chapter 17. To tackle the present problem, they must be included.

The Burger's vector, molecular volume, and moduli and the strength at 0 K are adequately described by:

$$b(p) = b_0 \exp - \frac{p}{3K}$$

$$\Omega(p) = \Omega_0 \exp - \frac{p}{K} \qquad (19.8)$$

$$\mu(T,p) = \mu_0 \left[ 1 + \frac{T_M \, d\mu}{\mu_0 \, dT} \left( \frac{T - 300}{T_M} \right) + \frac{p \, d\mu}{\mu_0 \, dp} \right]$$

$$K(T,p) = K_0 \left[ 1 + \frac{T_M \, dK}{K_0 \, dT} \left( \frac{T - 300}{T_M} \right) + \frac{p \, dK}{K_0 \, dp} \right]$$

$$\hat{\tau} = \hat{\tau}_0 \left( 1 + \frac{5p}{K} \right)$$

Here, $K$ is the bulk modulus and $p$ the pressure, and the subscript "0" means the value at room temperature and pressure. The influence of pressure on diffusion and creep is included by making the activation energy depend linearly on pressure:

**TABLE 19.9**   **Summary of limits on conditions in the upper mantle**

| Material | Range of $\sigma_s/\mu$ | Range of $T/T_M$ | Range of $\dot{\gamma}$ $(s^{-1})$ | Range of grain size (mm) |
|---|---|---|---|---|
| Impure olivine | $10^{-6} \rightarrow 6 \times 10^{-4}$ | $0.5 \rightarrow \cdot95$ | $10^{-15} \rightarrow 10^{-12}$ | $0.1 \rightarrow 10$ |

$$Q_v = Q_{v0} + pV_v^* \quad \text{(lattice diffusion)}$$
$$Q_b = Q_{b0} + pV_b^* \quad \text{(boundary diffusion)}$$
$$(19.9)$$

where $V^*$ is the activation volume for diffusion. The diffusion coefficients depend strongly on these activation volumes, the values of which are not well established. Because of this, the maps for olivine at pressures other than atmospheric may be seriously in error.

The data listed in Table 19.10 are the best we are able to deduce from the information at present available. Maps computed from these and the parameters given in Chapter 15 are shown in Figs. 19.27 and 19.28. The first is for atmospheric pressure, the second is for a pressure of $0.1\ K_0$.

### Conclusions of the case study

Guidance in identifying the dominant flow mechanism in the upper mantle is obtained by plotting the range of stress, temperature and strain rate given in Table 19.9 onto deformation maps for olivine. The reader will quickly convince himself that olivine, at atmospheric pressure, and with a grain-size of 1 mm or larger (Figs. 15.1 and 15.3), will flow, under these conditions, by power-law creep. Pressure displaces the contours and field boundaries in such a way as to make this even more certain, as will be illustrated below.

The smallest possible grain size of olivine in the mantle is, perhaps, $0.1$ mm. Fig. 19.27 shows a map for this material at atmospheric pressure. The

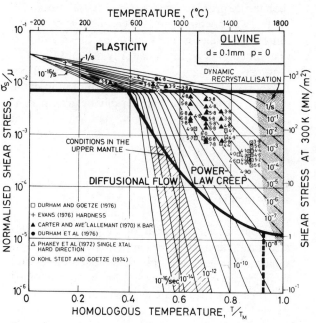

Fig. 19.27. A map for olivine with a grain size of $0.1$ mm at atmospheric pressure.

conditions of Table 19.9 are shown on it as a shaded band: they lie entirely in the field of diffusional flow. This, however, is misleading: the pressure in the mantle is far higher than that of the atmosphere. Fig. 19.28 shows a map for olivine of the same grain size, but at a pressure $p = 0.1\ K_0$, corresponding to a depth of 380 km. The pressure has raised the yield strength and reduced (by a factor of about 1000) the rate of creep. The shaded area now straddles two fields: power-law creep and diffusional flow.

**TABLE 19.10**   **Additional data for olivine**

| Bulk modulus | $K$ | $(MN/m^2)$ | $1.27 \times 10^5$ | (a) |
|---|---|---|---|---|
| Pressure dependence | $d\mu/dp$ | | $1.8$ | (b) |
| Pressure dependence | $dK/dp$ | | $5.1$ | (b) |
| Temperature dependence | $\dfrac{T_M}{K_0}\dfrac{dK}{dT}$ | | $0.26$ | (a) |
| Activation volume | $V_v^*$ | $(m^3/mole)$ | $6.9 \times 10^{-6}$ | (c) |
| Activation volume | $V_b^*$ | $(m^3/mole)$ | $6.9 \times 10^{-6}$ | |

(a) Huntington (1958).
(b) Graham and Barsch (1969); Kamazama and Anderson (1969).
(c) Ashby and Verrall (1977).

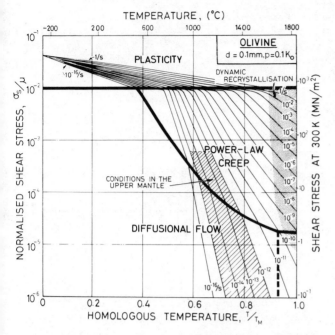

Fig. 19.28. A map for olivine based on the same data as Fig. 19.27, but for a pressure, $p$, of $0.1\ K_0$, ($1.3 \times 10^4$ MN/m²), corresponding to a depth of 380 km. The flow strength at low temperatures has increased by a factor of $1.5$, and the creep-rates have decreased by a factor of about $10^3$.

We conclude that, if the grain size in the upper mantle is 1 mm or larger, the dominant flow mechanism is power-law creep. But if the grain size is 0·1 mm or smaller, an important contribution is made also by diffusional flow. We cannot at present be sure that some new mechanism, such as fluid-phase transport (Stocker and Ashby, 1973; Ashby and Verrall, 1977) does not contribute also. But—even allowing for the wide range of uncertainty in the imposed conditions and in the material properties —the presumption of power-law creep seems the most reasonable one.

### References for Section 19.8

Ashby, M. F. and Verrall, R. A. (1977) *Phil. Trans. R. Soc. Lond.* **A288**, 59.

Graham, E. K. and Barsch, G. R. (1969) *J. Geophys. Res.* **64**, 5949.

Huntington, H. B. (1958) *Solid State Physics* **7**, 213.

Kamazama, M. and Anderson, O. L. (1969) *J. Geophys. Res.* **74**, 5961.

McKenzie, D. P. (1968) *The History of the Earth's Crust* (ed. Phinney, R. A.). Princeton University Press, Princeton, N.J.

Stocker, R. L. and Ashby, M. F. (1973) *Rev. Geophys. and Space Physics* **11**, 391.

# MATERIALS INDEX

Alkali halides   75 *et seq.*, 135 *et seq.*; Figs. 10.1 to 10.6, 18.4 to 18.6; Tables 10.1, 18.1
Alkali metals   30; Table 18.1
Alpha-alumina   98 *et seq.*; Figs. 14.1 to 14.4; Tables 14.1, 18.1
Aluminium   26 *et seq.*, 118; Figs. 4.13, 4.14, 4.15, 4.16; Tables 4.1, 17.4, 17.7
alloys   Table 19.8
oxide   *see* Alpha-alumina
Austenite   60 *et seq.*; Table 8.1
Austenitic steels   *see* Steels

Barium oxide   84
Bismuth   Table 18.1
B.C.C. metals   30 *et seq.*, 60 *et seq.*; Figs. 5.1 to 5.23, 8.1, 8.7; Tables 5.1, 8.1, 18.1
Borides   80

Cadmium   47 *et seq.*; Figs. 6.4 to 6.6; Tables 6.1, 17.4, 17.7
oxide   84
Caesium chloride   Table 18.1
Calcium oxide   84
Calcogenides   75
Carbides   80 *et seq.*; Figs. 11.1, 11.2, 18.4 to 18.6; Tables 11.1, 18.1
Cerium dioxide   93, 95
Chromium   36 *et seq.*; Figs. 5.11 to 5.13; Table 5.1
sesquioxide   98 *et seq.*; Figs. 14.5, 14.6; Table 14.1
Cobalt   Table 17.4
oxide   84 *et seq.*; Figs. 12.7, 12.8; Table 12.1
Columbium   38 *et seq.*, 118; Figs. 5.14 to 5.17; Table 5.1
Copper   24 *et seq.*, 118, 156; Figs. 4.7, 4.8, 4.9, 19.23; Tables 4.1, 17.4
alloys   Tables 17.5, 19.5
Corundum   *see* Alpha-alumina

Diamond 71; Table 18.1
Diamond-cubic elements   71 *et seq.*; Figs. 9.1 to 9.5, 18.4 to 18.6; Tables 9.1, 18.1, 18.3
Doped tungsten   150; Fig. 19.16

F.C.C. metals   20 *et seq.*; Figs. 4.1 to 4.20, 18.4 to 18.6; Tables 4.1, 18.1, 18.3
Ferric oxide   98 *et seq.*; Figs. 14.7, 14.8; Table 14.1
Ferrite   60 *et seq.*
Ferrites   105 *et seq.*
Ferritic steels   *see* Steels
Ferrous alloys   60 *et seq.*; Figs. 8.7 to 8.10; Table 8.1
Ferrous oxide   84 *et seq.*; Figs. 12.9, 12.10; Table 12.1
Fluorite-structured oxides   93 *et seq.*; Figs. 13.1 to 13.6; Table 13.1
Forsterite   *see* Olivines

Germanium   71 *et seq.*; Figs. 9.3, 9.4; Table 9.1

Hafnium   43
carbide   80
Hematite   98 *et seq.*; Figs. 14.7, 14.8; Table 14.1
H.C.P. metals   43 *et seq.*; Figs. 6.1 to 6.13, 18.4 to 18.6; Tables 6.1, 18.1
Hydrides   80

Ice   111 *et seq.*, 157; Figs. 16.1 to 16.6, 18.4 to 18.6; Table 16.1
Indium   131; Table 17.7
Iridium   20
Iron   29, 60 *et seq.*, 118; Figs. 8.1 to 8.6; Tables 4.1, 8.1, 17.4

Lead   28 *et seq.*, 144; Figs. 4.17, 4.18, 4.19, 4.20; Tables 4.1, 17.7
Lead alloys   144; Figs. 19.3, 19.4; Table 19.8
Lead sulphide   Table 18.1
Lithium fluoride   75 *et seq.*, 135; Figs. 10.3, 10.4; Table 10.1
Low alloy steels   *see* Steels

Manganese oxide   84
Magnesia   *see* Magnesium oxide
Magnesium   48 *et seq.*; Figs. 6.7 to 6.9; Tables 6.1, 17.4
olivine   *see* Olivines
oxide   84 *et seq.*; Figs. 12.1 to 12.6; Table 12.1
spinel   *see* Spinels
MAR M200   53 *et seq.*, 148; Figs. 7.8, 7.9, 19.11, 19.12; Tables 7.1, 17.5
Metals   Chapters 4 to 8
b.c.c.   30 *et seq.*, 60 *et seq.*; Figs. 5.1 to 5.23, 8.1 to 8.6; Tables 5.1, 8.1
f.c.c.   20 *et seq.*; Figs. 4.1 to 4.20; Tables 4.1, 18.1, 18.3
h.c.p.   43 *et seq.*; Figs. 6.1 to 6.13; Tables 6.1, 18.1
Molybdenum   39 *et seq.*, 118; Figs 5.18 to 5.21; Table 5.1
Monel   58

Nichromes   53 *et seq.*; Figs. 7.1 to 7.3; Table 7.1
Nichrome-Thoria   58; Figs. 7.6 and 7.7; Tables 7.1, 17.5
Nickel   4, 20 *et seq.*, 53 *et seq.*, 118, 127 *et seq.*; Figs. 1.2 to 1.6, 4.1 to 4.6, 17.6 to 17.9; Tables 4.1, 17.4.
alloys   63 *et seq.*; Figs. 7.1 to 7.9; Tables 7.1 and 17.5
oxide   84
Nickel-Thoria   58; Figs. 7.4 and 7.5; Tables 7.1, 17.5
Nimonics   *see* Superalloys
Niobium   38 *et seq.*, 118; Figs. 5.14 to 5.17; Table 5.1
carbide   80
Non-ferrous alloys   53 *et seq.*; Table 7.1

Olivines   105 *et seq.*, 158; Figs. 15.1 to 15.4, 19.27, 19.28; Tables 15.1, 19.10
Oxides   Chapters 12 to 16
α-alumina structure   98 *et seq.*; Figs. 14.1 14.8; Tables 14.1, 18.1
fluorite structure   93 *et seq.*; Figs 13.1 to 13.4; Tables 13.1, 18.1
ice   111 *et seq.*, 156; Figs. 16.1 to 16.6, 19.26; Table 16.1

163

Oxides—*cont.*
   olivines   105 *et seq.*, 158; Figs. 15.1 to 15.4, 19.27, 19.28;
     Tables 15.1, 19.10
   rock-salt structure   84 *et seq.*; Figs. 12.1 to 12.10; Tables 12.1,
     18.1
   spinels   105 *et seq.*; Figs. 15.5, 15.6; Tables 15.1, 18.1

Phosphorus   Table 17.7
Platinum   118
Plutonium dioxide   93, 95
Potassium   Table 17.7
   bromide   135
   chloride   135
   fluoride   135
   iodide   135

Rare earths   Figs. 18.4 to 18.6; Table 18.1
Refractory metals   *see* B.C.C. metals
Refractory oxides   *see* Oxides
Rhodium   20
Rock-salt structured solids   75 *et seq.*, 80 *et seq.*, 84 *et seq.*;
     Table 18.1
Rutile   Table 18.1

Sapphire   *see* Alpha-alumina
Semiconductors   71 *et seq.*
Silicates   105 *et seq.*; Table 18.3
Silicon   71 *et seq.*; Figs. 9.1 and 9.2; Table 9.1
   carbide   81
   nitride   81
Silver   25 *et seq.*, 118; Figs. 4.10 to 4.12; Table 4.1
   bromide   Table 17.7
Sodium   Table 17.7
   bromide   135
   chloride   75 *et seq.*, 135; Figs. 10.1, 10.2, 18.1, 18.2; Table 10.1
   fluoride   135
   iodide   135
Spinels   105 *et seq.*; Figs 15.5 and 15.6; Table 15.1

Stainless steels   *see* Steels
Steels   Chapters 8, 17 (Section 17.1), 19 (Section 19.3)
   304 stainless   69 *et seq.*, 119 *et seq.*, 146; Figs 8.11 to 8.13;
     Table 18.1
   316 stainless   69 *et seq.*, 119 *et seq.*, 146; Figs. 8.8 to 8.10,
     17.1 to 17.3, 19.6 to 19.8; Tables 8.1, 17.1, 17.5
   1% Cr-Mo-V   68 *et seq.*; Fig. 8.7; Table 8.1
Strontium oxide   84
Superalloys   53 *et seq.*, 148 *et seq.*; Figs. 7.8, 7.9, 19.11, 19.12;
     Tables 7.1, 17.5

Tantalum   41 *et seq.*; Figs. 5.22 to 5.25; Table 5.1
   carbide   80
T–D nickel   53 *et seq.*; Figs. 7.4, 7.5; Tables 7.1, 17.5
Thorium dioxide   93 *et seq.*; Figs 13.3, 13.4; Table 13.1
Tin   71; Tables 17.7, 18.1
Titanium   49 *et seq.*, 123; Figs. 6.10 to 6.12, 17.4; Tables 6.1,
     17.3
   carbide   80 *et seq.*; Table 11.1
Transition metals   *see* B.C.C. metals
Transition metal carbides   80 *et seq.*; Figs. 11.1, 11.2; Table 11.1
Tungsten   9, 34 *et seq.*, 152; Figs. 2.3; 5.1 to 5.7, 19.16; Table 5.1
   alloys   152; Fig. 19.17

Uranium dioxide   93 *et seq.*; Figs. 13.1, 13.2; Tables 13.1, 17.5

Vanadium   35 *et seq.*; Figs. 5.8 to 5.10; Table 5.1
   carbide   80

Wurtzite structured solids   Table 18.1
Wüstite   84 *et seq.*; Figs. 12.9, 12.10; Table 12.1

Zinc   46 *et seq.*; Figs. 6.1 to 6.3; Tables 6.1, 17.7
Zinc-blende structured solids   Table 18.1
Zirconium   43
   carbide   80 *et seq.*; Figs. 11.1, 11.2; Table 11.1

# SUBJECT INDEX

Activation energy   7 *et seq.*; Tables 4.1 to 16.1
Activation volume   131 *et seq.*; Table 17.1
Adiabatic shear   120 *et seq.*, 153 *et seq.*; Fig. 17.4
Alloy design   141
Alloys   53 *et seq.*, 60 *et seq.*, 141 *et seq.*
   *See also* Materials index
Alloying, effects of   10, 14, 16, 53 *et seq.*, 60 *et seq.*, 141 *et seq.*
Anomalous diffusion   31, 50
Applications of deformation maps   141
   *See also* Case studies

Boundary diffusion   15; Tables 4.1 to 16.1
Boundary-diffusion creep   15 *et seq.*, 126 *et seq.*
Boundary dislocations   16, 56, 57, 126 *et seq.*

Case studies   Chapter 19
   lamp filaments   150
   lead pipes   142
   mantle rheology   158
   metal forming   153
   polar ice cap   157
   reactor components   145
   turbine blade   148
Ceramics   *see* Materials index
Climb-controlled creep   11 *et seq.*
Coble creep   15
Cold working   153
Constant structure   2 *et seq.*, 117 *et seq.*
Constitutive laws   3 *et seq.*, 117 *et seq.*, 133, 136 *et seq.*, 141, 148
Construction of maps   17 *et seq.*
Continental drift   158
Core diffusion   12, 75; Tables 4.1 to 15.1
Core diffusion creep   12
Covalent bond   71 *et seq.*
Creep   Chapters 2 and 17
   by boundary diffusion   15, 16, 126
   constants   Tables 4.1 to 17.1
   by climb   11
   by core diffusion   12
   by glide   11
   by lattice diffusion   11, 15, 126
   transient   117 *et seq.*
Cutting tools   80

Data   19, 134; Tables 4.1 to 17.5
Deformation mechanisms   1, 6 *et seq.*
Deviatoric stress   2
Diffusion
   anomalous   31, 50
   boundary   15, 126 *et seq.*; Tables 4.1 to 16.1
   core   12, 75; Tables 4.1 to 15.1
   data   Tables 4.1 to 16.1
   extrinsic   75, 84 *et seq.*, 101; Table 12.1
   intrinsic   75, 84, 101; Table 12.1
   lattice   11, 15; tables 4.1 to 16.1
Diffusional flow   15 *et seq.*, 119 *et seq.*, 126, 130

Dimensionless material properties   136 *et seq.*; Figs. 18.4 to 18.6; Table 18.3
Discrete obstacles   7, 8, 15, 16, 128
Dislocations
   boundary   15, 16, 126 *et seq.*
   Burgers' vector   Tables 4.1 to 16.1
   charge   75
   climb   11 *et seq.*, 124
   drag   9, 16, 120
   effect of pressure   130
   glide   6 *et seq.*, 117 *et seq.*, 130
   mobility   7, 9, 120
Dispersion hardening   8, 14, 15, 53 *et seq.*, 64, 124 *et seq.*; Figs. 7.4 to 7.7, 8.7
Dominant mechanisms   18
Drag coefficient   9, 120
Drag-controlled plasticity   9, 16, 120
Dynamic recrystallization   14, 19, 47, 48, 49, 58, 114, 153

Effective diffusion coefficient   12, 15
Elastic collapse   6
Elastic moduli   22, 31, 67, 136; Tables 4.1 to 16.1
Electron drag   120
Explosive forming   153 *et seq.*
Extrinsic diffusion   75, 84, 101; Table 12.1
Extrusion   153 *et seq.*

Field boundaries   17, 18
Flow strength at 0 K   7, 8; Tables 4.1 to 16.1

Glide-controlled creep   11
Glide-controlled plasticity   6 *et seq.*, 117, 130
Grain boundary
   creep   15 *et seq.*, 126 *et seq.*
   diffusion   15, 126 *et seq.*; Tables 4.1 to 16.1
   dislocations   15, 16, 126 *et seq.*
   melting   114
   sliding   15
Grain size, influence of   15, 22

Hardness data   73, 74
Harper-Dorn creep   13, 15, 22, 26, 28; Figs. 4.14, 4.18
High strain-rates   10, 122 *et seq.*, 153
High temperature plasticity   11 *et seq.*, 117 *et seq.*
Hot working   153 *et seq.*

Ideal strength   6
Impurities, effect of   19, 51, 75, 127
Intrinsic diffusion   75, 84 *et seq.*, 101; Table 12.1
Ionic bond   75 *et seq.*
Isoelectric temperature   75
Isomechanical groups   133 *et seq.*; Table 18.1

Lamp filaments, case study   150
Lattice diffusion   11, 15; Tables 4.1 to 16.1
     controlled creep   11 et seq., 15 et seq.
Lattice resistance   8 et seq., 20, 43, 71, 80; Tables 4.1 to 16.1
Lead pipes, case study   142
Low strain rates   124 et seq.
Low temperature plasticity   6 et seq., 117

Machining, case study   153 et seq.
Macroscopic variables   2 et seq., 117 et seq.
Magnetic phase change, influence of   61, 64, 67 et seq.; Figs.
     8.4 to 8.6; Table 8.1
Mantle rheology, case study   158
Material properties   3, 136; Tables 4.1 to 16.1
Materials   see Materials index
Mechanism fields   17, 18
Mechanism of deformation   6 et seq., 117 et seq.
Metal forming, case study   153
Metals   see Materials index
Microscopic variables   2 et seq., 117 et seq.
Minerals   see Materials index
Moduli   22, 31, 67, 136; Tables 4.1 to 16.1
Multiaxial stress and strain   2 et seq., 141 et seq.

Nabarro-Herring creep   15, 16
Normalization of mechanical properties   133 et seq.
Normalizing parameters   133 et seq.; Table 18.2
Norton law creep   see Power-law creep
Nuclear fuels   93

Obstacles   7 et seq., 57, 124 et seq.
Order, effect of   98
Orowan stress   10, 57, 124
Oxides   see Materials index

Peierls stress   8 et seq., 20, 71, 80; Tables 4.1 to 16.1
Phase changes, influence of   43, 49, 60 et seq., 72
Phonon drag   9, 16, 120 et seq.
Polar ice cap, case study   157
Power-law breakdown   13, 20, 24, 26, 51, 121 et seq.; Figs. 4.7,
     4.13, 4.14, 6.10; Table 17.2
Power-law creep   11 et seq., 118 et seq., 124, 130
     core diffusion control   11 et seq.
     lattice diffusion control   11 et seq.
     data for   Tables 4.1 to 16.1
Primary creep   see Transient creep
Prismatic slip   43, 98
Pyramidal slip   43, 98

Rate equations   1 et seq., 6 et seq., 117 et seq.
     adiabatic shear   122
     diffusional flow   15, 119
     discrete obstacles   8
     dominant   18
     drag-limited glide   10, 120
     elastic collapse   6
     Harper Dorn creep   13
     lattice-resistance control   9
     Norton-law creep   12, 118, 132
     power-law breakdown   13, 121
     power-law creep   12, 118, 132
     relativistic limit   121
     superposition of   18
     twinning   10
Reactor components, case study   145
Relativistic drag   120

Salts   see Materials index
Scaling laws   133 et seq.
Shear strain rate   2
Shear stress   2
Steady state approximation   2, 117 et seq.
Strain hardening   117 et seq.
Strengthening methods   7 et seq., 149
Solid solutions   10, 12, 14, 16, 75, 126 et seq.
Stoichiometry   80, 84, 93, 95, 99
Structural ceramics   see Materials index
Structure variables   3, 117 et seq.
Superplasticity   128
Superposition of rate equations   18
Symbols   ix; Table 1.1

Taylor factors   9, 19, 34
Three dimensional maps   5; Fig. 1.6
Threshold stress   56, 124 et seq., 126 et seq.; Figs. 7.5, 17.5, 17.6;
     Tables 17.4, 17.5
Time-hardening law   118
Transient creep   2, 71, 117 et seq.
Transient maps   117 et seq.; Figs. 17.1 to 17.3
Turbine blades, case study   148
Twinning   10 et seq., 37, 43
Types of map   3 et seq., 177 et seq.

Warm working, case study   153 et seq.
Wire drawing, case study   153 et seq.
Work hardening   117 et seq.